RE viewing thinking turning

Essays on Life, Ecology and Design

alan wittbecker

Books by Alan Wittbecker
Global Emergency Actions
Eutopias Or Outopias
One Earth Many Worlds
Good Forestry
    from Good Theories
    and Good Practices
Topopoetics
Poetic Archaeology of
    the Flesh

RE viewing thinking turning

Essays on Life, Ecology and Design

alan *wittbecker*

Cambridge Books Urania Science Press
2006

Cover Design: 2002, Ebooksonthe.net
Author drawing & photograph: Merissa dePasse
Other graphics and photographs: Alan Wittbecker
Inspiration and support: Precious Woulfe

Published by Cambridge Books
& Urania Science Press
USP at SynGeo ArchiGraph
8051 North Tamiami Trail, No. 32
Sarasota, Florida 34343

For more information on sites and projects:
SynGeo ArchiGraph Co.: www.syngeo.net
Ecoforestry Institute: www.ecoforestry.us
G. P. Marsh Institute: www.gpmi.us
Rian Garcia Calusa: www.re-design.us
Eutopian Ecologists: www.eutopias.net

Publisher's Cataloging in Publication Data

Alan Wittbecker 1946 —

REviewing REthinking REturning / Alan Wittbecker

Includes bibliographical references and index.
ISBN 0-911385-13-4

1. Human Ecology. 2. Ecological Design. 3. Deep ecology.
I. Title.

GF51.W46 2006

Book Design by Rian Garcia Calusa
Printed in the United States of America

Acknowledgments

Chapter 1. Logical Fallacy and Computer Life. *Computing Center News*, April 1985, pp. 6-8.

Chapter 2. The First Self: Animals, Computers, and the Human Mind. *Computing Center News*, Jan 1985, pp. 19-20.

Chapter 3. Literacy and Computers. *ACM Conference Proceedings*, 1984, pp. 113-116. Rep in *Electric Exchange* 3(1):4.

Chapter 4. Letters and Talks on Water, Wilderness and Animals. All letters published in *The Idahonian, Lewiston Morning Tribune, Pullman Herald*, Spokane *Spokesman-Review, Seattle Times, Aprilci Commentator*, or *Sofia Echo*, on dates indicated.

Chapter 5. Wolves in Bulgaria. Wolves in Bulgaria. *Eko-Planeta* (Bulgarian/English edition) No. 11, 2001.

Chapter 6. Two Utopias. (unpublished).

Chapter 7. The Tragic Species. Article rejected by *Inquiry*, 1986. Printed in *Pan Ecology* 10(1):2-6.

Chapter 8. Reverence for Life. *Proc. Marsh Inst.* 7:4-22.

Chapter 9. A Middle Way of Eating: The Conscientious Omnivore. *Pan Ecology* 1(1): 3-5, 1986.

Chapter 10. Mythical Dreaming and Ecological Advertising. Lecture, *Humane Society of the United States*, Washington, DC, 1989.

Chapter 11. Ecologists Want to Rule the Earth? *Pan Ecology* 7(2):4-7.

Chapter 12. Invincible Ignorance: Hunting and Morals. Editorial, *Women in Natural Resources*, 9(2):2-3, 1988.

Chapter 13. Aesthetic Education, Organic Dialectics, and Deep Ecology. Lecture, National Association for Advancement of Environmental Education, Eugene, OR, 1986.

Chapter 14. Radical Education. Closing talk at graduation, International College, 1983).

Chapter 15. Minimum Wilderness Areas. Contributed paper, 3rd World Wilderness Congress, Findhorn, 1983.

Chapter 16. A Maximum Population. Contributed paper, Ecol. Soc. Am. annual meeting, Grand Forks, 1983.

Chapter 17. Ecological Design and Planning. GPMI Occasional Papers No. 6 (NS).

Chapter 18. A Proposal for Moscow 2020. Letter, *Idahonian*,1992.

Chapter 19. Transportation and the Pullman Plan. Letter, *Idahonian*, 1992.

Chapter 20. An Ecological Impact Statement for Rainforest Lotion. Invited Proposal, 1992.

Chapter 21. A Proposal for the Potlatch Corporation. Letter, *Idahonian*, 1992.

Chapter 22. Ecological Community Design at Nieman Ryan. News Review Publishing, 1991.

Contents

Preface: REdoing Everything

IN THIS book, I review numerous articles and books on ecology and forestry, on animals and computers, and on conservation and wealth. A theme runs through many of the works reviewed. This theme is that humans are doing really well at managing themselves and the planet, despite a small number who voice the loss of wilderness, the hundreds of millions of starving children, and the dangers of thoughtless globalization.

Too often, most of the authors of these books and articles hammer home one thin point that is too weak to penetrate the topic, but sharp enough to damage the faculties of some of their readers. Too often their simple arguments are transparent attempts to reconcile inappropriate actions with half-forgotten virtues, or to justify economic gains at the expensive of others or even of the life-support system we call earth. They, or at least the topics they address, need reviewing. Not because they are well-argued or correct, but because people long for simplicity and want to believe in the simple solutions most of these authors present.

Viewing and Reviewing

Most of the authors of the texts reviewed here seem to be writing behind glass, insulated in their own homes and offices, pampered by having a surfeit of resources and time (make no mistake, these are wonderful times for some, with miraculous technologies, longer healthier lives, more toys, and more options). They do not need to look very far or very wide, and they do not. They do not need to question their old blinders or filters, and they do not. But, it makes their analyses and conclusions irrelevant and weak.

Their world is simple because their image is simple, so they think there are simple solutions to simple problems. Many of these authors believe that energy and food increase automatically as people multiply. That simplifying ecosystems can increase their productivity. This exemplifies the first of a series of failures, the failure of imagination. We should not confuse the limits of our mind with the limits of the world, as the philosopher Schopenhauer warned. We seem not to have the ability to see what we have lost, in our rush to be civilized and big. Even in Concord in 1856, H. D. Thoreau mourned for the loss of the lynx, beaver, and other animals — even his relative wilderness he considered a "maimed and imperfect nature."

Rather than viewing the problems of humanity as things tossed at us by an angry planet, we might consider them as human shortcomings, that can be corrected by a wider perspective or better lenses. Reseeing nature with new metaphors, new glasses, and yes, new filters, such as nature not "red with tooth and claw," Alfred Tennyson's famous image, but "peaceful" and cooperative, in Stephan Lackner's newer image — or from seeing humanity in an ecological perspective.

Human behavior depends on constructed images. Images of the environment structure reality. We create an image, then see nature work that way. But we create nature by seeing what the brain filters. We are not only shaped by our buildings, as Winston Churchill recognized, but by our wildernesses and gardens, that is, by all of our images. Images may be of events, places or ideas. Images may come from experience or the arts. The arts manipulate images by their very presentation; they may be clear or vague, certain or uncertain, local or global, rigid or flexible.

The human desire to refine the focus has neglected the frame of reference. We need to place human values within a global framework, attaining a balance of human and ambihuman nature in a field of being. The frame of reference cannot be neglected or discounted.

The topics need to be reviewed, with new images and the whole frame of reference. Reviewing means viewing again (as well as 'examining thoughtfully' or 'evaluating critically'). Things need to be viewed again and again for many reasons. Because things change, because we change, because our understandings and needs change. It is interesting that words that begin with "re" change the meaning of the root word to include doing again or doing anew, whether seeing (viewing), thinking, locating, designing, adjusting, animating, vegetating, inhabiting, valuing, packaging, cycling , storing, specting, newing, generating, fitting, grounding, thinking, or turning (and sometimes it is preferable to use the present continuous tense to describe these activities because we need to keep doing them over time, and because they cannot ever be completed). Fortunately, this is a unique human strength, to remold patterns of thought and behavior; we are capable of undoing and redoing patterns any number of times. After reviewing things, it is important to rethink them, that is, to consider them again, from different perspectives, using different models or metaphors.

Thinking and Rethinking

We live on a planet where most people do not have enough security (of home, food, health, or work). Those people, who are secure are overfed (but still malnourished) and incredibly wasteful, and fool themselves that they can help the others by getting even richer and being more generous with leftovers. And those people feel the need to protect themselves from others who are different (whether in religion, wealth, customs, or land). Fearing the large destructive wars, people tolerate many small permanent wars (which are just as destructive of ecosystems and cultures). We think that peace might be stultifying or boring. Yet, there are some small trends. Some kinds of fighting, such as duels with guns or duels with dreadnoughts have fallen into disrepute, because they were too effective and hence of no use in establishing a clear victory. The Scandinavian countries have lived in peace for centuries. For a slightly less time, all of North America has also. Certainly if we tried a global peace and it was boring, then we could start a few small wars later.

We think that we have to address things one at a time, that we cannot see ourselves or our actions in the whole ecological system, partly because we are interested in continuing our immediate pleasure (even if over a short-term human lifetime) without regard to the indirect and shared costs to the system and to others within it, and partly due to a failure of imagination.

We simplify ecosystems such as forests or cornfields, because they are youthful and quickly productive, but like youthful systems they are also unpredictable and unstable. Since the natural limits are not conserved, we must supplement them with artificial ones. Since the systems are not resistant to pests and diseases, we have to protect them. But, our supplements and protections cause other effects in the larger system. It might be better to work with nature by using a mature system to fulfill our needs. But, a mature system has less available resources that can be taken from it, which means we would have to reduce our demands or population as well as increase our efficiencies and recycling.

For some reason, there have been many negative reactions to suggestions that we humans plan our populations. There are even more negative reactions to proposals for birth control or licensing parenthood, or for reducing consumption. The cornucopians and technophiles always seem to include several unchallenged or unconsidered assumptions: That all human beings, and their nations, have the irrevocable right to produce as many children as they want; that technology will always be able to correct our thoughtlessness; that we must use technology and speed; anything can be substituted for anything else; and that the value of human life in the present is absolute and everyone must protect it, at any cost (this last does not apparently apply to lives over 12 years old — adults are discounted).

Our thoughtlessness, and inability to plan or think, is imposing broad-scale changes on the planetary ecosystem. Forests, especially tropical forests, are being mowed and burned; wetlands filled in; oceans harvested. We have little idea what changes these activities will cause, as critical "environmental services" are disrupted or altered.

Rather than plan for even a maximum use of resources, we have become engaged in a thoughtless grand-scale experiment. We are increasing our populations exponentially, while using resources and altering natural ecosystems for our use. It is a bad experiment, not only because it is large scale, but because we have neglected to have a "control" planet.

Because, our current tenure on the planet is such a thoughtless experiment, I want to propose thought experiments to allow us to try to make decisions about planning our societies and their relationships and impacts. Similar to the thought experiments proposed by Einstein and Infield to discover the laws of physics, we would make thought experiments to test ideas about our living ecosystems. Although ecological systems, especially with humans in them, are orders of magnitude more complex than physical systems, perhaps we could imagine and use such experiments to help us understand what is happening with our complex environment, which is composed of many interlocking ecological systems. By asking all kinds of questions and then imagining the answers, given what we know about our history and ourselves, we could discuss things that are often taken for granted, or not even thought about (see contents for two examples). The best response to a question may be a hypothesis, that is, a thought experiment. Through that, we can create explanations and discover answers in a dialogue with others.

In practice, this means being cautious with new approaches to conservation and being properly skeptical about claims for sustainability. But it means erring on the side of preservation. Maybe after the thought experiments, we may want to drastically reduce our demand for natural products, through conservation, reuse, recycling, and human population control, so that the greatest possible amount of natural habitat can be left wild and degraded lands have time to be restored to health. I also want to propose one grand thought experiment for our consideration: Eutopias. Eutopias are approaches to building appropriate places, using the cultural wisdom that we have distilled from thousands of failures and successes at living, to live within the ecological constraints of places. To go back to a conversation or practice of these things means to return to it.

Turning and Returning

Everything in nature moves and turns. It is seen as a result of its moving and turning. Turning describes the change from visible to invisible (as when you try to follow a squirrel around a walnut tree). As a result of its seeing and being seen, it becomes more. Proteins attain a biological active form by a process of turning. This is the secret: each time something turns we see a new, unexpected universe (and the word universe itself is from the Latin, meaning "one-turning").

The world is constantly changing, but also returning to previous forms and states. This may be observed in the states of water, or life and death. Things follow in succession, then turn back at a limit. They revert. The world is something to observe, to learn from and to employ respectfully to meet our needs. The world changes but returns to previous forms and states. This movement is a series of unfoldings, observed in seasons, water and life and death. Things are transformed by cycles. Much of the environmental crisis is caused by failure to understand cycling. This is especially true with industrial agriculture. Regenerative agriculture (organic gardening) follows natural cycles (this nonintervention can be considered as a form of control, to allow a portion of crops to insects and birds and deer, that they would try to take anyway).

The universe unfolds but does not unravel. It becomes richer and more complex, so it does not know its own future states. Bits or information are being generated. The universe produces genuine novelty through its unfolding. Biological processes also generate complexity (or information). Consciousness allows us to experience the complexity directly.

Having a dialogue with things that are turning is not a linear path leading to a definite conclusion. It is a dialectical spiral, that requires the creative effort of many people, writers and readers, scientists and farmers, researchers and animal activists, to put together. If it asks more questions than it answers, it is for the purpose of stimulating even more questions. I find myself returning to the same topics and relating them to more and more things, using wider and wider perspectives. This is sort of a phenomenological spiral, that gathers up old things and puts a new turn on them.

This is in fact the operation of science, the return to things that we have seen before, but with more understanding gained from looking in a different way (perhaps more patient and less demanding). Science does reasonably well as a conscientiously objective system, but it is not really objective — it just pretends. So, it is not an adequate basis for a world view. Perhaps something better than science may develop: Ethosophics, that is, a fusion of knowledge plus value plus ethics, a process with daring exploration tempered with caution and understanding but done peacefully without harm to other species.

Economics does reasonably well as a consciously objective system, but it is the same pretense. A mature economy will resemble a mature ecosystem. It will be "steady state" in Herman Daly's words. The mature system does not immediately freeze in a state and then decay. It changes and develops. Limits are observed and interactions become appropriate within the limits. For our economy to act like this, we will need new institutions. For instance, one that licenses parenthood or limits the number of children in the society (not necessarily to every couple or group, just the whole society). Or another institution to control the stock of physical resources (through a quota system or some other device). Or another institution to distribute the wealth of society more equitably, through taxes or other limits on acquisition.

Language itself is both a needle and a screen. It can let us weave dialogues, but it can also separate us from the topics under consideration. Ideas can never be totally divorced from the flesh, or from the history of turning and unfolding.

Since we cannot foresee the problems, we do not know what kinds of questions to ask to devise solutions. Therefore, we must always wait until some system fails before trying to correct it. Therefore, we will always be reacting to natural and artificial situations. Therefore, there will always be unavoidable costs in lives and resources before some new working situation is found. There are many examples: We did not clean air until the pollution reduced visibility to a few meters; we did not clean up lakes until all the fish were dead and we could not drink water from them or swim in them — we could not force ourselves to pay for the ounce of prevention, no matter how unhappy we were with the tons of cure.

Finally, it is important to mention limits to our rethinking and redoing. Although we remake our individual and cultural worlds, we do not need to remake the planet and larger cultural images, because these are already made, and were made long before us. We would have to destroy them first in order to recreate them, and we do not have the ability or knowledge to do that. We should not diminish the otherness of nonhuman life or the regenerative capacity of the living planet. So, we must temper our remaking with large amounts of conservation, of nature as well as of traditional cultures.

We know what we have to do, really. But, it seems that only in times of war or great catastrophe do we have the nerve to do something, although we do not seem to have the nerve to avoid those catastrophes or wars that could be avoided by planning. Perhaps it is fear of pain or change. Perhaps it is a character flaw in the species, the failure of nerve at critical junctures. Perhaps it is just the lazy habit of mob thinking. But, we must overcome it and plan dramatic changes.

Remembering

When I started promoting the ideas of radical ecology and making good places ecologically in the late 1960s and early 1970s, as part of a Eutopian Framework, it was very difficult to get published or even listened to. Now, with successful activist groups, such as Earth First! , Sea Shepherd, Greenpeace, and Deep Ecology, it is still difficult to get published, because now I am considered a promoter of old ideas, or a compromiser, who is not radical enough! Nevertheless, I have continued to develop my own ideas of "Eutopias," that is, that good places can result from good cultures. And this theme runs through my reviews of these other works: That with ecological planning, respectful of ecological and cultural limitations, we can make good places to live better lives, while keeping vital, wild ecosystems.

In thirty years working as an ecologist, I have been attacked by: angry loggers, who were upset when I found a breeding pair of spotted owls on federal land, by farmers who resented the inclusion of a two-tractor-wide buffer between fields and a road, by academic ecologists who suggested that I was "confused" because I worked on a variety of projects across academic boundaries, by conservationists because I found merit in a logging plan for a secondary forest, by landowners when I suggested that they set aside small reserves for buffers, by hunters who suggested I was an animal lover for suggesting guard dogs to control coyote predation of sheep, by animal lovers when I suggested hunting as an interim solution to deer overpopulation, by newspaper columnists who suggested that I was as cruel and hard as Garrett Hardin in my articles advocating birth control (Professor Hardin sent me a nice postcard, congratulating me for being in his company, and I am honored to be so). By the early 1990s I was being insulted and castigated on a regular basis. But, I preferred it to being ignored during the 1970s and 1980s — at least it meant that I was heard a few times. Unable to find employment at universities or with research or development companies, I started my own ecological design group, SynGeo ArchiGraph, as well as helped my friends establish Institutes of their own, notably, The G. P. Marsh Institute (Cambridge Massachusetts) and the Ecoforestry Institute (Portland Oregon) — for each I was a director for three years. Some universities and institutes, such as the University of Delaware or Arcosanti (the ecological city in Arizona), have generously allowed me to present my ideas as a visiting scholar.

When I tried to design Eutopian societies it was a challenging intellectual exercise, but when I tried to apply some of the ideas to my home community, it was much more challenging. For instance, the community was willing to plant trees and recycle bottles, but not to enforce buffer zones or to impose population limits or size constraints on the city itself. When I was able to design several forest ecosystems and smaller landscapes, with the input of the owners, the designs became exercises in compromise, especially when the owners had their own favorite exotic species of plants. Applying Eutopian ecology and political ideas has proved to be difficult. Applications of ecological designs are treated with suspicion, especially since there is no academic or political support for it. The attempt for real-world conservation means accepting the opinion of everyone, regardless of their motives or level of understanding. The attempt therefore becomes the beginning of reeducation, which may be a long process, longer than single human lifetimes. And so, I offer these thoughts and reviews, which will hopefully outlast my voice.

Dedication (Chronological Order)

Special thanks to Dr. Michael W. Fox, for asking me to write several articles in the early 1980s and then for showing me how to write. After he edited my dissertation on ecological cosmology, he invited me to be on the Board of Advisors of the Center for Respect for Environment and Life (at the Human Society of the United States). Dr. Fox has been a tireless worker on behalf of animals on several continents, as well as a prolific writer and original thinker. I owe him more as a mentor and a friend than I can squeeze into one small paragraph.

Special thanks to Professor John B. Cobb, Jr. for his help with economic and philosophical topics. Professor Cobb, a classic polymath, has written books, taught classes, and lectured on ecology, religion, philosophy, and economics—I urge him to address politics next. He has been unselfish with his time, reminding me that I need to read things and look at things more carefully before talking or arguing about them.

Special Thanks to Professor Arne Naess, who at 90, is still able to outbox me. Professor Naess was kind enough not to correct me, but to suggest that I keep reading and thinking about philosophy and ecology until I could have a meaningful conversation with him. Fortunately, I did and we have had several long conversations. As long as I have a photograph of him trying to play tennis, however, I will be assured of a fair hearing.

And, as always, thanks to Mike Barnes, Norman Bowie, Alan Drengson, Buckminster Fuller, Twila Jacobsen, Neil Keefe, Devorah Levi, Boyd Martin, Linda Martin, Johanna Metzger, Eugene Odum, David Parker, David Perry, Theodore Roszak, Paul Shepard, Paolo Soleri, and Precious Woulfe, for their conversations, criticism, suggestions, and support, even if I did not pay sufficient attention to all of it.

Life, Computers & Animals

Chapter 1

Logical Fallacy and Computer Life

In his book of the same title,[1] Goeff Simons asks, "*Are computers alive?*" He answers "yes" and presents a series of arguments for his position throughout the book. Yet, many of us readers may feel uneasy. If mainframes and micros were as lively and tenacious as viruses, businesses and universities would be far more dangerous places to work.

Simons' arguments tend to be counter arguments for arguments against the hypothesis of computer life. He tries to present his arguments logically, but the arguments fail as a result of his semantic and pragmatic fallacies. For example, Simons considers the possibility of the acquisition of rights by computers, predicting that rights will accumulate as machine life-forms evolve.[2] This is a semantic fallacy, affirming the consequent: 'If computers are alive, they will have rights; they will have rights, therefore they are alive.' Life is not a prerequisite for rights; in fact, human law assigns rights to fictional, nonliving beings such as corporations. Even being human is not a guarantee of rights — many human groups have just recently been awarded rights by some cultures.

Simons implies that the recognition of "generations" of computers is a proof of living.[3] This fallacy of equivocation is semantic: 'Living beings produce new generations; there is a new generation of computers, therefore computers are alive.' Furthermore, he says tagging computers as "high-speed idiots" is tantamount to admitting that computers are likened to human idiots, who are still relatively intelligent.[4] Simons writes as if he is oblivious to metaphor, taking every cliché or hackneyed metaphor as a flat identity.

The fallacy of division is also a semantic fallacy: 'Simons is alive; his hair is part of him, therefore his hair is alive.' The conclusion is based on an argument that the whole necessarily relates to each part. Simons contends that since humans are alive and since computers arose from human development, computers are alive. He says, "computer life is growing out of humankind" and means it in much the way that new species emerge from existing ones.[5] However, all parts of a living being do not have to be alive. Trace elements are not alive; hair is not alive. Consider this argument instead: If machines are extensions of human being, as tools, then they are no more alive then an artificial heart or hip prosthesis.

The appeal to authority (*argumentum ad vericundiam*) is a pragmatic fallacy; for example: 'Michael Crichton believes that computers are alive, therefore they are.' Simons indulges in this argument regularly, quoting Lycan, Sloman, and others (passim). He even presents a Latinate classification of machines by genus and species — Walter's *Machina labyrinthea* and Ashby's *Machina sopora.*[6]

Then, there is the fallacy of accident, where it is argued that a specific case be subsumed under a general principle. The form of this fallacy is: 'Living beings can move and calculate; computers can move and calculate, so they are alive.' Simons argues that computers have superior mental abilities to many life forms; "frogs cannot weld car bodies as well as robots" Simons says.[7] Let's try that argument ourselves: Since cars are faster than cheetahs, are cars alive? Movement, chess, and welding are not true tests of life or intelligence. It is true that you can get more out of computers than you put into them, but that is a result of synergy, not life. Auto mechanics have known this about cars for decades.

Likewise, the fallacy of begging the question (*Petitio principii*) leads to premises that are insufficient to establish the conclusion: 'When something is moving, it is alive.' Simons spots behavior as a feature of many life forms and considers that the physical movement of machines makes them more lifelike. Computer behavior may be observable and even purposive, but it is lacking self-organization and self-preservation.

The fallacy of ignorance of purpose (*ignoratio elenchi*) uses an argument to support a conclusion that is not the proper conclusion of the original argument. Thus, Simons sets up and demolishes the argument of mimicry against computer life, by limiting his comparison to the distinction between the words 'mimicry' and duplication.'[8] Let's look at the definitions. Mimicry is a superficial biological resemblance of organisms to other organisms or objects that may result in a survival advantage for the first organism. Duplication is an exact resemblance or doubling. Computers can duplicate certain animal behaviors, but the purpose is the exactness of the duplication. Mimicry, on the other hand, is both selective and ambiguous.

The fallacy of complexity appears throughout his book. Here arguments have multiple assumptions, and attacking a weak assumption has the appearance of attacking all the others, which may be worthwhile. Thus, Simons considers an argument against computer life to be an attack against all good computer qualities. Many people argue that computers have good and useful functions and applications, but do not believe that they are alive.

Simons concludes that hostility to computer life results from having high expectations. He points out[9] that we do not make such demands of bacteria or primitive plants, yet we say that such beings satisfy the "necessary life criteria." Yet what are those criteria? Nowhere does Simons list them. His arguments may have been more persuasive had he done so, unless of course, they are unfavorable to his hypothesis. Indeed it is possible to present a definition of life that does not exclude the possibility of computer life or alien, nonterrestrial life forms.

Is there any unique quality of life that distinguishes it from nonlife? Living beings move, whereas nonliving ones tend to be inert; but, rocks and machines can move quite rapidly. Plants and animals increase in size as they age; most nonliving things do not, although glaciers and mountains can. Plants and animals reproduce, creating offspring; few nonliving beings do, except for crystals and fires. Plants and animals are 'irritable,' that is, sensitive to stimulation; yet, many chemical compounds are irritable. Plants and animals have definite forms; so do molecules and stars.

There are several approaches to defining life: Intuitive, functional, structural, or synthetic. The intuitive definition assumes that we can recognize living beings: A pine marten is aggressively alive; salt is not. One functional definition states: A living being effects a local decrease in entropy. This definition is not exclusive, however, since crystals also decrease entropy. A sample structural definition is: A living being is made of cells. This is inadequate, also, since dead organisms are made of cells. A synthetic definition combines the latter two: A living organisms is made of cells, grows and reproduces, and causes a local decrease in entropy.

The synthetic definition leaves out computers, as well as viruses. A virus is alive, but it has no cells. Computers, as Simons admits, do not reproduce.[10] Cells contain the reactions that are necessary for life, but the cell structure may not be absolutely necessary. What is necessary are the enzyme reactions that control metabolism. A molecular biological definition of life does not need to mention cells: Life is a structural replication of enzymes based on a close (but imperfect) reproduction of nucleic acid molecules. By this definition, a virus is alive, even though it uses the host's enzymes. Enzymes are tools, like hammers or computers. Life can be characterized by the possession of at least one nucleic acid capable of replication.

There are other important attributes of life. Life performs activity by itself, that is, a virus penetrates the cell for a purpose. Life modifies its surroundings to suit itself. Life has the ability to make choices (of mates, places, or whatever). But, the specialty of life is historical organization. It is not possible to think of the development of a living organism without taking into account its deep path of genetics and ecology.

Organic life fits the unique properties of carbon and water. But, life could possibly fit ammonia, which is almost as good a solvent as water. Silicon can form chains analogous to carbon; the chains are stable and can combine with oxygen. Of course, in this terrestrial environment, ammonia and silicon are not as efficient as carbon and water. Nucleic acid is the basis of organic life, but it is not the only means of carrying information for a self-replicative pattern. Consider this final, more abstract definition, that could include the possibility of computer life: A living being is complex, reactive, and able to maintain itself under changing conditions; it has a stable structure that carries information on its pattern, and it can reproduce its pattern in a separate being.

Early theologians argued that living beings came into existence, like William Paley's watch, through creation. Evolutionists have argued that living beings came into existence through self-production. Perhaps life can exist through human fabrication. Remembering the philosopher Hegel's warning that we are "more often wrong in what we deny than in what we affirm," computers may someday become a silicon-based life form, but for now, despite Mr. Simon's hopeful but fallacious arguments, they only represent a prebiotic level of complexity, perhaps similar to the organic molecules necessary for the existence of carbon-based life.

CHAPTER 2

The First Self: Animals, Computers, and the Human Mind

In her book, *The Second Self*: Computers and the Human Spirit,[1] Sherry Turkle concludes that: "Before the computer, the animals, mortal though not sentient, seemed our nearest neighbors in the known universe." In that one sentence, she makes two glaringly erroneous assumptions: That computers are sentient, and that animals are not. Her book has been styled (by the publisher's blurb on the dustcover anyway) as "an urgent philosophic inquiry." Her inquiry, however, does nothing to mitigate the urgency that surrounds our relationship with nature, ourselves and our tools, and it leaves philosophy untouched. She does touch unwittingly on the two greatest challenges facing humanity: Its relationship with nature and its relationship with machines.

Professor Turkle presents the computer as a "new mirror"[2] and as a "perfect mirror,"[3] which could allow us to see ourselves from the outside. She recognizes the possibility of narcissism, but distorts the classical interpretation of Narcissus as the embodiment of proud, ignorant self-love— she instead interprets him as "loving an objectified image."[4] If Narcissism is a metaphor for the human condition, as Christopher Lasch contends, then it is unlikely that we can construct an "objectified image" of ourselves, although perhaps the metaphor applies to computers.

Narcissus, remember, was a handsome youth who harshly scorned the woman (Echo) who loved him. His nemesis was to fall in love with his own reflection mirrored in the water, then slowly pine away and die (as did Echo, until only her voice was left). Lasch characterizes narcissism[5] as "to live for the moment, for yourself, not predecessors or posterity." As a metaphor, it is to use the mirror as a source of pseudo self-insight, as a support for fantasies, and in the service of evasion, but not for self-discovery or even objectification.

Turkle wonders whether computers are good or bad, but concludes neither; instead, she writes, "they are powerful."[6] She avers that computers offer a new and powerful way to see ourselves, a "Rorschach," she says, reinterpreting this term as well. (A Rorschach is an impressionistic, free association from an inkblot. It is a projective test used, for instance, to identify the pathognomic response of mixing parts of animals and humans, or machines and humans, in the same figure.) Nevertheless, she offers the "Computer as Rorschach" as a metaphor.[7] Perhaps she really means to say that computers are a test for personality disorders, but probably not. Perhaps she means a different metaphor, "computers as gestalt therapy." If so, Turkle metaphorically stumbles again. Gestalt therapy is a phenomenological approach to perception, based on an understanding of meaningful patterns, which stresses awareness of the world as a whole. Computers cannot deal with perceptual wholes, so that metaphor, too, is inappropriate.

Computers construct fascinating virtual spaces, as dimensions of human spaces. The computer promises the power of an informationally rich world. But, that world is delimited by the type of information that can be digitized. Meaning, context, and sensory texture are lost through digitization. The computer simply makes facts out of that which had been in-form-ed digitally. As Goethe recognized, all fact is a blend of theory, perception, imagination, and needs. The computer filters perception so that only a quantifiable dimension remains. The myth, spirit, and history of humanity on the earth are ignored.

In spite of these limitations (and others ably noted by Paul Shepard, C. Lasch, A. Koestler, and many more), Professor Turkle continues to use the man-as machine metaphor to understand human psychology. She addresses people who see themselves "as a machine"[8] or who turn to machines for relationships. Although that is a proper use of a metaphor, it may not be a good one. After Shakespeare saw our bodies as gardens tended by will, Descartes saw them as machines. From the mechanical to the cybernetic machine, this metaphor did successfully explain detailed processes without answering fundamental questions of meaning.

Machines are the new "totems," according to Paul Shepard,[9] replacing animals as metaphors for the human condition in industrial cultures. Machines are seen as models of human experience. In Turkle's interviews, people reveal that they lose themselves in their roles so completely that their independent identity seems lost; they become like automatons, lacking any feeling or empathy with other beings. Yet, she argues that machines are the proper metaphor for understanding the human condition. If the metaphor we choose to describe ourselves is characterized by detachment rather than engagement with the world, then we have likely made a poor choice. Perhaps the computer is not a good metaphor for human feeling.

At the other side of the "man-as-machine" argument lies the impulse of humans to endow their machines with anthropomorphic or biological characteristics. Perhaps because of the identity crises in technological societies, we have come to see machines as humanoid and possibly, desirably, "friendly."

Machine metaphors for the body, animals, the earth, and the solar system were illuminating for a while, but, they have been extended too far. The body is not a machine. Animals are certainly not devoid of sentience — elephants, like many other animals, are aware of death and will not abandon a wounded companion (as noted by George Schaller and others); wolves, like many other animals, are loyal and devoted parents (as revealed by the research of M. W. Fox and others). Not only are animals sentient, but they respond to sentience. Accepting the sentience of others provides kinds of knowledge not available to machines or to machinomorphs.

Animals present us with "related otherness;" the human mind benefits from exposure to animals, whose existence helps us to develop our own self-identity — the *first self*. We human beings are biological organisms who have co-evolved with other species in nature over deep time.

Pushed from concern by industrial culture, nature is viewed now with detachment and suspicion, dangerous attitudes for a dependent species. The computer sometimes adds to the separation, especially in its role as a useful tool that fosters a utilitarian attitude. Although computers can permit the simulation of fragile ecosystems, helping us to understand the complexity and intrinsic values of wild places, they have also enhanced our ability to destroy such places.

As Turkle implies, nature is no longer the mirror for culture, as it was in the eighteenth century; that function has been, and is being, replaced by human technology more and more. It should be remembered, however, that humanity mirrors merely one aspect of nature, and that the computer mirrors only one facet of the human mind. To be whole, we cannot dismiss or ignore our emotional and economic dependence on animals and the earth to which we have been fitted over many tens of thousands of years.

The nineteenth-century poet Novalis recognized that one necessary step to full consciousness is introspection, the exclusive contemplation of the self. Computers can be a useful tool for this step, although psychologists like Turkle tend to emphasize this stage and stay there. The other step must be a genuine outward observation of others, not just other people and a few animals but the entire ultrahuman world. Computers can assist with this step also. But, only with both steps will the reflection in the mirror be whole, and not just a shadowy second self.

CHAPTER 3

Literacy and Computer Illiteracy

Computers are marvelous devices that have simplified data processing and stimulated new applications in many fields. As tools, the machines have become so useful and attractive that educational authorities are calling for computer literacy and ordering large numbers of microcomputers and full-screen terminals. The computer is becoming a dominant all-purpose tool. But, like any tool, it is not a panacea for the difficulties of modern civilization; it may not even be a critical part of literacy. In fact, the computer may have distinct educational and biological disadvantages for our species. This article places computers in a broad educational perspective.

The Good Computers Do

Computers are powerful tools. They can process immense quantities of data and solve incredibly complex equations, reducing the time for answers from human life-times to mere seconds or minutes. They can correlate literary texts, reducing the efforts of some scholars from lifetimes to years. Books can be written, edited, and typeset from a microcomputer, using sophisticated programs. The variety of software for applications is increasing daily. Systems are becoming more reliable monthly. Costs are decreasing logarithmically.

Computers can extend human abilities. Consider some of the progress in the last three years: Very large-scale integrated circuits, voice-recognition, and scanning. Consider what is coming soon: Teachable computers, learning, adaptive, distributed computers that can synthesize patterns and process gestalts. The computer is almost like a brain itself. Like the brain, the workings of the computer are invisible to normal eyes, but the effects are visible. Perhaps the design of computers is limited by the biological structure of the human brain. In a recent article in *Datamation*,[1] Vincent Rauzino argues that computers are moving from a linear, left-brain orientation into the right-brain realm of inference and intuition. So, perhaps not.

Computers can facilitate education by permitting self-paced learning, much like tutor-texts used to do in the 1960s. They can simplify business procedures for people. They can assist with repetitive writing tasks and text-editing for writers and editors. Computer simulations are cost-effective and literally life-saving in the fields of medicine and veterinary medicine. In ecology, computers can be used to demonstrate differences in productivity of ground-covers or estimate the effects of hydrological cycles from clear-cutting the Amazonian rainforest. The only cost of a simulation is in time or dollars. This is not true in real systems, where acid rain kills fish, or where nuclear radiation kills humans. Computers are used for far more than simple computing; they are used for games, word processing — even status.

In another recent issue of *Datamation*,[2] Ted Nelson speculates on the possibility of a hyperworld, a vast realm of information and graphics available to anyone instantly — something like a giant word processor with links, where literature propagates like electronic wildfire through the system (perhaps avoiding the immense waste of news and articles published), where people are brought together by the computer, rather than driven apart by television. This wild transformation could make education come alive with graphics and answers to immediate questions (of course, it could just as easily encourage separation and shallow relationships).

The Machines of Desire

The universities are caught up in the excitement. They are promoting the "important and strategic role" of computing, according to Nicholas Lee,[3] "Computer literacy and its definition are issues in the new computing environment." At Princeton,[4] EDUCOM announces seminar series on Faculty Computer Literacy. The sessions emphasize planning, courseware selection, budgeting, and team building, to facilitate the management of "information technology." Sheila Widnall at MIT[5] questions the basic value of a bachelor's degree in science if computers are not used to enhance the undergraduate program. She bemoans the fact that the computer is not viewed as a full partner in natural science. There is a push for student computer literacy by administrations and computer communities (including manufacturers). The faculty at Ohio State[6] argue over who needs computer literacy and who is qualified to teach it. The Computer Science Departments are then pitted against departments that offer unique applications in their own areas.

Desire and Disease

Happy salespeople tell everyone that the computers save time, increase efficiency, and solve problems. Sometimes they do. And a disease—computer illiteracy—is invented for those unable or unwilling to use computers for every purpose. Much of this emphasis on computer illiteracy is self-serving; the computer companies wish to sell computers, the universities need funding, professors need employment. So, the university is sucked into offering computers at cost—something never done for books or supplies—instead of setting up programs, users groups, and networks. However, there is a real disease present: Computotosis. Computers are too much with us, like telephones or televisions or automobiles. Ivan Illich[7] even goes so far as to say that computers are forcing some kinds of communications and eliminating those parts of culture that do not "fit the logic of machines."

This hysteria has everyone hoping for miracles. People want to be spoiled by computers. Many users who come into our consulting office want keys with pictures on them, like fast-food registers, so that they will not have to learn commands. Users seem to be wanting a technological dependency. They seem too willing to delegate their responsibility to experts. Excessive reliance on any tool may have unfortunate consequences that we may not foresee.

Although the computer is a magic tool, like the loom or radio, sometimes the tool determines the limitations of a project. And metaphors of tools, especially mechanical metaphors, determine the limits of thought. For classical humanity, the mind was a clay pot or clay tablet; for early moderns, it was a steam engine or clock. Now, it is a computer. The computer is becoming a master technology, capable of dominating other machines, as the regulator dominated steam and the clock, as the wheel dominated transportation and agriculture (even now when mag-lev trains are possible). The computer is also capable of imposing a machinomorphic[8] view of reality on its users, which could suppress other intellectual and spiritual abilities.

Computers are being used unthinkingly for any application. But, as the radio was not very good for comprehensive communication, computers are often not very good for creative paperwork or efficient short communications, such as a single memo. Much computer work relies on the use of well-structured, linear language skills, in short, on linear thinking. This linearity limits most current applications. In fact, it may not be advantageous always to use a computer for every possible application. Pat Wagner[9] concludes that well-organized projects do not need computers to improve the projects; the problems of the projects will be magnified by computer installation.

The success of computers is partly due to the unending flow of cash by patrons of the art, influenced by fad as much as any patron is. Furthermore, investment in programs is built on the lives of thousands of programmers and users (wage slaves), who have a vested interest in having their programs become the standard, regardless of ease of use. There is no doubt that computers have brought about many good changes, as have the telephone or automobile. But the phone and auto had undesirable effects; the invasion of privacy and pollution, for example. What are the hidden costs of computers? What undesirable effects will they have? Loss of freedom? Less liberal education? Furthermore, not everyone has a car or phone, or wants one. Yet, auto and phone illiteracy are not considered problems.

Types of Literacy
The thrust of artificial intelligence in theoretical computer science is to create a field of inquiry to understand the nature of mental processes. The direction of computer evolution is for people to be able to deal with machines and programs in an intuitive manner. Programmers are aiming to simplify the representation and retrieval of knowledge, especially in the "Fifth generation" of computers being created. As computers become more sophisticated, the programming becomes less visible, easier to use, and so literacy is not a factor. Computer literacy has addressed shallow questions about how much we know. But, literacy occurs on more than one level. Beyond familiarity with the terminology and applications, computer literacy is competence at solving problems using a computer, or even the professional design of new machines. Complete computer literacy is the knowledge of how the computer operates, as well as its design, manufacture, and programming—few have that complete kind of literacy. Classical literacy can survive the collapse of printers and newspapers. But, could computer literacy survive the collapse of computers? Perhaps this question hinges on the definition of literacy.

Literacy is the quality of being literate. Specifically, it is the ability to read a short passage and answer questions about what was read.[10] The word comes from the Latin word *litera*, meaning letter, and *Litterae*, meaning

25

writing (from *Linere*, to smear or daub). Being literate is being characterized by learning, cultured, educated. A person who is educated has been "lead" from ignorance, out of the self in other words, by fostering the growth and expansion of knowledge through a course of formal study. Knowledge is a condition of knowing, an acquaintance with theoretical or practical understanding. There are no limits on what can be known. But, most knowledge is concerned with survival first. It is important to know what plants to eat, where to find shelter, how to make clothing. This was, and still is, the most basic level of literacy.

Literacy is taken as synonymous with education, but it is not. Martin Mathes, a biologist at the College of William and Mary, writes of the importance of Botanical Literacy[11] as a key ingredient of scientific literacy. Garret Hardin points out that the standard literacy is not enough.[12] It needs to be supplemented by "numeracy," the ability to think in numbers — computers, remember, use numbers for everything — and, on another level, by "ecolacy," the understanding of the complexity of the world, that things are interconnected and affect one another.

Gandhi put literacy in a similar perspective,[13] "Literacy is not the end of education, not even the beginning, it is only one of the means." Certainly, computers can be valuable for certain aspects of education, but we must not forget what function the computer is assisting. In other words, computer use should not displace the skills themselves. Education should include a core of mathematics as a liberal education always has. And poetry and narratives should still be memorized, as well as written or examined on a computer.

Adaptation and Computers

The use of computers may actually have long-term psychological effects that should be considered as a problem of education. If we consider our machines as "energy slaves,"[14] then Americans each use the equivalent of fifty slaves per day. The computer is an information slave and must be the equivalent of at least three other human slaves. Slave owners love the privilege. And the exploitation of inanimate slaves is more easily justified. In fact, to some extent, the culture of arts and sciences may not be possible without machine slaves. But, at some point, slavery corrupts the owners, making them physically or mentally soft. What is the exchange for summoning these information slaves so easily? The failure to develop intellectual ingenuity? Loss of imagination? Lack of trust in intuition? Let us consider an earlier exosomatic adaptation that helped humanity. Knives permitted hunting larger game animals or deeper roots, but the long-term anatomical result was partial degeneration of the human jaw. Garrett Hardin notes that all exosomatic adaptations bring about "a corresponding degeneration in the endosomatic function," i.e., knives change the function of teeth.[15] The species is then more vulnerable to accidents, since external adaptations, like pacemakers or artificial kidneys, become required for the health and maintenance of civilization. Perhaps writing has changed the function of human memory. Perhaps computers will change the structure of thought.

The computer has allowed us to keep track of inventory without memory and to add without effort. But, when these mechanical aids fail, people have more difficulty adding up their grocery bills themselves or keeping track of all the widgets made or sold. Computers successfully augment human intellectual abilities, but we must take care that they do not allow the abilities to degenerate. The ultimate importance of education is how to live harmoniously on earth, with the wealth of living beings and interrelated systems. Computers can be an important part of education, but not necessarily a critical one or one requiring special literacy.

Figure 1. Winter Thinking
Photograph by Merissa dePasse

CHAPTER 4

Letters on Water, Humans, Animals, & Wilderness

June 28, 1979

Dear *Idahonian* Editor:

Although Jim Sterling, Latah County Weed Supervisor, might enjoy his 2,4-D cocktails (*Idahonian*, 6-25-79), it is doubtful the "mixed as directed" caveat includes the addition of Banvel or Tordon, the other chemicals added routinely to 2,4-D. It is doubtful, indeed, if he means what he says at all. For example, 2, 4-D has been shown experimentally to disturb the physiological process of respiration in the cell and to imitate x-ray damage of chromosomes. The comments made by Mr. Sterling and Mr. Lee are suspect, calculated at a third-grade level to "teach people that these chemicals are safe." The chemicals in fact are not safe, nor are their methods of production, the problems of their disposal, or (shudder) Mr. Sterling' s serene "faith" in them.

It is becoming clear, as one reads and observes, that because of the extraordinary stresses and demands already placed by man on himself and the environment, caution is our most important and beneficial resource. This caution is certainly advisable in the use of chemicals, particularly because they have the potential to be pervasive and persistent in the environment and to boomerang, forming more toxic and carcinogenic compounds than the original substances. The routine use of chemicals, sprayed, dusted, injected, or otherwise forced into the complex network of the interdependency of living things, seems likely to carry with it the penalty of addiction: the madcap acceleration of deterioration in environmental quality and safety.

As pointed out, not of course by Mr. Sterling or Mr. Lee, but in a recent (April 1978) College of Agriculture publication at the next university over, one disadvantage of chemical controls is that "their cost is high, in both initial investment and subsequent application." I do not argue that handpicking weeds would not be tiresome on a county scale, although certainly as effectual as poisons. There also are biological methods, including cultivation of desirable plant species or mowing, that might be, in the long run, less costly alternatives.

The *idee fix* that chemicals are the best and most efficient means of altering the environment (they can never control it) is incalculably more noxious than the weeds the chemicals purport to control. So, rather than "bottoms up" Mr. Sterling, try to be "heads up" about the proper use of dangerous chemicals.

Dear *Idahonian* Editor:

I was amused to read J. Shelledy' s dismissal of D. Hughes as nonsensical; it seemed inconsistent with your *Idahonian* policy of headlining its own nonsense. For example, last Saturday' s (19 September) paper had an article on Maynard Miller, Dean at the University, riddled with absurdities far more frightening for their influence. Examining some of them, we find that pure water isn't 'good for you;' but pollution is; although pollution control laws are part of what 'killed' Bunker Hill, according to Miller.

    I disagree. Pure water is good for you. Water is a universal solvent; it is such a great solvent that absolute purity is only a theoretical goal; even highly distilled water contains gases and solids. The fresh water we drink contains about 1% solution of carbonates, with various nitrates, silicates, minerals, compounds, and trace elements. These elements in normally pure water are absorbed by plants that we eat.

    By contrast, the Webster's definition of pollution is 'contamination.' To contaminate something, like water, is to make it unfit for use by introducing undesirable elements, such as sulfur, phosphorus or lead compounds. What makes these elements undesirable is their excess. The generation by human beings of excess wastes aggravates the capacity of biological systems of the earth to absorb and recycle wastes. So pollution, or excess waste, ruins forests, crops, fisheries; destroys whole species and the productivity of local biological systems; and impairs the health of ecosystems and children. Perhaps the lawyers of the children of Kellogg would like to hear of the benefits of a 'little pollution?'

    One origin of pollution is industrial waste, especially new compounds and heavy metals. Until recently we imagined that waste disposal was provided for free by the environment. Clean-up was an operating expense that we ignored. Now that we have the bill, we complain — an old bill does not contribute to productivity or the balance sheet. But a 'good economy' is one that pays all its debts, and a good environment is the highest cost, not the lowest and last. Several years ago, relatively minor and inexpensive improvements to Bunker Hill would have resulted in a 50% improvement. Harsh laws didn't help kill Bunker Hill; bad economics did.

    Since air and water are cycled so often, over large areas, places like Bunker Hill are a global problem. Our 'cowboy' economy, with its exponential growth of products and pollution, advances the time of some catastrophe. Hughes' letters may stimulate a little participation, but Miller's views spread misinformation; let him contain himself to his area of expertise, faculty salaries.

March 1, 1982

Dear *Idahonian* Editor:
Having read Professor Gier's discussion of the new right, it is tempting to view the new right's position, as presented by Mr. Thomas, as much like Satan's: they have power without charity. Unfortunately, Mr. Gier has charity without knowledge.

The humanism he describes has been a dirty word for a long time. Rightfully so — it is concerned only with humans and only humans are considered to have intrinsic value. Valuing only humans causes animal and plant deaths by the millions, far more lives than those lost in the name of Christianity. Perhaps one *species* a day becomes extinct through human action — one thousand times the natural rate. Ironically the word animal is derived from the Greek word for soul. Devaluing animals leads to devaluing human souls. Devaluing nature leads eventually to devaluing human life. Many modern philosophers, such as Merleau-Ponty in his last program, have rejected humanism.

Other thinkers have been more concerned with creating a model of thought in which nature is valued. This ultrahumanism is described in ecophilosophy, phenomenology and process philosophy. It is important that more people than just Gier and Thomas are aware of this trend. It has knowledge with charity.

14 January 1985

Dear Editor:
Regarding the recent article in *Environmental Ethics* on eco-feminism by A. K. Salleh, there are numerous shortcomings and inconsistencies. The summary of Deep Ecology by Arne Naess was not advanced as an exhaustive position. But its anti-class posture can in no way be considered anti-feminine.

It is true that patriarchal cultures have suppressed women for centuries. Balance needs to be restored by radical revision. Deep ecology offers more possibility for such revision than logical positivism or academic feminism. And revision is necessary for the existence of healthy human civilizations in diverse habitats.

Certainly women are different from men, but it is silly to believe that women are grounded more firmly in nature and genetically 'flow with the system,' or that men are more dependent on women as mothers, any more than women are dependent on men as providers, or that biological egalitarianism is a 'grab' at women's special potency. Salleh might consider what it is like to carry the seeds, a unique masculine experience, as well as to bear the fruit.

Deep ecologists attempt to enlarge human consciousness; certainly no one worthy of the name would accept the annihilation of feminine identity or ignore the inventiveness of women. Perhaps the human sense of self-worth in academic circles, exhibited by Salleh herself, is exaggerated and requires bombastic terminologies, but that is not an exclusively masculine trait. Nor do women seem less inhibited by the constraints of 'status validation.' Salleh herself lists her own work in six of nine references to her article.

Salleh's narrow filter distorts cultures and religions. In taoism, for instance, there is no inherent 'Man/Woman hierarchy,' as she contends. Nor is human limitation at odds with life-affirming values, as she writes. In fact, limitation is necessary. This is known in many aboriginal cultures, such as the Campa or Arunta. The earliest myths, for example, the Sumerian myth of 'Enke and the World Order,' show the destructive results of trying to go beyond the limits of 'Mother Earth.' Furthermore, the suppression of the feminine is not 'universal,' as she concludes. Certainly not in neolithic matriarchal societies, nor in many archaic cultures, such as the Haida, Coeur d' Alene or Pawnee.

Hysterical hyperbolism is a perilous path to consciousness; fortunately, it is not the only path. Many men permit expression for their feminine aspects. But Salleh seems to think, in concluding, that women have no masculine aspects. They do. And those aspects need expression for full consciousness.

Salleh's arguments rely on a deductive, reductive logic. Feminine experience is treated as the organic basis for all philosophy. It is only a part. A synthetic logic would be more appropriate. She even implies that the alienated 'Other' is woman. It is not. The Other that we need to recognize is ultrahuman nature. Feminine experience is implicit in human experience.

In summary, Salleh's critique seems 'facile and arrogant,' as she herself labels others. Having uncovered the foundations of deep ecology, she offers only muddy invective in its place. Any critique is lost in a welter of loxocrative framis, unsupported by insights from ecology, cultural anthropology, or philosophy.

23 January 1986

Dear Editor:
Living Conscience
While not wishing to address the inconsistencies or errors of Mr. Arnzen's letter in Thursday's *Idahonian*, or dismiss it automatically as resulting from ignorance, the question of abortion must be considered. There are alternative birth control measures, and these may work in some cases, but not nearly all; abortion is therefore necessary. For all of us, taking part actively or silently, the conscience is wounded, but not killed. The quality of life, however, is thereby improved by the reluctance to cast more lives in doubt. Even now, the number of children born from ignorance, or fear of Christianity, or cementing a shaky marriage, or commercial gain (i.e., welfare), or revenge, or selfishness, or indifference, vastly exceeds the supply of competent, loving parents in the world. Ideally, abortion would never be necessary, in loving communities.

Since these are few, or nonexistent, the death of an unfinished organism is preferable to an adult crippled by the absence of love. Abortion, the prenatal termination of life, is painful enough for physicians and husbands, not to mention the woman whose creation is destroyed. But death also is a natural process, as necessary and majestic as life. Death and life are intertwined inextricably. Without death, the quality of life would be a meaningless phrase. In nature it is the young that perish in the greatest numbers; in fact some animals spontaneously abort if conditions are not correct. Even peoples as sophisticated as the early Greeks regularly exposed babies on hillsides, if conditions for survival were unfavorable (i.e., famine).

If we are to survive as fully human (as wise animals), we must learn from the natural processes of the world, without reducing ourselves to unthinking, unfeeling robots. We must learn to balance all of our values wisely. Lately we have worshipped the individual at the expense of society; the quality of life refers to its social context. Truly, if a fetus has rights, then so must the legions of the unconceived and unconsidered — those who might someday inherit a barren plain of writhing limbs and moaning faces. People turn nasty and brutish without love. Mechanically forcing unwanted children into unloving and unsatisfying circumstances is far more destructive in the long run. Trying to improve the quality of life, partly through abortion, is no crime. The real crime, in India and here, is bearing unwanted children into a callous world, who only repeat the cycle, and bestow their undying resentment on their own unloved children.

Abortion is not a solution of death, but a reluctant willingness to make a difficult decisions and accept great pain. True reverence for life, more than the blind maintenance of vital signs, means to undertake the responsibility to deny life to one's own unborn offspring, until that life would be wanted and loved, appreciated and cared for, in a society planned as carefully.

February 4 1986

Dear Editor:
Society is not cruel. It is unfortunate that Ms. Tartoue has nightmares over the possibility of life denied; apparently, she is less concerned for starving children or the voiceless poor who are freed by their starvation or other disasters to appreciate the same sunrise that Ms Tartoue appreciates when she is distressed.

That there is so much waste, that the wretched are not helped, is not because society is cruel — if it were the arguments for abortion would be even more undeniable — it is because society is like nature: dispassionate. Waste is unavoidable in natural processes; seeds are lost, the young die in incredible numbers, and the unborn and inexperienced are cut loose prematurely, before they can suffer.

The two examples offered earlier were meant as illustrations. Most animals, such as deer or rats, are genetically disposed to abort under unfavorable environmental conditions, as in the instance of a strange male prowling the nesting area. Species that exhibit this preventative behavior are far from extinct; they are thriving and may outlive humanity. Similarly, by their mimesis of the ecological economy of nature, the Greeks were wise. They recognized the necessity of balancing their values (as some civilizations have not), even if it meant not carrying babies to birth when they could not provide for them. They mourned them nevertheless, and Greek society, while not nearly perfect, has lasted longer than ours has thus far.

It would be noble to save every baby conceived—especially if it meant that the resulting plague of hungry mouths would deny the enjoyment of life to everyone. Perhaps humanity will perish yet for its nobility, as dinosaurs perished for their strength (and lack of flexibility). But, perhaps not; the automatic reaction of a species faced with the crisis of survival may render all of our arguments irrelevant. Human groups faced with extinction have resorted to cannibalism and worse behaviors.

In the world, abortion is the most widely-used form of population control. Most people who have had abortions are not indifferent to the suffering involved; if anything, they are more compassionate afterwards. When I worked as a psychologist at a clinic, one of my responsibilities was to provide follow-up care for university students who had abortions. It may seem idealistic to attribute their motives to concern with improving the quality of life in society, rather than to their personal limits, but their actions do have that effect, even if the motive is a healthy selfishness.

Jonas Salk, who has saved as many lives as any one person, has written that the human race must become wise, or act as if it were wise, to survive the next several generations. This wisdom must entail recognition and submission to the limits and ways of nature, in the same manner prescribed by Lao Tse over two thousand years ago. It is foolish and unconscionable to conceive children who cannot be supported by their parents and who may even be a burden to the ecosystems that support all of us. Until we learn to reproduce thoughtfully, abortion remains a necessary corrective.

March 1 1986

Dear Editor:
I heard President Reagan's pledge to save the unborn from government intervention. How muddled and inconsistent!

As an ecologist, I am aware that a fetus is as uniquely human at conception as at birth and at death, and has as much potential for a soul as any human or animal. And, as a radical humanist I am aware that we have a moral responsibility to aid all human beings. But, we ignore our responsibilities. Over ten million people starved to death last year (five million of them were children). Another ten million more died from nutrition-related diseases. We are not acting responsibly. We cannot be aware of this, resting in front of our television fantasies, and still claim to be a moral community. The President and his supporters are hypocrites.

No one should cry out against abortion until they have done something to help save lives elsewhere. People are limited by their own needs; they give to organizations who work heroically to feed, clothe and house the poor and disadvantaged.

If Americans cannot control their sexual encounters and conceptions, use birth control, plan their families, or restrain their actions, then abortion is necessary. If the starving born can be helped by killing the helpless unborn, then it is necessary and inevitable. We have numerous precedents for such preventative killing — we have killed living people to protect the property and resources of other living people, especially in oil and resource-rich areas. We even label and kill wild animals to protect our domestic animals that we later kill for food or parts. Killing may be spiritually depressing, but it seems inevitable. And humans, throughout most of history, on every continent, know this. We make decisions economically. And from an economic basis it makes sense to kill those with the least experience, who will suffer the least, and who benefit society the least, that is, unborn children.

The Reagans and the Falwells, and their loud, fearful supporters, are locked behind glass, in their immoral television world. They cry to look to heaven for answers — there is even a bumper sticker that claims that Jesus does not guarantee the ride, only a soft landing. But, it was Jesus who also warned us that "What you loose on earth is loosed on heaven."

Letters on Wilderness & Animals

11 October 1981

Dear Editor:

I am writing to express qualified support for the Institute of Resource Management at the University of Idaho, as founded by Robert Redford, and to describe a radical (meaning deeply rooted) view that the Institute should consider to guide its direction. Although the institute is a step in the right direction, it is based on many bad assumptions. It is difficult to believe that the groups from Southern Idaho fear Redford and distrust the university. The university is just an assembly line for special interests, and Redford, in spite of his concern, shares the same flawed understanding as all resource groups.

Redford states that he has been seeing two "dire" positions develop that are false. And he says that there is a need for a balanced approach to our environment that recognizes the need for development and our responsibility to protect our resources. He claims the institute will provide balanced education in resource management. Environmental debate is often reduced to opposite views. This polarity is appealing in its simplicity. But the positions mentioned are not really opposite. The position of the group from Southern Idaho, the vandal position, is merely a narrower perspective of Redford's conservative position; and his position is just a much narrower perspective of the radical position.

The very title of the Institute shows a bias toward the utility of nature, for development and resource protection (for future development). Natural resources were originally defined as objects provided by nature for human use. This concept has been expanded to include minerals, wildlife and people. The idea that everything should be managed is based on an extreme belief that nature is a resource to be processed. Furthermore, management self-perpetuating and self-justifying. The objective of resource management is to increase the measure of quality of life for affluent people in overdeveloped countries.

I n short, conservation management is based on economic objectives. And as Leopold pointed out, the weakness of relying on economic motives is that most members of the earth's community, such as wildflowers and songbirds, have no economic value. Yet all the members of the community contribute to the integrity of the whole, which is vital to maintaining what we do consider important. Those beings with no economic value are ignored, or worse, labeled as weeds or vermin and destroyed so that crops and animals with short-term advantages for human ends can be substituted. The goal of this institute is that kind of temporary control.

The impulse to manage nature is an expression of the judgment that we know how the world should be run. But we are finding that we do not know at all. We did not know about the effects of DDT or radiation or chemical dumps or special drugs. The whole approach of the conservative position ignores the physical and ecological dimensions of resources. The vandal position is even more basically ignorant. Laws of ecology must be obeyed for these laws determine our existence and that of "resources." There can be no "balance" between obeying some laws and disobeying others.

History records some civilizations that tried to manage their resources and failed; they existed in the Americas, the Middle East, Africa, Asia, the Pacific, and Europe. Even a modern, balanced exploitation may destroy forests and fisheries. Currently, many resource managers espouse the ideas of equilibrium maintenance and maximum sustainable yield. These ideas are poor guides to management. By trying to maintain habitats in equilibrium, we often set them up for catastrophic decline, for instance, in fire-climax pine forests, or destroy resident species, e.g., the condor. The use of maximum sustainable yield in wildlife management has resulted in the degradation of the populations involved, whales and salmon, for instance. A carrying capacity is not constant; species that live near the limit of capacity cannot be killed at a maximum. Even small numbers, e.g., less that 6% in Sandhill crane, hunted could result in extinction. This is also true of wolves, bears and mountain lions.

Some managers, like Japanese whalers, are far worse. They do not try to manage for a continued maximum yield; they try to maximize the economic value of a resource, in spite of an awareness of extinction – the rape of one "resource" provides the capital for the rape of the next. We are unbalanced. Our whole industrial world view is unbalanced. Balanced resource management will still unbalance nature, though perhaps at a slower rate. The balance of nature has to come before the balance of resources. We will continue to be unbalanced until we enlarge our understanding of nature and let ecological limits suggest new technologies and techniques.

A radical view emphasizes the value of all nonhuman beings and tries to use as few individuals as possible from as many species as possible – moderate yields, within the limits of the food chain. Many species should not even be used, for instance, whales and wolves. Ecosystems cannot be forced into stability for our convenience.

The radical view refuses to compromise, to use cost-benefit analysis and management of resources, to accede to shallow economic domination. Radical ecologists choose their own terms of reference and consider passion and feeling in planning a human role on the earth. There is a much validity to loving a forest as to measuring its net boardfeet. Plants and animals are as much of our heritage as art, history and tools. Ignorance of one is just as sad as ignorance of another. The fight for radical conservation is the fight for ourselves and our homes and places, not for scientific hallucinations of energy flows and biomasses. If we fight to conquer the world, we may end up stupefied by the television image of a polluted desert. We need a wild universe to live fully. When we understand our roles and relationships in nature, then we will not be managers or stewards, but participants and sharers in experience.

The radical view still uses technologies to be part of the food chain, to feed and clothe humanity. This view can support a high human culture, within the limits of our knowledge of nature, by emphasizing differences, diversity and change. It shows how to care for the earth, using what is needed, but letting the rest be.

8 April 1984

Dear Senator McClure:
The Marsh Institute has been working since 1976 to scientifically determine the minimum amount of wilderness necessary for Idaho — as part of our efforts to determine minimum world wilderness, which is necessary to support human civilizations. This work is as arduous as it is pioneering. It will not be finished for another decade.

Our preliminary indications, however, are that Idaho should have from six to nine million acres of wilderness. Even nine million acres is not a very large percentage of land, compared to that dedicated to mining and timber interests, or for agricultural or civic purposes.

Therefore, we request that you increase the number of acres in your proposal to nine million acres, which is probably a safe estimate for the future good health of Idaho wild and human populations.

3 May 1984

An Open Letter to Senators McClure and Symms and Representative Craig
Thank you for your form letters and reports. However, we do not accept your rationalizations. The act introduced by the Idaho delegation is frivolous and irresponsible. We suspect that you subtracted the minimum recommendation from the median acreage, instead of trying to reach a compromise. Furthermore, your responses indicate an alarming lack of understanding of ecology and economics. Here are some of our objections.

Rep. Craig thinks that he creates wilderness. We do not create wilderness, we set it aside. God created wilderness, for wild animals and plants as well as for people, and there is not much left. Seven percent of the total area of Idaho is not very significant. There are countries in South America, Europe and Africa that set aside larger percentages at a greater sacrifice.

Sen. McClure boasts of how much wilderness Idaho has. Idaho does have more wilderness than most states, but that was because the land was passed over by pioneers looking for Oregon and Utah, who misunderstood its value. Still, the people of Idaho should be proud of wilderness and want to increase and preserve it. Let other states be known for their smog and freeways. Idaho could be known as the wilderness state, but only if it has enough wilderness.

Rep. Craig considers that advocates of larger wilderness areas are not affected by wilderness. We, and other Idahoans, are already affected by wilderness, which cleans our air and water, supports healthy ecosystems, and provides recreational opportunities. We also vote.

Rep. Craig calculates that only 20 percent of Idaho's land base is available to support Idahoans. Yet statistics put out by the University of Idaho (1980) indicate that, of Idaho's over 52 million acres, almost 22 million (41.5 %) are rangelands, used for grazing, over 9.5 million (18 %) are croplands, and 2.5 million are paved, built over, or wasted. In fact, Idahoans are using over 83% of the land, not including the free services of wilderness.

Idaho's nonwilderness land base is more than adequate to support its small human population. Monies for education and public services depend on industry and incomes, which depend on the tax structure, not wilderness, on human ingenuity, intelligence, and creativity, rather than on once-through grabs. If Idaho suffers as the result of having wilderness, it will be because of stupidity and laziness.

Sen. Symms regards roadless areas as posing a long-term threat to Idaho jobs and prosperity. Yet, the jobs he describes are "easy" jobs, the simple hunting and gathering of timber and minerals. Idaho should be developing "hard" jobs, permanent jobs that depend on stewardship and invention. High-technology jobs or service jobs. Mining and forestry could still form an important part of the economy, but be limited to the 46 million acres (88%) of nonwilderness.

Not all Idahoans view mountains as resources; certainly, we do not. Nor do they regard all trees as crops, as does Rep. Craig. Most Idahoans respect a decision to set aside more wilderness. In fact, as you well know, most Idahoans want more wilderness put aside than the small amount proposed. What the delegation does not seem to understand is that when wilderness is open to development, it is no longer wilderness, regardless of any future plans for re-evaluation.

Sen. McClure recognizes the misunderstanding and confusion in his proposal. We understand his confusion, but there is no confusion or misunderstanding on our part. We realize that wilderness is restrictive, by definition. The Wilderness Act, enacted to set aside wilderness for present and future generations, is quite generous. It permits hiking, camping, rafting, canoeing, skiing, fishing, hunting, research, and many other activities. It prohibits only off-road vehicles and permanent building.

Wilderness is not waste, as Sen. Symms implies in his report. The importance of wilderness is the health of the land, not the amusement of the people. Genetic reserves, not matchsticks. Places for nonhuman life, not for off-road vehicles and snowmobiles, which damage land. Fires and insects, on the other hand, perform natural functions in wilderness, as they have for millions of years. In fact, some Idaho forests are dependent on fire for existence.

Sen. McClure emphasizes the difficulty of the process, but obviously not enough of his work has concerned the ecological functions and importance of wilderness. More research by the delegation would reveal the faults in their proposal, which is more concerned with fast action than good representation.

You can be of assistance to us in the future, as you generously offered. You can increase your recommendation to 3.1 million acres for the final bill. We hope that you will reconsider.

Alan Wittbecker, W. A. Raider and J. R. Bays

3 September 1984

Dear Senator McClure:
Please support the Idaho Wildlife Federation proposal of 4.5 million acres. Northern Idaho and Eastern Idaho both have ecologically significant biomes that are not being protected.

There are ethical reasons to leave more wilderness. Humanity has taken its own opportunities. These opportunities have been codified for centuries as rights. Now, we must allow other beings equal opportunities. The interrelatedness of life dictates the interrelatedness of rights. And these rights are necessary to the integrity of the whole planet. Humanity developed in a community of animals and plants, as part of a clade on the same tree of life. The quality of human life has always depended on the quality of animal life. Animals have sensations and feelings, as important to them as ours are to us. The extension of rights to animals and plants does not deny any traditional human rights. Animals should be accorded higher moral regard and legal standing to reflect the intrinsic worth afforded by their existence and sentience. Welfare laws to conserve species and to guarantee humane treatment in research, transportation, and slaughter indicate a growing concern among people. A new ethic can keep animals free from human intervention, prejudice, or overuse. Animals should be preserved because they are as they are; their existence is moral justification. Their intrinsic worth is independent of the instrumental values imposed on them by humanity.

The extension of ethics to animals and land is an ecological necessity. Extended ethics defines a social conduct that is a mode of cooperation and, ultimately, symbiosis. Leopold argued that ethics are voluntary limitations of freedom, necessary in a complex world of which we remain incredibly ignorant. Ethics are developed in response to problems that arise from increasing knowledge. Science has phenomenally increased our knowledge of physical and biological processes. It has now become the basis of our moral code, but it cannot very long be a science divorced from feeling and art if that code is to help us survive. To do this it requires aesthetic perception as well as disciplined thinking and feeling. As there is a rational component to ethical judgments, so there is an intuitive and emotional one, also. An ethics can appeal to religious, philosophical or scientific reasoning; or to all three to form a coherent whole.

Humans need to recognize that they automatically participate in everything, and that they cannot unparticipate by choice. The disenchantment of the world is only another name for the hushing of mediating voices between nature and human. Without these resonances, Levy-Bruhl's mystical participation in nature is impossible. The ecological health of civilization depends on an environment that is healthy, that is, that has sufficient wilderness to renew air, water, genetic resources. The survival of society now depends on an expanded ecological consciousness, an awareness of the global system in its complexity and connectedness. The spirit of humanity depends on an ecological consciousness that would place humanity in a proper relation to the wild places of the earth, taking what it needs, but letting the rest be.

9 October 1984

Dear Senator McClure:
I want more wilderness for Idaho. It is annoying to me to hear that only rich backpackers benefit from it. Wilderness is already to everyone's benefit. Just like it is to their benefit to have a heart or kidneys. In fact, the wilderness is like the body of our species. We can do without some of it, but not without all of it. So we can live without one kidney, some of the liver, or arms or legs. Consider, there is a large market for organs to transplant, so everyone could sell a few of their organs if they wanted. To not sell them, in fact, is to not take advantage of the resources of our bodies in an economic sense. Most of us don't sell because feeling whole and healthy is more important than temporary income. That's the way it should be with wilderness.

Wilderness does not have to be destroyed or used to be of benefit. Or visited. I have never visited a wilderness, but I want more of them. It is valuable as wilderness, because it is there. Aldo Leopold said that "to those devoid of imagination, a blank space on the map is a useless place, while to others, it is the most valuable part." There should be more blank spaces, that we can just leave alone. There should be wolves and bear and martins and lichen somewhere, without human company.

What is the human relationship to the earth or wild beings? Pandominant species, lord and master, good steward, or fully conscious and self-limiting beings? I prefer the last. What are humans without animals? Without plants and flowers? Little more than machines, I doubt. Variety is important. Even Aquinas spoke in favor of variety, saying that although an angel is better than a stone, it does not follow that two angels are better than one angel and one stone. Wilderness provides variety and depth. Set aside more wilderness now!

11 December 1984

An open letter to Senators Symms and McClure:
I received Senator Symms letter of 6 December 1984, in which he struggled to present his compromise on wilderness, by supporting an acreage below the lowest recommended by the preservationists, the people, or even timber interests. Claiming to represent foresters, biologists, and mainstream Idahoans, he characterized the preservationist community as "hard-line" and "unreasonable." Affirming his commitment to "recreational and economic" interests, he signed his letter "Yours for a free society," without understanding freedom, economics or wilderness.
In 1984, only two percent of the lower United States has been left in natural cover. As much as possible of that needs to be designated wilderness. In Orwell's book, *1984*, the rulers of the police state abolished wilderness, because it supported freedom of thought and action. Wilderness is not just a resource, a recreation area; it is a freedom. To diminish our wilderness is to diminish our freedom.

As Mark Sagoff recognized, wilderness is the central symbol of American nationality. Wilderness shaped the American character. We are obligated to preserve it, as a national value and cultural tradition. As we have the right to vote, we have the right to participate in our culture, a culture that depends on wilderness for its very life and uniqueness.

As an economic resource, wilderness is worth far more as wilderness than as logs or bike trails. Each acre of wilderness cleans air and water that would cost millions to do chemically. But wilderness is more. It is the source, not a resource, of creativity, wonder, and all living beings. It is a church, where the sacred may be sought.
The Idaho Forest Management Act was a travesty of the wishes of Idahoans. Despite the efforts of Governor Evans and the major conservation groups, Symms and McClure refused to compromise. As a voter, a partner in a sawmill, and an ecologist, I urge them to consider the interests of the land and the people and compromise on a larger acreage.

30 December 1984

Dear Senator McClure:
There are two kinds of people on earth; the ones that love the earth and those who fuck it. What is the difference? you ask.

Love is more than an expression of desire — it includes self-restraint as well as self-expression, reverence for life as well as self-respect, humane knowledge as well as self-interest, and caring as well as self-preservation.

Loving the earth means leaving parts of it untouched by human interests. It means shaping the human adventure to the cosmic adventure and its limitations. One can live simply and nobly, and love well.

Fucking the earth means, as you know, since you have been told how by your rich puppeteers, using it without regard to any ethical or ecological consideration.

Please learn this important difference and then act accordingly.

May 11, 1982

Dear Editor:

Eugene Farley Heap's execrable May 11th letter to the editor deserves a response. Heap's sarcastic reply to the concerns expressed by Jeff Dahl regarding animal experimentation and exploitation shows unconsciousness and ignorance. Medical and soft-science experimentation routinely involves the use of animals in situations for which there are alternatives, or where the value of the research or procedure is questionable.

Animal experimentation has, in the American way, become an industry. A nicely planned research project will be funded, even if its symmetry is all that it has going for it. Companies and independent entrepreneurs make their fortunes by supplying animals to researchers. One notorious sort, called a "buncher" in the trade, acquires dogs and cats from city pounds, and sells them to research groups.

Your family pet may wind up strapped to a table and mutilated, made sick with induced cancer or heart disease, or with its nerves severed — as in a recent research project which entailed the severing of the nerves in various places on each of 20 different monkeys to discover what effect this procedure had on the animals' ability to compensate. As was expected the researcher *found* that the animals practiced self-mutilization on those limbs without sensation. In addition, the researcher ignored the medical and dietary needs of the animals, withholding veterinary attention and feeding them moldy and unfit "food."

For Heap's information, medical and other research long ago graduated from using only rats in their laboratories. Some of these even Mr. Heap might feel a bit of sympathy for. A Boston surgeon performs head transplants on dogs. Other animals, including primates, are used in frivolous, repetitive, spurious research projects to gain trivial bits of information. Scores of eminent scientists, authors, and others have openly criticized the exploitation of animals in the laboratory. These and other proponents of animal rights — the right of an animal to its own destiny — argue that a society's ethics are observable in the way in which it regards all forms of life, not just human. This is an oversimplification of a profoundly sensitive view of life, but it places an emphasis on reason, responsibility, and respect.

For every thoughtful Dahl there are dozens of Heaps, no doubt. This country lags behind Great Britain, which has adopted a rather rigorous set of laws regarding animal experimentation. Great Britain also has been a leader in using computer simulations to replace actual replications of experiments in the laboratory.

Laboratory animal experimentation may not even be essential in medical research. Notwithstanding the latest estimates that 40% of the drugs on the market are useless, although they have been animal-tested, the one drug that turned medicine around in this century — penicillin — was never first tested on animals. Penicillin may be one of many substances that, while lethal to most animals, is of use to humankind. Heap's carelessness with his facts and the absence of sensibility regarding this issue illustrates the clumsy heavy-handedness, which so far characterizes our inability to understand and work with natural processes.

Certainly there are enough human subjects for smoking and alcohol tests, without capturing primates and forcing them to acquire our deadly habits. Certainly enough women and men have self-tested cosmetics for many thousands of years, without forcing rabbits to look made-up against their will. The Draize test, used by the FDA and the hundreds of corporation's which manufacture products for human use, from eyewash to floor wax to pesticides, cruelly subjects rabbits to a brief life of utmost discomfort. The rabbits so unfortunate as to end up in a Draize test are permanently harnessed so that they are entirely immobile. Caustic dyes, wax, mascara, detergents, and all the other accouterments of our daily lives are dripped over the rabbits' eyes repeatedly over a period of several days. The rabbits are unable to blink, to wash themselves, or to move while this process is going on. Ask yourself if you need 20 different shaving creams, 100 different liquid foundations, or 1000 kinds of eye creams. We cannot justify the short-sighted and infantile premise that anything we want ought to be gratified at whatever cost to the nonhuman environment. We take it for granted that it is our privilege to extol this rent from other creatures for sharing the earth; we have the mistaken notion that we are the landlords or simply lords.

Whether or not you want to argue whether all this is necessary for the sane and healthy survival of mankind, it is a fact that animals deserve our respect. There is nothing sacrosanct about human life when it is characterized by mindless activities and irresponsibility. Humankind is still very short on ethics, which it likes to claim for itself alone. Studies have shown that some nonhuman animals may in fact be more advanced than we are with respect to inter special respect. It is interesting to note that the animal with the most bad press, Heap's apparent favorite the rat, shares with humans many characteristics, including belligerence and an ostensible social order.

Hardly the right object for Heap's inane jocularity, animal exploitation by humans is symptomatic of a graceless and dangerous anthropocentrism in which all values have a human source. Writer John Fowles states, in a recent *Harper's Magazine*, that we are fools to think our relationship with nonhuman creatures has no bearing on our relationship with each other. It does. We should not do to animals anything that we would not want done to ourselves.

Alan Wittbecker and Carolyn Hagen

January 3,1977

The Editor *The Lewiston Tribune*
On page 15a of the January 3, 1977 edition of your newspaper, I noticed the item, "Some people wouldn't pay two cents for a coyote; George Fliger would," written by Sylvia Harrell. I believe this article is a piece of irresponsible filler for the following reasons.

The title as well as the substance of the article devalues the intrinsic value of the coyote as wildlife, assuming without justification that coyotes are "pests." This is a false assumption. Federal and numerous wildlife organization studies, including one by the Marsh Institute, where I work, have shown that the coyote is a valuable natural predator which helps maintain an ecological balance. Coyotes are extremely sensitive and perceptive; if you have ever watched one follow an airplane across the sky, as I have, you would believe this. Coyotes are caring and playful; if you have ever watched two parents play with pups, you would believe this. Coyotes prey on animals we do not want in our orchards or homes; if you have ever watch a coyote pounce on mice, one after another, you would believe this.

The threat of the coyote to livestock is a myth promulgated and greatly exaggerated, especially by the ranchers themselves, who eagerly seize upon the reputation of the coyote for their own sporting perversions and profit. The article written by Ms. Harrell does nothing to educate the reader to the natural stature of the coyote, but blindly supports the ignorant view that they are pests to be controlled and killed.

In addition, the article serves as an impetus for the wanton killing of coyotes for the sole reason that their pelts will be bought, depending apparently, on the number of bullet holes, for somewhere between 5 and 55 dollars. Companies in Spokane buy them for coats. The inference is that it is perfectly acceptable for humans to use nature in their usual thoughtless way for profit. Those who could wear the skin of a creature so killed and not be aware of the excruciating vanity of it must themselves be spiritually dead.

The publishing credo that 'the public shall be informed' should be taken by the *Tribune* for its educational and positive values rather than cheap sensationalism. Give people balanced stories, especially about wildlife, such as coyotes, that directly benefit human communities.

Talk at the Aprilci Community Forum, 11/21/2000 (English Translation)

I am not, like the rest of you, an expert. I am just an ecologist. Being an ecologist is quite the opposite of being an expert. I do not know any one thing very well. In fact, I have kind of a professional myopia that prevents me from getting mired in details, which always seem fuzzy, contradictory, and really confusing. Because of this I look for large patterns. I see many patterns in a system and many connections between systems — the systems I mean are ecosystems, in size from a stream bed to our city of Aprilci nestled in the mountains here or to the entire montane region.

So, I do not have any solutions to the problems presented here today.

I do not know the best way to rebuild the infrastructure of water delivery or how to deal with trash collection. I cannot match your expertise with plans for a water treatment plant or strategies to increase tourism from other countries.

But, I can look at some of the things that are happening and connect them with other things. I can suggest some relationships between animal care and health, between trash collection and tourism, between advertising and market sales. For instance, I notice that many of you are coughing on this fine sunny day. I suspect, well, I know because I was with some of you, that you stayed up drinking and smoking all hours, then went to the disco until dawn—neglecting your sleep—then went home to uninsulated houses heated with wood, and finally, to get here, you walked on roads traveled by all your domestic sheep and cattle (breathing in pulverized manure thrown up from cars driving over it). All of these things contribute to respiratory distress. Therefore, I would suggest a few changes in lifestyle, as well as new routes for domestic animals.

There are many other links between our actions and health, as well as between domestic animals and wildlife, clean water and technology. We can see these links by examining the history of the area, as well as our human attitudes. We can think about how different models of thought or behavior could result in different, and hopefully better, lives for our inhabitants and communities.

I was asked to describe the ten most pressing problems facing the city right now. Let me start with a few observations about what I have seen here in the past three months. First, however, let me say how happy we are to be living in a town as beautiful and prosperous as Aprilci—in a beautiful setting between two beautiful rivers. We are still observing and learning about Aprilci and Bulgaria—and I trust you will forgive my slow and faltering Bulgarian language. Dr. Elena Stefanova will help me with the more difficult phrases.

We recognize many of the same problems that we have in America, and not all of them have been solved there, either. Let me talk about some of these problems that we share and suggest some low-cost, labor-intensive solutions. I know we are all concerned with money, but much can be done with effort and ingenuity.

1. The worst problem is the indifference of people to wildlife and to their natural surroundings, which are more than just surroundings but a complete ecological support network that provides air water, food, and even the deepest of enjoyments.
   The solution here is to have ecological education in school—I'm sure the children would love to get out near the rivers and forests! This participating in the wild teaches us real values as much as anything.
2. The next is the disappearance of wildlife and wild places to live. This is a problem for the whole planet. I know that many of you do not miss wolves, but they are very important to keeping the deer populations adjusted and healthy. Pets and domestic animals also need to be

considered as competitors for wildlife. But, they also need care, even when it means sterilizing them to keep them from breeding freely.

The solutions to this are to save some habitat, that is, do not convert it to tree farms, farms, or roads. Zone the valley, using an ecological plan. Limit stray domestic animals. Then, limit the takes of all wild species, not just wolves but also deer.

3. The pollution of water. I know now that we have no treatment and that the water goes back directly into the river. I know because I traced the pipes from our building. This is far worse for those living downstream, such as those in the town of Zla Reka. If we cannot build a treatment plant, we could start with lagoons and biological drain fields using native plants that thrive on our wastes and clean the water.

4. The pollution of air. Some of this is from smoke from the fires that heat all the buildings, from cars and their exhausts, and from animal manure in the streets. We could start by having separate animal paths, which would take them off the streets. Yes, I know, it sounds funny, but most countries, such as Switzerland and America, have exactly such laws to protect the health of the people. Of course, burning carefully and tuning the car engines would reduce some pollution. For the rest, there have to be technical improvements on stoves and engines.

5. Trash in the streets, rivers, and in the forest. I know that there is no trash in your homes, or in your yards or even in the sidewalk or street in front of your homes, but every place else seems to be free for dumping.

I am not unaware of your long history under the Turks and Soviets, but people must keep their common areas free from trash. The simplest thing, besides education, would be to have trash cans everywhere. Then recycling is being done in larger cities now. There is no reason we could not start recycling here.

6. Trash and industrial waste. On my visits to the dump, I noticed large amounts of medical waste and pharmaceutical wrappings, as well as broken bricks and old clothing (rags actually).

Some of these materials could be reclaimed and recycled. But, if we could try to develop an industrial ecology between our industries then they may not be produced and reach the dumps and land in the first place.

7. Overcutting of trees. Bad forestry practices lead to stream and hill erosion. I have seen a lot of it in the forestry enterprises.

New forestry practices, such as leaving snags (dead standing trees) and downed woody debris (fallen tree trunks), would improve the health of the forests. Ecological planning and restoration would also help.

8. Nonefficient use of energy. I know that the infrastructure is old and needs to be updated. I know that is expensive. This includes trends in transportation, where cars are increasing and train travel is decreasing.

But, there are new technologies for some things; I am thinking of solar power for lights, which could be very efficient here in sunny Aprilci. Maybe for hot water also, but probably not for heating. The efficiency of the technology in place could also be improved. Trains are more efficient and Bulgaria has a good train network. Cars and fuel could be taxed to pay for new trains and buses.

9. General ugliness. Parts of Aprilci are not pleasing. The park is trashed, the streets are filthy, the river is a dump and open sewer. We have twenty hotels in this area, and at least eight are working. To get tourists to come here, the place has to *look* well-kept; we need to be proud of the beauty here, not drown it in our wastes.

   We can start by planting more trees, by cleaning up our common areas, such as the park and river. Then we need to make everything beautiful. I know that we know how — I see it in our homes, with their profusion of plants and home-made fabrics and rugs.
10. Taking care of ourselves. We are dying too young here, getting too sick too often. We are as special as wild animals and wild forests, so we need to treat ourselves with equal respect.

   This should not need education. We need to stay warm, eat right, get enough sleep, and smoke and drink less. I know we get enough exercise. I have seen you all walking and working.

The most important solution to any of these problems is an ecological education. That includes listening, not only to what people really need and want, but to the voices of all existence, from insects to cows. Television can help with that, by showing where things, like water, come from and where they go.

   Another thing that can be done is to have an ecologist on the staff of the city, and just let her monitor the actions of the city and its residents. An ecologist could measure the exact conditions of the water and air, forests and fields, animals and people. For instance, I have seen dead animals in the river. Who puts them there?

   Ecological planning would be a great beginning. We need a complete inventory for the whole large community. We need to think of land use and maps. We need to develop partnerships in a complete community ecology, where the wastes of one group become the resources of another.

   Another thing is to encourage cooperation between villages and cities. For instance the City of Troyan has a problem disposing of pharmaceutical wastes also. Perhaps, there could be a joint project. As you know, many things, such as ozone depletion and air pollution are international problems.

   Thank you all for your consideration in listening today. Thank you, Dr. Stefanova, for translating my incorrect or new words. Thank you, Madame Chairman, for allowing me to speak today. I will be living here for the next two years and would like to participate in the life of this community. Please invite me to discuss things.

We would like to thank you all for being here today. We would like to thank you for your hospitality and for help with our language and projects for the past two years.

You have thanked us, with this award, for working on behalf of the City of Aprilci. But, all we did was learn about Bulgaria and her people. You have thanked us for getting money for projects to help people and businesses. But, all we did was make friends and work on projects with them. You have thanked us for connecting Aprilci to the world. But, all we did most of the time was talk about what our lives were like in America and listen to what your lives have been like in Bulgaria, while consuming good rakia and wine.

We want to thank you for hosting us, for making us a part of your community, for forming friendships with us, and for helping us learn about life here (especially how to make and drink rakia and wine). We have learned a lot about Bulgaria, the nature, the wildlife, and the people.

In the years ahead, we will tell people in America about Bulgaria, what your lives are like, what your ambitions are, and what you want to communicate with them.

In the years ahead, you will tell people in Bulgaria about what our lives have been like, about what friends we have made, and about what projects we have all worked on.

And, thus, all the goals of the Peace Corps have been fulfilled: To learn about you, to let you know about us, and to work together to improve some things.

And, thus, the goals of Bulgaria were also addressed: To make things better for people, to encourage her unique cultural traditions, and to protect the unique and wild nature, while letting people understand it and participate in it.

Thank you for your hospitality. We want to continue our friendships and conversations. We want to return to visit you again as friends and to invite you to visit us. Thank you.

Alan Wittbecker and Marcella Crider

Chapter 5

Wolves in Bulgaria

When people first came to the Balkans, they learned to share the region and game with wolves. For thousands of years they lived together in relative harmony. There may have been as many as 6000 wolves in Bulgaria then (based on habitat requirements). But, as people increased, they took over habitats, then started hunting and poisoning wolves. By 1980, it was estimated that only 100 wolves remained in the remote areas of Bulgaria.

Then, after poison was banned, and people had their own difficulties with economic competition, wolves started increasing again. Game was plentiful and there were many open habitats. But, after a while, game started decreasing, for many reasons, including poaching. Wolves started finding easier prey, such as sheep and cattle, which often were not protected. Now, the wolves may be decreasing again. Many people suggest that we have too many wolves and we must kill more of them, while others say that we should protect them and let them increase. Perhaps we do not know enough to do anything, yet.

Wolves in Bulgaria are excellent hunters and prey on large hoofed animals, such as roe deer and red deer. Since wolves weigh far less than their prey, they must hunt in packs. Although a single wolf could kill a large deer, hunting in a pack is safer and more reliable. Except under rare conditions (such as heavy snow), wolves do not determine deer populations, whose numbers are limited by food supply. Instead wolves cull weak and diseased individuals that lag behind their herds. Wolves help strengthen herds by killing such animals. Wolves are also efficient scavengers of domestic animals that are sick or have died. Old or unhealthy animals can be a burden on a herd, for example, by eating browse that a healthy animal might need or by infecting young deer. By eliminating such animals, wolves perform an important natural function in wild ecosystems.

Their persecution could reduce the complete functioning of ecosystems in immeasurable ways. Wolves are keystone species and an indicator of the quality of wildlife habitat. Their actions are crucial for maintaining the long-term viability of ecosystems. Other predatory species are also supported by wolf kills. These species include ravens, foxes, wolverines, vultures, bear, and eagles. Wolves contribute to an ecological 'balance' and prevent overpopulation in deer and other grazers. Wolves are evidence of the diversity that we value so much, from genetic diversity to the full spectrum of an ecosystem.

Biologists monitor wolves using a combination of techniques. The primary methods are surveys for sign (tracks, scat, fur, and snow urinations), incidental observations, aerial tracking by airplane, and radio-telemetry. Because of the expense of the last two methods, wolves in Bulgaria have been studied primarily using the first two methods. Because of the unknown size and unknown distribution of the wolf population in the country, we needs to supplement these techniques with newer techniques, such as scent stations, radio-tracking, and mathematical modeling.

Wolf research is used to determine habitat preferences, prey selection, movement corridors, mortality causes, and the location of critical areas such as den sites and trails. Unfortunately, wolves, like many carnivores, are secretive and (occur at low density), especially in Bulgaria, where they have learned to hunt at night. Accurate estimates of abundance are difficult to obtain. Yet, this information can have an important bearing on the management of protected areas, and may influence selection of locations and levels of human use of trails, facility location, and seasonal closures in critical areas, such as den and rendezvous sites.

Although some studies of wolves have been done in Bulgaria, and monitoring has been proposed for some parks, it has not been enough to determine or monitor current populations of wolves. This work needs to be done to pursue a rational wildlife policy and to create species management plans.

So, how many wolves are in Bulgaria? Geko Spiridonov and Nikolai Spassov estimate the springtime wolf population (before young are born) in 1998 at 600-700 individuals ("Large Mammals of Bulgaria"). A year later, Nikolai Spassov et al. calculate the number in Bulgaria at 1000 individuals. The National Forestry Board estimates the wolf numbers for the year 2000 at 1796 and for 2001 as 2160, based on a census in forestry units. The Ministry of Agriculture and Forestry states that there are also 2160 wolves (Snezhana Paskaleva, Ministry Press Center, August 2001). Alexander Dutsov of the Balkani Wildlife Society suggests that there are less than 1300. Working in and outside the National Parks, my own estimate (admittedly crude and unfinished) is 750 wolves in 2001.

Obviously, the number is not well-known. Taking a mean number of 1200, and accepting forestry and hunting incidental sightings of pups and killed wolves, and assuming normal reproduction and change rates, it is possible to make a simple model of the wolf populations in Bulgaria. Observations indicate that there are only about 1-2 pups seen with adults. Hunting statistics state that about 300 wolves per year have been killed for the last four years. Using simple mathematical calculations based on these observations, and on birth and death rates (and immigration and emigration rates), we can project that the wolf population, which has just begun to return from a minimum less than 30 years ago, could be extirpated from Bulgaria in less than three years. This is a grim forecast. But, it is not fate. If hunting and poaching were banned, the population would rise again.

Public attitudes towards wolves, especially in rural areas, are still very negative. Many people fear wolves, because of their reputation in myth and folklore, and also because they believe wolves attack human beings (the howl can be frightening). But, wolves avoid people as much as possible. Many people hate wolves because they kill domestic animals, such as sheep and cows, but wolves prefer to eat deer. Wolves are blamed for many livestock losses, which may result from poor health, accidents, or feral dogs. In fact, poaching causes more losses of game animals than wolves. Many hunters dislike wolves because they kill game animals, such as deer, but wolves usually only take sick or young individuals and not healthy trophy animals.

Many attitudes and problems can be overcome by research and education. Research needs to be performed not only to understand the status and ecology of wolves in the ecosystems, but also to understand the extent of the threats to the species. Social and political measures cannot be taken until there is a sufficient knowledge base for the decisions. Neither wolf policy nor wolf management can be addressed until wolf ecology, numbers, (damages,) are known definitely. Studying wolves will do much to preserve sufficient habitat for wolves and benefit wildlife conservation.

Education can demonstrate that wildlife is important to people and communities for four main reasons: (1) aesthetic value (beauty), (2) economic value (especially tourist income), (3) scientific value (knowledge), and (4) survival value (the health and diversity of ecosystems). If people cannot learn the need for wildlife conservation, the threatened and endangered species will become extinct. Then, many others will die out. If this happens, human beings will lose much of great value that cannot be replaced.

Although education and research about wolf ecology is essential, wolf policies may not be successful unless the human population is healthy and stable, and is related to the carrying capacity of the land, which would include also the populations of wild animals. Then, wolves and people might be able to live in harmony again.

Figure 2. Bulgarian wolf (*Canis lupus lupus*)
Photograph by Balkani Wildlife Society

Tragedy & Ecological Design

Chapter 6

Two Utopias

Utopia and War. Nuclear war is unthinkable. Limited disarmament seems to be unworkable. Human and environmental degradation are unconscionable. Utopias, as plans for better societies, seem to be unimplementable. Is anything workable or thinkable without being hopeless? The problems of war, aggression, nationalism, disarmament, degradation, goodness and peace are problems of human nature, that is, of symbol, culture, politics, and ecology. They are not simple problems and not easily solved. They are interrelated as human groups and natural ecosystems are interdependent ecologically, politically, economically, and technologically. Human interactions are dominated by symbols. The most powerful set of symbols, embodied as nationalism, has a direct relationship to war. Large-scale war has the potential of creating utopias on earth, that is, no places, with nothing and nowhere. But, it is possible that an ecological politics can create eutopias, that is, good places that are in harmony.

Symbol and Culture. For most of human history, the habitable earth has formed a mosaic of separate territories and peoples. Human ecology was a part of local ecology, which was part of planetary cycles and systems. Human life, depended, as it still does, on the productivity of natural systems. The most common form of social organizations, at subsistence and pastoral kinds of existence, was the tribe. Different groups in unique areas developed habitual ways of dealing with the characteristics and limitations of their surroundings. Each way can be referred to as a culture, a pattern of behavior based on shared images and behaviors adapted to the local environment. Lewis Mumford stated that culture turns into symbolic language. The particular symbolic idiom that is concerned with cultural institutions as manipulative objects was defined by Morton Wheeler as political theory. Politics deals with words, and words are arbitrary symbols for things or events.

Misguided politics can arise from the wrong relationships of words, events and symbols. For example, things are sometimes regarded as symbols for words in totalitarian states. This reduces individuals to stereotypes, that can be tortured or disposed of, without personal involvement. Such semantic traps confine and warp thought, as well as increase suffering. People can become prisoners of a cultural order that rejects new knowledge and solutions.

The desire for order without a rational search for the nature of order can have disastrous political consequence. The Aztecs, for instance, used the wrong symbols to interpret the universe; they believed that the sun needed human blood to survive, so they sacrificed great numbers of lives to ensure the sun's life. Their political policies were based on raids for victims, and this policy contributed to their overthrow and decline with the arrival of the Spanish. Some modern countries have equated nuclear weapons with defensive strength. The use of wrong symbols, from bloodthirsty gods to nuclear weapons, can destroy cultures and natural habitats.

Conflict and Nations. Many human societies advanced by fighting and expanding. Fighting is a common form of human behavior. It occurs in children, at least as a ritual limited by pain, and in adults as a ritual limited by symbols. The growth of the brain, and its capacity for abstract thought, seems to have bypassed the ritualization of social conflicts common to other mammals. Human fighting is not as formalized; this is what permits humans to objectify and slaughter one another. Many human groups follow two standards of morality, one for insiders and one for outsiders. Most aggression was directed outwards; the losers were often exterminated.

Many tribal groups confederated into larger groups or nations. A nation by definition has a single central government representing people who occupy contingent lands and are conscious of a common identity, as Aztec, Iroquois, Chinese or Babylonian. The professional ruling class is free of kinship bonds. The structure is stratified but internally diversified.

Almost the entire land surface of the globe is divided into centrally governed nations. Human affairs are managed within the framework of these autonomous units. Decisions are made on narrow political and economic grounds, rather than on environmentally sound principles or on international concerns. Isolated problems stir only mild attempts at reform. But, few problems are isolated. Modern government is the assumption of responsibility for problems without adequate knowledge. Modern citizenship is the abandonment of responsibility, on the assumption that others know how to manage problems. The frightening aspect of government and citizenship is the total failure of world leaders to learn or understand the simplest facts of science or technology.

Lord Acton observed that nationalism aims solely at making a nation, the abstract idea of a political state, and not at liberty or prosperity for the people. He predicted that the result would be moral and material ruin eventually. Nations are basically exploitative, within their spheres of influence: The United States concentrates on Latin America, Europe on Africa, Russia over eastern Europe and western Asia, China over Tibet and eastern Asia, and Japan over the Pacific. The reasons for this domination are: The rapaciousness of society; the acceptance of war; and the economic advantages of large-scale operations. The cultures of industrial nations are based on the unethical accumulations of materials and energy. Inequality is maintained by symbols and power, not by persuasion.

Nations and War. Power politics, in the context of nationalism, creates problems that cannot be solved, except by war. Questions about recognizing the best nation or the truest religion lead to organized slaughter as the answer. "War is not merely a political act," said Clausewitz, " but also a political instrument, a continuation of political relationships ..." As long as human institutions were large and brittle, war was an effective way of disassembling them. This form of renewal, however, was very expensive. Karl Marx may have been right in thinking that war was necessary under certain conditions, as a last resort. But, those conditions no longer exist. His observation may have been true from the 1830s to the 1940s, but it has been rendered false by modern weaponry. Nuclear war has the capacity to destroy the parts as well as the connections, whole cultures and ecosystems, as well as senile political structures.

The economic rationalization for war also seems to be crucial for nations. Armament piling has become a vital part of American, Chinese, Russian, Iranian, and Korean economies, among many others. The recovery from the American depressions of the 1930s was not complete until the rearmament surge to combat Axis powers. The Korean and Vietnamese wars also spurred the American economy. Thus, American prosperity has its basis in the preparations for death. The fear of other countries ensures government expenditures of billions of dollars for deterrence or protection. The vested interests in this system are almost insurmountable. The concrete companies and bomb –makers put up their own puppet politicians to guarantee their part of the spoils.

The cost of war, alas, excludes the social distribution of wealth, as the American and European poor can attest. The situation is much worse in disadvantaged countries. The war industry, as measured by expenditures, is on the order of $200 billion a year (1978). Kenneth Boulding says that it is an unrecognized paradox that the cost of maintaining the war industry is greater than any possible damage that could be inflicted by an enemy. Furthermore, war intensifies the depletion of resources; therefore, it is counterproductive to fight to steal another nation's resources. For centuries, warfare has resulted in incredible wastes of resources. The latest multinational conflict, which began in 1914, and has been hot and cold during this time, has been the most wasteful.

Those who believe in the theology of nationalism are committed to fight with one another. The war ethos has been expanded and reduced to absurdity. War has become so big that there can be no victories or victors, and possibly no survivors. The only remaining purpose for war is the complete destruction of combatants, defined as nations, and most of the natural systems on the planet. Sadly, the only people who do not know this — or admit it — are those people in decision-making positions, who are compelled to prepare for what they subconsciously must know would be a terrible catastrophe. Their power has made them prisoners of their symbols of nationalism, afraid to be caught in any criticism or sign of weakness. Yet, they direct the money, skill, and knowledge of their citizens into projects leading to destruction, misery, servitude, and perhaps meaningless death — not life, liberty, and the pursuit of happiness.

The intellectual rationalization for the continuous preparation for war is the old Roman adage: "If you want peace, prepare for war." This adage has been so completely taken into the modern heart that most of the larger nations have spent half of every century in war, according to Pitirim Sorokin. Preparation for war has always lead easily into war. There seems to be no reason that the present preparation will lead anywhere else.

Perhaps it is the fate of humanity to die a glorious, radiant death, taking most other living beings with them. Who could resist the glory of complete annihilation? Perhaps that is our fate. But, what is fate? The hand of God, or the idea of Tolstoy: That historical events are determined by the summation of innumerable small decisions by the anonymous masses that add up to a tendency? In particular terms, the decisions to buy automobiles, poison coyotes, plant trees, or live simply, add up to fate. And, fate concentrates power in corporate, military, and political hands, which still belong to human beings, who cannot profit or find consolation in death or total ruin.

Disarmament. These leaders are pushing towards catastrophe. But, people cannot accept the possibility of complete destruction. Arms reductions are only tentative motions in the right direction, under the old rules. A transformation to peace must be complete; it cannot be partial. Someone has to heal the schizophrenic breach between human welfare and total destruction. It is the responsibility of citizens to tell their leaders to stop making and selling weapons, to tell manufacturers to stop making weapons and toys of grotesque weapons.

Furthermore, complete disarmament could be accomplished in a week. This is not a new idea. Earl Osborn, founder of the Institute for World Order, proposed the concept of sudden disarmament in response to the tedious phase-out envisioned by most plans. An agreement to disarm would not involve much negotiation. Taking the first step would add to the prestige of the first nation bold enough to agree. The United Nations could post a police force to disable all military ordinance (even hammers and welders are effective). In fact, "rapid disarmament would not be difficult. A thousand planes each carrying one hundred trained inspectors ... could distribute ... these men at all major centers in Russia and the United States within twenty-four hours," according to Osborn. This Police Force could be staggered by nationality to avoid cheating. And, the same force could be used to settle international disputes. There is precedent in using unarmed peace-keepers to mediate between hostile groups in Kashmir, Cyprus, and elsewhere.

How could disarmament work? Humans are one of the most pacific of large animal species; they have a tremendous capacity for kindness and decency. Violence results from fear and ignorance. Violence is learned and so can be reeducated. Morals can change according to conditions, and not be defined for all time. The atavistic obsession with violence as a solution to problems is inimical to an ecological outlook. The image of society can change from competition and aggression to cooperation and mutual respect.

Ecological Politics. The United Nations is a natural choice for a global agency with the responsibility to monitor human nations and supporting ecosystems. It is the only body with the machinery for constructing a world system. The beginnings of an ecological politics can be found in the special services of the UN — UNESCO, FAO, WHO. As long as political problems are addressed in a frame of nationalism and military power, these organizations are treated as peripheral. In an ecological frame, they become central. The politics of health and fulfillment need to replace those of power and misery. The United States, China, Russia, and others need to relinquish world leadership; resign from the security council; cease their propaganda; renounce foreign policy objectives; call back their soldiers; and sell no more weapons.

With an ecological consciousness, the weak and disadvantaged could get food and shelter by work or appeal, without plundering, without war. Once, war was judged to breed strength and nobility, which may have been suitable for hunting societies, but not for urban civilization, where it results in ruinous competition. War would be allowed to die with the protection of all people by a global governing body. Fighting would still occur, between individuals and small groups as interests conflicted and communications faltered. Nonviolence is possible only between rational individuals. Sometimes force is necessary against irrational objectives. Therefore, armed security would be necessary. A shift from armies to police forces could provide that security. This police force would be used for humanitarian intervention, assisting local jurisdiction to establish the dignity of all peoples.

Nations are only organized — most often by force — communities confined to a geographical area, and possessing the means to wage war — these have been the criteria for nationhood by the old UN. Many independent cultures in established countries, like California in America, Scotland in Britain, or the Nyiha in Tanzania, are organized communities, but they are not permitted to join the UN at present because they do not possess armies. A new global order would permit autonomous cultures to join the UN according to cultural or linguistic affinities and not merely by force of arms. Many nations could break up into preferred natural groups.

The UN could also work to equalize the wealth of the planet, taxing luxuries and offering educational and technological benefits to all. Technology provides wonderful opportunities for communication. Radio and television reveal the concern for peace virtually anywhere. But, they also aggravate the images of inequality. It is unrealistic to expect cooperation without some fair redistribution of resources and manufactures. Equalization would allow trust. Trust would allow customs and prejudice barriers to fade away. In designing a global system, everyone can participate. A social debate can happen in many places at many levels. Resolving conflict through social exchanges would free unprecedented resources for satisfying basic needs. That which has hitherto been left unsaid — the goals of humanity — could become explicit.

With the removal of war capabilities, the equalization of opportunity and wealth, and the respect for cultures, the remaining issues are less likely to incite violent passions. Disagreements over the best way to raise wheat or restore a forest may be more easily resolved than recognizing the best nation or finest religion. Hopefully, the death of large-scale dogmatic ideology and national idolatry could lead to the end of organized slaughter.

If America is to be a world leader, let her lead in tolerance or trust. Let her people lead in voluntary simplicity and not consumption. Let them be the first to give their allegiance to a global organizing body, the UN, and the first to divest themselves of nuclear weapons. If they fear for their safety, let them remember the success of nonviolence in India or the success of guerilla actions in Vietnam.

Two Utopias. This alternative will certainly be regarded as utopian, but, then the arms race is incredible utopian, with its ominous policy of complete destruction. After a nuclear war, there may be no place left on earth, and that is the original meaning of utopia, "no place." The word utopia was originally meant to be a word play on the Greek words "outopia" and "eutopia." Rather than risk creating no places, we could strive to create good places, "eutopias," peacefully and in good spirit.

Chapter 7

The Tragic Species
(Or Why Humans will not Change — Even to
Save Themselves or Their Wilderness Home)

Human beings have been very successful at monopolizing nature for
themselves and for their own purposes. This very success, however, may
lead eventually to failure, as the tactics that worked so well in an uncrowded
world rich in resources are applied to a humanized world with immediate
ecological limits. As the Greeks recognized, success can lead to failure, that
is, to tragedy.

Theatrical Tragedy

In defining tragedy in terms of its formal characteristics and emotional effect,
Aristotle characterized it as "the imitation of an action that is serious and
also, as having magnitude, complete in itself; in language with pleasurable
accessories [i.e., rhythm and harmony], each kind brought in separately
in the parts of the work [i.e., some in verse, others in song], in a dramatic,
not a narrative form; with incidents arousing pity and fear, wherewith to
accomplish its catharsis of such emotions."[1] For Aristotle, tragedy was the
mimesis of a good or noble action.

In a theatrical play, the tragic hero triumphs at first, and incorporates the
successful behavior that lead to the triumph. This behavior, employed in new
circumstances, then leads to disaster. Hence, the failure to give up a chosen
role or pattern of behavior leads to great loss later.

Disaster results from breaking of the principle of temperance
(*sophrosyne*), or "nothing in excess." Reversal (*peripeteia* in Greek), the change
of a situation to its opposite, is an important plot mechanism. When the
principle of temperance fails, the actor sees "the transformation of his action
into its opposite." This antinomy illustrates the Heraklitean notion of the
return swing of the pendulum (*enantiodromia*); human ideals seem to swing
back and forth in dynamic opposition. In Greek tragedy, a single element
of value grows cancerously and destroys the whole. The Greek tragedy
showed the process by which absolute values swing through reversal to their
opposites. But the anomalies in tragedy called for a creative response — a
higher level of moral awareness born from the dialectical union of opposites,
especially for the wise observer.

Virtue is steering between the intellect and impulse. Folly is a
misjudgment in combining values, missing the just proportions. The tragic
error in Greek theater is a mistake of being in an evil predicament. Oedipus
did not sin — he was caught in contradictory circumstances.

Two fundamental notions of tragedy are hubris and nemesis: Hubris
is an arrogance that arises from a form of blindness, and nemesis is the
eventual consequence. The hero builds up an interpretation of events that
conforms to her ideal. Hubris leads to a decision made on the basis of the
hero's mistaken interpretation of circumstances; the decision leads to ruin.

The quality of irony brings with recognition of the situation a decree of necessity. The hero recognizes herself and her ideas in terms of laws. This is purgation. The Greek tragic hero was a typical man or woman isolated and projected onto a larger background of fate. Theatrical tragedy stresses the price of consciousness.

Joseph Meeker judges that the tragic view of life, as embodied in Greek tragedy, is based on a deep conviction that humanity has no part in nature, that human behavior must conform to moral laws that are extranatural.[2] This view also assumes human superiority over nature. Obviously, an assumption of human superiority would cause conflict, as would separation.

Humanity can still be a part of nature and still be tragic, however. Even if moral laws are natural. Tragedy is more than an ethical conflict. Tragedy can imply conflicts larger than the individual or even society. Becker claims that the tragedy of evolution is that evolution produced a limited animal with unlimited horizons. For him, tragedy is wanting an earth that is perfect, a heaven abstracted from imperfection. Of course, the swing of the pendulum ensures that humans cannot create a perfect place. In the sense of employing a successful strategy in all circumstances, perhaps natural selection in evolution is tragic (the tragedy of reality) — M. W. Fox and Garrett Hardin hold this view. Modern science is one-sided (or single-visioned), in denying poetic knowledge. The attempt to avoid the pendulum (enantiodromia), without understanding the operation of nature, ends in tragedy.

The definitions of tragedy can be linked to cosmology, the image of the place of humanity in the universe. Tragedy challenges the external order of things (the cosmos), even if the cosmos is only the size of a city (polis). The fatal flaw of the individual is the fatal flaw of the world-view of the individual. This is the root of Hardin's[3] Tragedy of the Commons — people are locked into a system of self-interest through economic gain without being bound by traditional rules for sharing or cooperating. Hardin's own definition of tragedy is "the working of fate." But, this kind of tragedy results from a failure of cosmology; humans are responsible ultimately, not fate or chance. Humans are tragic because they are responsible for their actions. They can choose a tragedy of the commons or of Leviathan — or they can expand or alter their cosmology.

Comedy

In *The Birth of Tragedy*, Nietzsche suggests that tragedy is dead. Joseph Meeker suggests that the moment after tragedy is a comic moment, where the actor steps out of tragic action and observes herself and her former universe, and laughs. The actor is then free of evil, restraint, body, and death. Laughter involves detachment, which is a fundamental form of freedom; freedom, according to Meeker, is the central value of comedy.

The comic actor tries to create a new and better universe, Meeker argues. The attempt is a celebration of human freedom. Farce, in opposing aging and death, affirms life through comic recognition of the impossible. Comedy strives for freedom. The farcical hero demands the universe to be changed; the comic hero demands that society accommodate itself to his will;

and the absurd hero sees the universe as a hoax and makes no demands at all. Comic heroes lack self-awareness—if they had it, the collision of will and act would be tragic. Comedy, however does not trivialize what is important, although it deflates what is overinflated.

The comic mode of behavior, as defined by Meeker, is a genuine affirmation of the instinctive patterns necessary for biological survival. Comedy is concerned with muddling through, not progress or perfection. Meeker argues that evolution shows all the flexibility of comic drama. Evolution is a shameful, opportunist comedy whose object is the proliferation of life without regard to morals. The participants must adapt, diversify, and accommodate. He contends that events in tragic literature could not occur if comic principles are observed. Comedy encourages necessity and acceptance; tragedy avoids necessity to try to accomplish the impossible. Comedy assumes that all choice is likely to be in error. To comedy, as in evolution, nothing is sacred but life itself.

Western images are tragic because death does not imply rebirth or living continuity. By comparison, Indian images of life are comic because death does imply rebirth, as rebirth implies further death. This comedy depends on surprise—death is really an illusion. Immortality is the shift of identity from ego to universe, the universal rhythm that exists in all beings of all sizes. This is the Indian rhythm of the juggler, balancing worlds; and if one seems to fall and shatter, hundreds of new jugglers are born.

Cultural heroes, by contrast, represent cultural ideals. Cultural heroes are often mythical beings, of human birth, who restore the balance to the people. Unlike theatrical heroes, who precipitate a tragedy by not changing, cultural heroes change to avoid tragedy (and they can do so without being comic). For example, many Japanese heroes change their roles as the play/action develops. The hero becomes multifaceted— ferocious, then poetic, after success followed by loss. Many Japanese legendary heroes, like Yamako Takeru, have a time of success followed by failure from the application of the strategy that made them successful in the first place, quite similar to Greek tragic characters. This behavior is called the Nobility of Failure, *mono no arvare*, the pathos of things.

Summary
Despite the arguments of Hardin and Meeker, evolution and ecology cannot be compared to comedy or tragedy, which are descriptions of the fitting of human images and behaviors to a changing environment. In the end, both comedy and tragedy encourage the acceptance of human limitation. Meeker emphasizes that comedy is immoral, like evolution, but he is wrong. Comedies of manners and types have a lesson to teach: Moderation and control are affirmed; rascality and immorality are denied. European High Comedies show that the world is good and that evil is introduced by those who lack control or moderation.

Tragedy proceeds by analogy and homogeneous substitution in the rationalization of the hero. Events are controlled to be consistent with an idea; the direction of expansion is integration; it ends in cumulative catastrophe and purgation. Comedy proceeds by wide variation and heterogeneous substitution. Every action discovers inconsistency. Expansion is a phase of discrimination. Ideas have continual purgation. Comedy plays with the ideas that tragedy discovers. But, the play of ideas is hedged with the mystery of tragic issues. Reason moves between tragic pain and comic disillusion. Comedy offers a rebirth. Tragedy also offers a way out, but it is through death. Tragedy confronts evil; comedy avoids it. In comedy, frames of reference collide and shatter harmlessly; in tragedy, the frames conflict.

Furthermore, the comic mode is an incomplete strategy for living. If the comic attitude is necessary for survival, then it must be moral (at least in the etymological sense of having rules for living together). The comic mode, as described by Meeker, cannot be reconciled with natural patterns of reproduction and death. Life is no more sacred than rock or air. What Meeker has tried to describe is not comedy, which is an analog of complete freedom, which cannot ever occur, but compliance, that is, the acceptance of partial freedom and partial conflict. The reduction of behavior to a comic mode would be tragic, in the Socratic sense, because it would refuse just proportion; it would be single-sided.

A complete strategy must include ecosystems and other species, those systems and beings on which our lives depend. We have already made a start in considering them, as when we suggest that each species has a characteristic face. And, each face is capable of a variety of expressions: Fear, anger, humor, indifference. Thus, a bear may appear curious, a wolf noble, a chimpanzee sad. These stereotypic expressions become as masks (and with which we risk fooling ourselves).

Human faces seem more dramatic to us, as evidenced in the Greek standards for tragedy and comedy — the two masks often displayed outside theaters. The mask is seen by an audience as tragic or comic. The mask offers a surprise. The surprise catches the hero or the audience, and the audience cries or laughs. The play is tragic or comic only from perspective — the difference between comedy and tragedy is in the point of view, not the subject matter. The mask of tragedy and comedy reflects this duality: The refusal or acceptance of the inevitable, detached indifference (comedy) or compassionate concern (tragedy). Both faces are combined in one mask.

From a cosmological perspective, there is no tragedy or comedy: Everything simply is. Thus, the human situation is tragic or comic to humans only; other species, ecosystems, geological cycles will continue or not. Responsible human behavior may avoid human tragedy, however. As de Montaigne wrote: "Our great and glorious masterpiece is to live appropriately."[4]

CHAPTER 8

Reverence for Life

Most philosophies are not adequate to deal with nature and ecological relationships. Many religions are too narrow to consider nature as more than a dominion. Albert Schweitzer examined many religions and assimilated some of their thought. Schweitzer, a musician, musical historian, theologian, and philosopher, chose to live in Africa, putting his Christian ethic into practice as a medical doctor. Under those circumstances, he formulated a new principle, the reverence for life. Ethical thought had been developing since prehuman history, he said, and it culminated in the principle of reverence for life.

Schweitzer began by examining Brahminism in India. Within its very character, Brahminism is world-denying. While its adherents praised *ahimsa* as the highest virtue, Schweitzer held that their compassion was not natural. It was only a derivative of egotistical metaphysical theories, causing adherents only to refrain from doing evil and not act toward the good. The reaction of the Bhagavad-Gita, in the Mahabharata, was to reconcile that severe nonactivity with human activity. But Schweitzer noted that since the Gita accepted the same world view, it also failed to meet the requirements of ethics, not doing more than providing a "phantom place for activist ethics within the philosophy of world-negation."[1] Buddha was also swept aside by Schweitzer, for setting limits to compassionate activity.

The Chinese sages, and Zarathustra, formed philosophies that were inclined toward affirmation of the world. Lao Tse and Confucius concluded that the fundamental virtue and common aim of all ethical conduct was good will toward men. Jesus, although misled into promulgating an attitude of renunciation of the world, nevertheless, in his ethical code, permitted unlimited action on the behalf of good. Humans were free to seek to achieve all that they regarded as requisite. Love became the supreme commandment, including all others in itself.

Unfortunately, people in the Middle Ages were concerned with the perfection of the world through renunciation. During the Renaissance, Christianity became affected by the spirit of affirmation, especially with the discovery and popularization of the teachings of the Stoic and Epicurean schools: Love of man was the virtue above all.

In the philosophies of the 18th century, where reason supported the commandment of love, a natural ethics was constructed. Bentham argued for the utility of love of neighbor; altruism was presented as a function of social welfare. Kant found a moral law immanent in human beings, so that good and evil were discernible by conscience. In Hume's analysis, ethical conduct flowed from the fount of natural sympathy — nature endowed man with the ability to share experientially the joys and sorrows of others.

While admitting the profundity of natural ethics, Schweitzer exposed its quandary: How are its responsibilities determined? How will concern for our own well-being be properly related to concern for the well-being of others? He felt that the questions were too imposing, that the answers were too individual and subjective for the clear formulation of commandments and prohibitions. But, he neglected that the function of science is to make common answers, and the function of rights is to ensure all can participate. Apparently, what led these thinkers into such a dilemma was the attempt to fit ethics into their world view according to the revealed nature of the universal will to life and to describe it in terms of human judging. Schweitzer considered this an error. Any thoughtful person, reflecting on the quality of altruism, could not help but enlarge the scope of ethical activity until it included all nonhuman life. "We perceive that ethics deals not only with people, but also with creatures."

Schweitzer noted that during the evolution of humanity, the circle of responsibilities gradually widened, beginning with family, then tribe, nation, and humanity — working toward all of life. Similarly, the circle of knowledge widened, increasing the understanding of the laws of phenomena. He felt that the streams were divergent, that ethics could gain nothing from understanding the universe, that there was no hope of finding meaning in natural phenomena: "A philosophy that proceeds from truth has to confess that no spirit of loving-kindness is at work in the phenomenal world." For him, nature had no reverence for life. It produced life a thousand-fold in the most meaningful way and then destroyed it a thousand-fold in the most meaningless way. Spiders sucked the blood of their victims; wasps laid eggs in live caterpillars; wolves ran down young caribou. He said: "Nature is horrible . . . cruelty is so senseless . . ."

Creatures had the will to live, but no compassion, they suffered, but had no compassion. The most precious form could be sacrificed to the lowest. In the struggle for survival, nature was maintained only by being in contradiction with itself. This was a contradiction of the will to live, for Schweitzer; there was life against life, suffering and death. Nature was a dreary spectacle of the manifestations of the wills to live in opposition to each other; each preserved itself by fighting and destroying the other. Nature taught only cruel egotism, briefly interrupted by the urge to love one's offspring. A terrible ignorance lay over each creature, as in a dark valley whose floors were perpetually covered by the fog of ignorance and egotism. But man, using his intelligence, was able to climb some of the peaks and catch glimpses of light; truth and goodness appeared.

It was obvious to Schweitzer that the deduction of principles of conduct from the world led, at best to a naive optimism, and at worst to skepticism and pessimism. Pessimism was intolerable, and optimism was incomplete, being concerned only with human relationships. When humanity extends its concern to relationships with all life, when intelligence operates on the will to live within an affirmative philosophy of life and the universe, then the "Reverence for Life" arises. We already possess understanding of the conduct our own natures require. It is our duty to share and maintain life. It does not matter that we have imperfect knowledge of insoluble contradictions, there is an elemental fact present in every consciousness that guides the spirit in harmonious philosophy: "I am life that wills to live, in the midst of life that wills to live." Sympathetic concern toward all the wills to live is the basis of ethics, for Schweitzer. Reverence for life is the greatest commandment in its most elementary form. The negative statement of this occurs in the Bible, Exodus 20:13: "Thou shalt not kill." (Recent translations say "Thou shall not murder.") Other parts of the Bible, Joshua, for instance, display killing as an effective way of conversion.

Schweitzer stated that we must struggle to wipe out antihuman traditions and inhuman emotions, that we must struggle against our own insensitivity. It is inevitable that we kill some things, unknowingly or to survive, but we must never come to take killing lightly — plucking flowers and squashing ants indiscriminately, we must not become thoughtless and blind, because all this killing weighs against us; "everything takes its revenge." The very saving of lives often calls for the sacrifice of others, but even this action is sometimes arbitrary, since we impose our own values on situations. True reverence for life makes no distinction between higher and lower forms. If we were to act so, distinguishing between pests and pets, we must do so in the sorrow of the recognition that we are killing.

Schweitzer warned not to "let your hearts grow numb." Compassionate awareness would maintain the soul on the way to real goodness. One should do good in gratitude for all the benefits received, in order to balance the "books inside." There is a secret to gratitude, which being more than a virtue, is a mysterious law of existence: its kindness spreads like the roots of a plant, but "whenever we penetrate the heart of things, we always find a mystery. Life and all with it is unfathomable." Our knowledge of life is the admission of mystery. To act justly, then, we must obey the laws that follow from recognition of the mysterious.

But why, after we have discovered the good, are we still at odds? Why is there cleavage instead of harmony? How can a force like God rationally create life and irrationally destroy it at the same time? How is the God of nature to be reconciled with the God of love? In all respects the universe remains mysterious to humans. Schweitzer believed that even if we despaired of comprehending the phenomenal world or the plans of God, we would not need to confront the problem of life with utter perplexity, because the ethics of Jesus, reinforced by reason, lead to the reverence for life, whose edict is the rule of universal love. This same reason would find the bridge between love for God, love for man and love for all creatures, and express reverence for all being, however dissimilar to our own, reverence and compassion for all that is called life. Such a foundation for morality forces the realization that when we establish gradations of values between lives, we only judge them in relation to ourselves and that is wholly subjective. How are we to know the importance of each? The principle of reverence for life rejects relativism — "it recognizes as good only the preserving and benefiting of life: any injury to, and destruction of, life, unless it is imposed on us by fate, is regarded as evil."

Most men are educated with a set of superficial principles, which evaporate when tested; most are brutal, ignorant and heartless without being aware of it. Previously, there was no absolute scale of value, because there was no reverence for life. Schweitzer urged a new renaissance to liberate us from "the poverty-stricken pragmatism" with which we limp along. We need such a spiritual renewal that everyone will reflect on the nature of goodness; our own thought must lead us from naive optimism toward a profounder affirmation of life, helping us progress from ethical impulses toward a rational system of ethics. "In the universe, the will to live is in conflict with itself; in us, it seeks to be at peace with itself." This is an act of spiritual independence on our part, but it carries an element of responsibility to which we must submit: acting toward the good. "The essence of goodness is: Preserve life, promote life, help life to achieve its highest destiny. The essence of evil is: Destroy life, harm life, hamper the development of life."

Individuals must transform themselves from blind men into seeing ones by following the new commandment: Revere life. The quality of personal existence depends more on it than on laws and prophets; it comprises the whole ethic of love in the deepest sense; it is the source of constant renewal for humanity. Those who do not help others by profession must help them as an avocation, seeking out others in need and helping to alleviate their suffering, and in so doing paying off one's debt for happiness already received. "The secret hour does not require of us that we should be happy — to obey the call is the only thing that satisfies deeply."

Humans are to prove themselves in doing and suffering. We are headed right when we trust subjective thinking to yield the insights and truths we need. Kindness does much to make the world better, but sometimes that must be accepted on faith. "Where there is energy it will have effects. No ray of sunlight is lost; but the green growth that sunlight awakens needs time to sprout." Nor may the sower always witness the harvest. Schweitzer thought that the quest for ethics was a hard fight, but that right thinking would leave "room for the heart to add its word." Schweitzer challenges: "Ethics must plunge into the adventure of making its adjustment with nature philosophy . . . . Let it dare, then, to accept the thought that self-devotion must stretch out not simply to mankind but to all creation, and especially to all life in the world within reach of humanity. Let it rise to the conception that the relation of man to man is only an expression of the relation in which he stands to all being and to the world in general."

The Limits of Learning
In spite of a catholic learning, Schweitzer was never quite able to escape many of the limits of 19th century European culture. He showed disdain for other religions. He was limited by the utilitarian ethics and by the social Darwinism inflicted on nature. Schweitzer was very selective in his philosophy. He read Goethe, but did not incorporate any organicism into his own work. Although deeply indebted to the German romantic tradition, he owed some of his arrogance about nature to the utilitarianism of J. S. Mill. These limitations are discussed.

How well did Schweitzer understand meanings in Asian schools of thought? In the dialogue between Krishna and Arjuna, Krishna proclaims the truths of reverence and the indwelling of God in all beings and things in the world process. The Upanishads depicted universal joy flowing through the air, earth and water to animals, trees and grasses. This is most evident in the Bhagavata (III,2934): "Bow to all beings with great reverence in the thought and knowledge that God enters into them through fractionalizing Himself as living creatures."

Mahayana Buddhism realized the same universal principle that allowed everything to harmonize with everything else. "Without turning towards anything, always unobstructed in his wisdom / He goes along in the world of living beings boundless in space, acting for the weal of beings." In the Mahayana school of Buddhism, compassion ranks with wisdom, the heart is as valuable as the head: "The ideal man is the Bodhisattva, who, caring nothing for his own salvation is vowed to dedicate his being and his every act to the salvation of each form of life, until the last blade of grass shall enter into Buddhahood." Lao Tse and Confucius taught universal reverence and nonviolence as well. These ideas have been stressed in modern times by Vivekananda, Tagore and Gandhi; by St. Francis and St. Thomas, Emerson, Maritain and Tillich. A. N. Whitehead said,[2] "The love in the world passes into the love in heaven, and floods back again into the world. God is the great companion — the fellow sufferer who understands. God gives to suffering its swift insight into values which can issue from it. He has in his nature the knowledge of evil, of pain and of degradation but it is there as overcome with good."

Primitive religion and animism do not seem to fit in Schweitzer's evolutionary scheme of consciousness. E. B. Taylor traces the evolution of primitive religion, noting that foremost is the belief in souls and spiritual beings of trees and animals. Alfred Wallace cites numerous examples of animal sacrifice, ritual worship and animal burial in the Upper Paleolithic era. Various songs of the American Indians also serve to indicate the sacredness of the relationships of human and animal, human and plant.[3]

Nahuatl: "the people assumed the forms or characters of birds and animals."

Pueblo: " —to take the place of a spirit with my mother the bear"

Pawnee: "I killed an eagle I consecrate the eagle" "Listen, the song of the aged buffalo, my aged father"

Is this not reverence for life? To be one with the animal you hunt or with the corn you grow, to realize that it becomes part of you, living through you? Schweitzer's manifests a disinterested love of life, unlike the primitive attitude.

Schweitzer echoes a profound horror of nature, seeing it, perhaps, as Tennyson did, "red with tooth and claw." He believed the world of phenomena to be a Darwinian battleground where the fittest survived the murderous struggle. Yet, there were a number of studies available to Schweitzer at that time that emphasized the cooperation of nature. In 1910, Herman Reinheimer published *Evolution By Cooperation: A Study in Bioeconomics*, in which he characterized organisms as bioeconomic traders, who put cooperation before competition. P.A. Kropotkin, in his work, *Mutual Aid: A Factor in Evolution*, concluded that the element of cooperation in animal life, even between different species, was more impressive than instances of competition. He recounted instances of lapwings protecting other birds from a preying eagle; porpoises or elephants not abandoning a wounded companion; Impalas standing by while a troop of Baboons drove off a cheetah; cats and rats collaborating in a psychology laboratory experiment.

Many more recent studies stress the primary use of cooperation as opposed to struggle: Konrad Lorenz, in *King Solomon's Ring*, studied intraspecial social rituals in animals; V. C. Wynne-Edwards studied self-regulating systems in populations of animals; A. Tinbergen documented the sharing of space by birds; L. Schaller, living among mountain gorillas, observed instances of play, compassion and "altruism;" L. L. Whyte, in his book, *Internal Factors in Evolution*, considered even genetic inheritance to occur by rules of cooperation. Schweitzer was never aware of the spectrum of niche-finding, dominance rites, symbolism, sociability, and sacrifices (usually of leaders and lookouts) in animals.

As Schweitzer blamed Hume and others for the unwarranted anthropomorphization of nature, so he himself readily assigned cruelty to animals, as well as egotistical motives and terrible ignorance. These terms are applicable to human beings only. Man he had already placed above nature rather than in it.

For Schweitzer, nature was a cruel drama of the will-to-live divided against itself. He did not love nature. The enormous mortality in nature was an embodiment of evil, against which the will-to-live struggled. His omissions are the by-products of the pursuit of an abstract harmony with nature. Schweitzer was most concerned with the protection of animals useful to humanity, not the preservation of all animals. He killed wild predators to save domestic goats. The imposition of rational ethics on nature leaves one only in the same paradoxical situation that one attempted to solve. Schweitzer's view of the world as a consequence of evil was a denial of the world, not an affirmation. He was limited by his knowledge of the evil in humanity, by the bias of Social Darwinism and by the ignorance of his European culture. His ethic was a matter of faith.

He referred to life in nature as meaningful and death in nature as meaningless. But, death is not meaningless in nature. Death means life in nature. Life is not so much destroyed as used to further more life. There is no waste; most of every being is recycled. Death is only the renewal of life. Certainly what Jesus meant by giving up life to live in him again could be applied metaphorically to the natural world.

Schweitzer realized that life and its joys were subject to death: "Death reigns outside — it reigns over you —" but its rule ended where men inwardly overcame it. He believed that the contemplation of death could be comforting and produce true love for life; what a dreadful and intolerable burden if life continued forever. The death that Schweitzer confronted was Christian, from Corinthians 15:25-26: "The last enemy that shall be destroyed is death." The concept of death that he embraced was as the difficult passage to the Kingdom of Christ. Schweitzer believes that "Something within us shall not pass away . . . goes on working and living where the kingdom of the spirit is present . . . it is working because we are able to reach life by overcoming death."[4]

He does not describe whether the soul is saved or perhaps only a pattern, as related by the Buddha. Nor does he conclude that something from each form of life, human and ambihuman, will go on. Referring only to humans, he said that a man who lives with death in his eyes, who accepts life as a gift, believes in eternal life, because it is already his. Yet, there is a contradiction: if death is not a limit, as he intimated, then killing should not be evil. Schweitzer hinted that without death, there would be no life as we know it—no happiness, beauty or renewal.

The fact of life entails death. The living dying process is integral in nature for the continuity and renewal of life. If life and death were considered absolute limits, and not the relative continuity preached by Schweitzer, life would be no less sacred or meaningful.

There seem to be a number of instances where Schweitzer could have drawn alternate conclusions from his premises. He called the action arbitrary when we kill worms to save a baby bird; but it is not arbitrary, it is value oriented by human reason, the same reason employed by him when he bemoaned a man sacrificed to the lowliest of germs. It is true that we cannot be sure of the actual importance of each, but neither can we be sure of God's plan for both. Schweitzer stated that any injury or destruction of life was evil, unless it was imposed by fate. But fate is either a deistic plan or a course of events before which we are helpless or ignorant. Then death through ignorance or necessity is excusable, and if that is so, then the commandment for reverence is a very restricted one.

Although he believed that to the truly ethical, all life was sacred, he was forced to kill to save lives. When forced to decide which would be sacrificed to the other, Schweitzer made distinctions, from case to case, under the pressure of necessity, and held himself accountable for the lives sacrificed. But this was pragmatism of the kind he rejected. The more logical behavior following the principle of reverence might be the Jainist attempt to avoid injury to all living things, by sweeping a path through life and eating only dead plant matter. The most extreme and altruistic action would be suicide, since one's very life causes some unavoidable suffering in others. Schweitzer could not have helped anyone without assigning a hierarchy of values to the phenomena. For him, thoughtlessly picking a flower was sinning against life, or worse for not being under the compulsion of necessity. But what kind of existence is dictated only by necessity? Necessities are limits for play. And we are not even aware of all the limits.

Schweitzer did not describe the extent of responsibility for the ethical being. In the case of one person curing the dread and painful disease of another and thereby saving a life, how is the good calculated if the person is only saved for forty more years of suffering, sickness and hunger? Is the greater good to extend and complicate a life of suffering or to give meaning and dignity to a short life of suffering? If one is responsible for a life saved then one is responsible for how it is lived afterwards. Perhaps reverence is best reserved for personal meetings. It is difficult to relate personally to the fate of over four billion people and billions more ambihuman lives.

The principle of reverence for life could be applied to the mistreating and killing of animals for amusement and nonessential learning. It could lead to the realization that all life has much in common and should be approached thoughtfully.

Part of Schweitzer's dilemma came from the prevailing myth of the opposition of life and death, strife and love. But we know that life and death are necessary conditions for each other. Hinduism presumed that the animate preceded the inanimate; Freud based his theory of the evolution of culture on the assumption that the inanimate existed before the animate. Whichever is ultimately true, the two principles are mutually interwoven now. M. W. Fox judged that Schweitzer's ethic was flawed by having no ecological ethic.[5]

True reverence for life entails reverence for death, since life and death are inseparable. No pattern can survive death, when death is the destruction of individual patterns. No one would mourn the content, which is even more evanescent. All life is sacred, but this can never be a reason for not killing, because that is how lives are sustained. Since life is of the utmost importance to the living, it should only be taken in sorrow, used and shaped with respect, and experienced with awe, for underneath it is still unfathomable mystery. Life is its own meaning. Human thought is all that we have to guide decisions about lives. It is impossible to avoid some killing, no matter how conscientious. Even Jainists kill intestinal bacteria, cells, virus, and some plants or insects.

We may fool ourselves into thinking that animals, in a pantheistic communion, allow their life force to be transubstantiated upward into a creature self-advertised as capable of grasping God on behalf of himself and all other beings. Eating is inevitable. But what is inevitable may be spiritually unendurable. It may be justifiable. But what is justifiable may be atrocious. We do not do the best that we can, even if that best may be only organized butchery. Society, as well as nature, is an organization of deaths as well as lives. Realizing this, we should at least be more aware and gentle, with reverence for all lives. Spiritual maturity depends on this awareness. If the world would ever be a garden again, as it once was written to be, nonlife and ambihuman life will have to be shaped reverently, from the values humans have created from the experience of living.

Our attitudes are grounded in a belief system that constitutes a particular world view. The system constitutes a coherent whole. With Schweitzer, the system began to shift toward a biocentric outlook. . The concept of reverence, *Ehrfurcht*, meaning honor-fear in German, offered some respectability for nature through a proper attitude. He judges that "The great error of earlier ethics is that it concerned itself only with the relations of man to man. The real question is, however, one concerning man's relations to the world and to all life which comes within his reach ... Only the universal ethics of an ever-expanding sense of responsibility for all life can be grounded in thought."[6] So man is only ethical when all life is holy to him. Schweitzer proposed an ethics derived from Christian ethics (but really larger) that affirmed the world. But reverence for life sometimes conflicts with the Christian paradigm, which is just a particular manifestation. And it can lead to an instrumental ethic (take care of earth because it takes care of us — this is a problem with the Gaia hypothesis, also.) Man is not the ultimate reference. To preserve the human we must refer to the universe.

As the circle of ethics was enlarged to include the realm of all living things, it could be stretched to include all things — at worst, this is only a pantheistic Monadism. The religious argument for this extension would be that God created everything. A scientific argument would begin with the difficulty of defining life. A genuine affirmation of instinctive patterns is necessary for survival. Adaptive modes should conform to ecological patterns. An ecological ethics is based on attributes of ecosystems and human compliance with ecological laws. The concepts of rights for nature could be examined etymologically. Science could demand an ethic directed to the preservation of life in its mosaic setting. But only a religiously conceived ethic has done so. And Schweitzer's reverence for life is the only one visible in Western world. His reverence for life principle acquires a new aspect when it is restored to ontologically firm ground. Arguments against killing and the consideration of only human values become untenable. The world becomes a synthesis of a philosophy of values with the mysticism of religion, characterized by love, compassion and the reverence for all things.

CHAPTER 9

A Middle Way of Eating: Conscientious Omnivore
(A nominal review of *Food for The Future*)

Probably no consequence of human development has had a greater impact on the natural landscape and ecological processes than the production of food. Patterns of eating have influenced the constitution of species and the very contours of the earth. Throughout their history, humans have used animals and plants for food and clothing. Animals were followed, herded, corralled, tamed, and finally bred. Plants were domesticated later. As technologies developed, human relationships with animals and plants changed. Hunting, grazing, and agriculture provoked large ecological disturbances. Early domestic animals were revered, but nondomestic animals were considered competitors or nuisances. Now, animals are treated as commodities processed in factories and wildlife is regarded as useless. Hunting persists, but mainly as recreation. A few plants provide the bulk of human diet; the rest are considered ornamentals or weeds.

Although dietary habits were stable for long periods, they have been changing, for economic, personal, and social reasons. Many people limit their intake of animal products to milk, butter, and eggs. Some are vegetarians or vegans; others concentrate on fruits. Humans have been represented as omnivores, carnivores, or fructivores by different factions. In view of our control over animal and plant populations, a reexamination of our use of animals and plants is critical. There are advantages and disadvantages to strictly carnivorous or vegetarian approaches.

Carnivorism. Humanity has a long history of eating animals, and traditions of eating are important to the integrity of archaic cultures. Besides flesh, animals provide many high-quality materials that cannot be duplicated by an appropriate technology: wool, leather, lard, tallow, manure. Most animals eat food that humans cannot; they concentrate protein and convert low-quality protein to high-quality. Where animals are allowed free range, they graze in nonproductive places, such as steep, rocky slopes. In many places, wild animals are hunted; wild animals are adapted to range unsuitable for domestic animals and are usually twice as efficient in converting protein—an oryx, for example, needs a third as much water as a steer and is immune to many of the diseases. Insects are an abundant, but neglected, source of food.

There are many arguments against carnivorism, especially as practiced in industrial cultures. Often, animals compete with humans for the same food, corn and soybeans, for instance; in 1980, livestock ate enough food for 14 billion humans. Intensive meat production causes tremendous organic waste and water and air pollution; moreover, factory farming methods are inhumane—calves, pigs, and poultry are squeezed into the smallest possible spaces. Currier Holman[1] said that his business at Iowa Beef Processors "is very much like waging war." And, as in war, the innocent suffer. The inhumane treatment of food animals results in lower-quality protein, while drugs and chemical additives, beyond the toxins and saturated fats already in meat, increase the danger to consumers. Circulatory and heart diseases are linked to a diet based on animal foods.

Vegetarianism. Limiting the diet to plants avoids the suffering associated with food animals. The practice is more efficient overall; more people can be fed per acre. By eliminating the cereals fed to animals, the acreage in production could be reduced by 51 percent. Grain is an efficient food, having a high calorie to waste ratio. Furthermore, technology could reduce the area needed to produce edible plants by 95 percent, with greenhouses, hydroponics, and alternate sources, such as algae. Vegetables offer other, untapped, sources of protein, for which appropriate technology currently exists: leaf protein, which is abundant, efficient, inexpensive, and suitable to tropical and subtropical growing areas; algae, which is efficient and protein rich; single-cell protein, from yeast and fungi, which is fast, efficient, waste-free and pest-free, and can be grown on petroleum waste. These new microbial foods could lessen the burden of land use. Furthermore, the use of wild vegetables could encourage the exploration for and use of neglected and unknown plants.

But there are disadvantages to vegetarianism. Plants are living organisms, also, and many plants are living when they are eaten, although sentience is a more important criterion. Many of the alternative sources, such as single-cell protein and algae, are deficient in amino acids or are of poor nutritive quality. Cereal crops, by themselves, do not provide a good balance of proteins, so many kinds of plants would have to be used. In some areas, this would mean importing protein, at the risk of economic imbalances and threats to local self-sufficiency. Plants also have poisons, to discourage invertebrates; still, many plants must be cooked before being consumed. Much of the land pressed into production by industrial agriculture is not suitable for domestic plants; habitats are ruined by the development of land for special crops, such as chocolate, tea, coffee, and tobacco.

Omnivorism. It is obvious that humans are not pure carnivores, but it is less obvious that they are not pure herbivores or fructivores. Physiological evidence, such as the shape of teeth or length of the intestine, points to an omnivorous existence for many thousands of years. Although some human cultures concentrate on large animals or on roots, most cultures depend on a combination of sources of food.

Vegetarianism is an ethical response to the suffering promoted by factory farming. But vegetarianism is not a compartment separate from industrial agriculture, social mores, cultural traditions, the rights of wild beings, and the necessity of sufficient wilderness. Diets are part of the cultural traditions that provide individual identities for all people. Cultures maintain regional differences and emphasize the unique social aspects of consumption; meals often provide important social and psychological benefits.

Many of the problems associated with human patterns of consumption are problems of scale, efficiency of exploitation, and a universal, commercial diet. Our lust for food has resulted in a war against other species, less reported than human conflicts, but waged more constantly, viciously, and mostly out of sight. We cannot eat without killing animals or plants. Human cultures are based on killing. Often, our wants and charities result in deaths. If we have zoos, to save a few species after destroying their habitats, then we must kill for those animals that are carnivores (as we do for our pets).

Knowledge of our carnivorous history should not paralyze us with guilt. Rather, as Raymond Durgnat says,[2] "it's a reminder that what is inevitable may also be spiritually unendurable, that what is justifiable may be atrocious, that the best we can do will always be an organized butchery—and the possible best is itself light-years from fulfillment." Durgnat concludes that when we realize that society is "an organization of deaths as well as of lives, can we become more aware, gentle, alive, sad, free for a Schweitzerian reverence for life ... for all lives, and not just the next one."

Knowledge makes us aware of the costs of eating and, perhaps, inspires us to eat simply, as part of a simpler and more frugal pattern of life. To make choices, we need a way of calculating. Michael Fox[3] presents a scale of suffering of animals. For example, dairy cattle are the least intensively raised while veal calves suffer the most inhumane conditions; turkeys have better conditions than battery hens. A conscientious person minimizes animal suffering by limiting diet.

For understanding the ecology of beings, Aldo Leopold offers the image of a biotic pyramid; individuals at the top, fewer and usually larger, are more complex and capable of feeling. We argue about the extent that each species can suffer, think, or anticipate, but there is evidence of feeling on all the levels of being. A healthy biotic pyramid could be used as a basis for ranking value. For example: one human is considered more valuable than several wolves, or many deer, or millions of willow shoots, or billions of root bacteria. For species, the pyramid is reversed; bacteria are most important to a whole ecosystem, like the Amazon, which supports so many species and individuals, and humans are least important.

As John B. Cobb implies[4] the entire pyramid must be healthy and whole. Human health is rooted in the integrity of ecological systems. Physical, mental, and spiritual well-being are dependent on a good diet. Being a vegetarian may be the best option for the urban residents of industrial cultures. For archaic or post-industrial cultures, where human needs are kept simple, limits are respected, and all beings are revered for themselves first, being a conscientious omnivore (the term is from M. W. Fox, who is a vegetarian himself)* is a middle way that preserves the meaningful rituals of eating, yet uses animals and plants in an humane and optimal way.

Author's Note: *When I first met Dr. Fox I was a vegetarian and he was a conscientious omnivore. After days of discussion, he became a vegetarian (he is now a vegan) and I became a conscientious omnivore—as a part-time beggar (or volunteer) from time to time, I am dependent on the charity of other people for food, so I eat gratefully what I am given. At home I am a vegetarian."

CHAPTER 10

Mythical Dreaming and Ecological Advertising
(A nominal review of *Adbusters* journal)

Advertising creates the mythic images of our industrial cosmology. The myths are powerful, but trivial, and memorable, but inadequate to convey the meaning people need to live. Perhaps the myths are restricted by their content. If so, then ecologists and artists, as well as urban planners, historians, and politicians, need to use the strengths of advertising to convey ecological sense and traditional wisdom, the feelings of balance and the dreams of nature.

Our dream of nature, in modern Western culture, is the dream of order and beauty. But, as Aldous Huxley noted, the dream of order begets growth and tyranny, the dream of beauty, monsters and violence. Our dreams are nightmares because they are not complete. The nightmares are symptoms that reflect unbalanced and immature cosmologies, that is, images of the earth. A traditional cosmology evolves with people's needs, fears, and knowledge. But, if it is incomplete, or if it does not fit environmental conditions, it may fail. Many early cosmologies, primitive or advanced, failed to fit the earth.

The modern industrial image of nature as a resource has resulted in pollution, material shortages, and environmental degradation. A culture that degrades its ecosystem risks its own extinction.

Industrial cultures, however, are not the only cultures in existence. There are hundreds others, although at one time, around 1900, our species had over 1,000 different cultures and 3,000 languages (roughly equivalent to the number of natural biogeographical provinces and subprovinces on earth). Each culture exists in a particular location with a unique history. Later developments are not more adaptive than earlier; nor do they replace them. Ethnic groups are not evolutionary stages culminating in America, but are equally valid ways of life. Each culture is only one of many possibilities, a way. There is no single or correct way.

Each culture has a root metaphor. In the West, it is the machine. The advent of the machine made processes of order more amenable to description. Although only a closed system itself, the machine was a fruitful metaphor for living systems. The theory of the living organism as a mechanical contrivance explained biological phenomena from the physiology of an organism to the processes of cells. The cybernetic machine metaphor was successful at explaining detailed processes without answering fundamental questions of meaning.

Science makes extended use of the metaphorical process to construct its models. For example: "Man is a system" according to Ernst Laszlo or "Man is a computer" according to Michael Arbib. Kenneth Boulding offered the perfect machine metaphor for the operation of the earth: as a spaceship. As a metaphor, a spaceship suggests the limits of earth and the value of a limited life-support system (unfortunately, it also implies something of human creation that can be controlled and fixed by human intention).

The use of the word "ecology" by Ernst Haeckel implied that the natural world was a place to live, a house, rather than a machine to control. Making the earth into a house is fundamentally a poetic activity, according to Gaston Bachelard. Poetry also is a way of understanding the universe through metaphor, a literary device that transfers the characteristics of one term to another. As Picasso said of art, poetry also is a `lie that tells the truth.' For example, Shakespeare said "The body is a garden; and Harvey said "the body is a machine." The body is not a garden or a machine, but the metaphors extend our understanding of the body.

Poetry is communicative of the quality of things. Like science, it discriminates the unsuspected in the commonplace. It is not different from science, but more diffuse; not better than science, but more comprehensive. It accepts ontological parity, the equality of beings; aspects of the world are not negated or reduced by one another. As metaphorical knowledge, which may be prerational or metarational, poetry can avail itself still of scientific references. Poetry can measure a whole qualitatively and mimetically, a germ or the cosmos with its imagery. Poetry is a tool for comprehending partially what cannot be known totally. A poetic language could include a view of the interrelatedness of all existence in a sublime ecology.

People need to be made aware of the power of self-determination. People need to feel things, like the immensity and uniqueness of nature or the strangeness of a tick, before they can act. Poetry can help people feel themselves as part of the web of life or on an oasis in space. That feeling, more than laws or injunctions, can justify preserving the ecological systems of the earth on which we live. Humanity is a poetic species, as Richard Rorty noted, "one which can change its behavior by the words it uses." We need desperately to change our behavior.

Mythology can join science with feeling to help us change. Mythology is not limited by method. Mythic symbols store information concisely, which makes it possible for a person to assimilate the collective experiences of a culture. Myth combines us with other beings. Mythologies are in fact great poems that function to awaken the experience of awe and humility before mystery, create a cosmology, validate and maintain an established order, and bring the individual into harmony with the whole.

Unfortunately, the myths of the predominate industrial cosmology are inadequate. The myths are powerful, but trivial and misdirected. Poetry and art are undervalued as forms of communication, not to mention as ways of shaping and making. Business has transformed much of art and poetry into advertising, to match the style and attention span of the people in industrial cultures. Advertising, quite literally from the *Wall Street Journal* to college textbooks, refers to its activities as "shaping the American dream." Like art, advertising creates an image of a way of experiencing. Unlike art, it limits its focus for a specific goal—profit. Like art, it mirrors us. Unlike art, it intensifies and glorifies only the positive aspects of culture, ignoring the dark, negative aspects.

Its simplicity is irresistible. Our environment deteriorates according to ecologists, but gets better according to economists.

And their pictures are prettier. People want to hear that it is getting better. Advertising tells them it is. People want to act stupid, greedy, and selfish, and spend the inheritance of their children on themselves. Advertising tells them their actions are rewarded. The real issues of life and death, destruction and hope, make people feel helpless and anxious, so advertising draws their consciousness to comfortable trivia.

Despite the ugliness of the dreams of progress and growth, of waste and stylistic frenzy, advertising, using sophisticated techniques and narrowing the focus out of context, makes the dreams desirable and irresistible. People in agricultural and hunting cultures interiorize the abstract industrial vision. African farmers are convinced to buy inorganic fertilizers, even though it degrades the soil; women to buy powdered milk for their children, even if it kills them. Tractors replace draft animals in the paddies in the Philippines, even though they are costly and less energy-efficient; French winter fashions are found desirable in tropical Brazil, even if they can only be worn in air-conditioned villas. People in industrial societies are convinced that their children will be ruined without personal computers. Disposability is offered as a fix to a wanting in the temperament. Advertising fuels the acceleration of conspicuous and compulsive consumption.

Yet, advertising may be the most effective means to reshape desires and reform buying habits. Advertising presents the symbols of modern experience, even if they are just the trivial ones. It could present healthy symbols equally well. Advertising does incorporate traditional values, like family, friendship, and love, although to sell beer and cereal and, sometimes, churches and hospitals. And, like art, advertising lies (although Jules Henry thought it was instead a new kind of truth — "pecuniary pseudo-truth" — not intended to be believed, or certainly, not to be proved).

Advertising is beginning to support more informational functions, such as the dangers of drug abuse and smoking. Advertising creates values — fur coats, fast cars, dark beer, slim cigarettes are certainly recent and artificial values — but it could be used to create positive ecological values and new identities that show that our needs for prestige, esteem, and belonging can be met without stylistic waste at mindless speeds. Advertising could promote new attitudes about appropriate technology, the rights of other cultures, and the place of people in nature. Good advertising could be as subversive and conservative as ecology. It could avoid confrontation with people's values; emphasize positive aspects without negative ones. A good ad could capture and carry the most self-indulgent viewer; for the most part, ads don't require effort, literacy, or consciousness, just attention.

Advertising has been serving the dream of progress, but progress is leading to catastrophe, a long, slow, global catastrophe. When people experience local, sudden catastrophe, they usually respond immediately, with heroism and sacrifice, aiding the victims of earthquakes or floods, sometimes famine. Advertising could bring to consciousness the slow catastrophes of erosion and population overshoot, and, perhaps, invoke the same altruistic responses to them.

To work towards this service, conservation groups could define and promote an integrative mythology as the basis for the framework of diverse efforts to protect life and the environment. Conservation groups could provide a meaningful philosophical foundation, as well as coordination for other humane, social, and conservation programs. But, the approach must be egalitarian: Respect for life cannot neglect human life and suffering. The approach must be Eutopian: A new cosmology cannot ignore adaptive cultural traditions that arose in place over centuries. Furthermore, in addition to formal education, they could provide reeducation through the most effective means, such as advertising. Conservation groups could spend money advertising "humane consciousness," moderation, and the joy of living (instead of just consuming or winning). Ecological ads would be unique and compelling, simple and effective. They would advertise not a product, but a way; not for a profit, but for a dream.

Addendum 1997: Advertising and Euphemism

The other day, tired of writing, I went to visit friends, who were watching a car race. It occurred to me that only with entertainment industries is there so much technical fireworks and coordinated enthusiastic teamwork. Imagine all that energy and enthusiasm directed to appropriate technology for reforestation or the proper use of forests. Imagine television coverage of forest work with the same amount of attention and detail. Why not a competition for the most beautiful or productive forest or teams working to restore devastated city areas — broadcast by a major network as an important event.

It also occurred to me that this remorseless entertainment is an anesthetic against fear, emptiness, self-searching, or death. Continuous entertainment is a kind of guarantee of health, riches, and long-life. Everything that is pleasurable, thought George Orwell, seems to be an attempt to destroy consciousness. Television seems intent on proving him correct. Ecology cannot ever compete with entertainment if it raises troubling questions or difficult expectations. As long as the industry can guarantee many forests through the arithmetic of fantasy, we will always seem to be complainers and false prophets — until it's too late, then we will be blamed for not avoiding the catastrophes.

Maybe the situation is not that bad. Maybe we can present images that rival the industry images. Maybe we too can speak the languages of euphemism that large corporations use to conduct their businesses of larceny and fraud. Positive images and pleasing language skills are everything these days; no one really looks for substance. The devotion to money, beauty and youth is our focus.

I think one way to compete might be to present parts of ecology as a medical discipline, aimed at restoring ecosystems and forests to health — and advertise it that way. As with any medicine, the patient actually does most of the work to become healthy, although the doctor gets the credit and the payment. This would also lead to more respect for the practitioners, but also to more responsibility and more rules. The first rule, which we might take to be basic, is identical to the first vow of the Hippocratic oath, "Do no harm."

CHAPTER 11

Ecologists Want to Rule the Earth?

I had trouble recognizing Richard Watson's characterization of Deep Ecology in the winter 1985 issue of *Whole Earth Review*. He seems to have quoted the program correctly, but has drawn the wrong conclusions from it.

For example, he seems to find somewhere in the literature of deep ecology that the ideal human impact would be on the level of hunters and gatherers (which is Paul Shepard's ideal anyway), with a global population of five million, after quoting Naess's ideal of 200 million. However, limiting human impact is not the same as returning to hunting or gathering or to subsistence agriculture, which no deep ecologist has ever recommended. Watson presents his own ideal at 500 million, based on a concept of cultural flowering. That figure is the same as Daniel Kozlovsky arrived at intuitively and the one I calculated for an optimum population based on global net ecosystem productivity (in 1981).

Watson's criticism of ecological ethics is based on a confusion of value. A nonanthropocentric ethics argues that beings have self-value, although not necessarily human value. Saying that the generic term value is independent from 'man' as Watson does, is meaningless. An ecological ethics can address the limited relationships of all beings without becoming entangled in the fuddle of reciprocity or sentience. By basing an ethics on 'what is,' deep ecology avoids the 'absolute' that bothers Watson.

Deep ecology is not anti-anthropocentric. In fact, it accepts the necessity of an anthropomorphic, anthropocentric, and anthropometric logic. It does not accept the extremes of such a logic, however, which assigns to humanity all value and creativity. Furthermore, deep ecologists do not claim, as Watson accuses, to know what is right or good for humanity, let alone the ultrahumanity on which we all depend. Indeed, most deep ecologists urge caution and noninterference with primary cultures or wilderness areas.

I do not know what the ecosystem wants Watson to do, as he begs it to tell him. Does anyone else know? But I doubt if it wants him to die for the human species, as he complains. Deep ecology is not concerned with the good of the human species apart from the diversity of species and the health of the ecosystems in which we live.

When apocalyptic rhetoric is not heeded, human societies vanish, leaving behind their monuments and deserts; hundreds of human cultures have 'bit the dust' in the last 3000 years alone—keeping archaeologists in business. Contrary to what Watson would have us believe, deep ecology is not concerned with telling people what to do or with saying what is right. Moreover, it stresses that we should not always attempt to say what is right for all.

Watson's fear for the ambition of deep ecologists is unfounded. It is unlikely that Naess, Hardin, or Skolimowski want to 'rule humanity' (as Watson warns us), although it would not hurt to keep an eye on them.

CHAPTER 12

Invincible Ignorance: Hunting and Morals

What is it about sport hunting — as opposed to the subsistence activities of aboriginal cultures — that attracts hunters? The snazzy gear? The prehistoric mystique? 'Free' food or room 'decorations'? Communing with nature? Sport? Shooting? Competition? Killing? Instinct? Half-remembered, traditional rites of passage to manhood (and more frequently womanhood)? Peer pressure, sexual release, emotional expression, or misperception (anti-Bambi reaction)?

In this issue of *wiNR* (Women in Natural Resources) are two articles that glorify or endorse hunting. Understandably, these articles do not address all of the issues associated with hunting or with animals as "natural resources." (One problem with natural resources is that we tend to think that every resource is available for our use, if not direct use for food and clothing, then for recreation and entertainment.) These articles address a limited perspective of hunting, that associated with the individual pleasure at stalking animals and using technological advantages to kill them. There are other dimensions and many larger questions unasked. Personal pleasure is one dimension; another dimension of hunting is the shear number of hunters (over 25 million). At what point does hunting break down as a sport when overequipped hunters overkill millions of individuals of overestimated (by game managers) wild species?

As phylogenetic omnivores, humans have a long tradition of eating and using animals. Many early cultures revolved around hunting as a way of life. These cultures had traditions that revered their prey and many taboos to limit taking of prey animals; many prime game animals were taboo altogether. Partly as a result of successful competition, human groups expanded and came to enter and dominate virtually every ecosystem, such that scale, greed, and ignorance resulted in extermination, overgrazing, overcutting, and wanton destruction.

Modern hunting is founded on a number of social and ecological myths based on early human traditions. There are many fallacious arguments for hunting.

1. "Hunting is necessary to keep game species in check and healthy, even if original predators are reintroduced." No, the population rarely increases beyond the carrying capacity before it restabilizes. Food is always the final check. Predators stabilize a herd at a lower number and keep it on the move (which improves its health). Furthermore, we do not know enough about species to indulge in maximum sustainable yield.

2. "Hunting is necessary to protect human economic interests, especially crops, cows, sheep and chickens." Not so. Usually preventive measures are enough; good fences and enough attention by the farmer. Otherwise, why not accept a ten percent tithing as payment for the provisions of nature, the economic free goods so often touted in economics books.

3. "Hunters take the place of natural predators to maintain the balance of nature." Why bother, the natural predators are more efficient and cost-effective. Why create an imbalance by killing predators and then try to take their place, mostly unsuccessfully? Hunting as a system is very unnatural, in its timing, its targets, and its scale.

4. "Hunting has a minimal impact on game species and almost none on nongame ones." Except for the fact that many habitats have been artificially skewed towards game species. Hunters often shoot prime specimens, which is the opposite tact taken by predators, who weave the sick and old from herds. And furthermore, the type of game shot has great repercussions for the surviving animals; in many areas the same percentage of deer starve every winter after hunting season. Also, many animals, especially rare species — lynx, marten, panther — cannot survive in the presence of hunting.

5. "Hunters do more to benefit conservation than the rest of the people." Hunting licenses provide much of the revenue for fish and wildlife agencies to mismanage lands to game species maximums. Very little of these funds (less than 3 percent) goes for nongame species.

6. "Sport hunters are concerned about their image, and only a small percentage are slobs." Hunting takes less agility and conditioning than other "sports." The lack is made up for by motorized transport (snowmobiles, trucks, dirt-bikes) and by powerful, sighted weapons.

7. "Hunters are just following nature's law of kill or be killed." It's an old law, never properly understood and repealed by new knowledge. The new law is cooperation.

8. "Hunting is okay because animals have few feelings and emotions and feel little pain." New evidence shows differently; even fish have well-developed feelings, according to a recent scientific study in England.

Hunting is an ethical issue, now, more than a social and ecological one. What is most traditional about hunting? It is the concept of "fair sport." Fair sport excludes slaughtering trapped deer with automatic weapons or bombing trout, although such things have happened in the name of sport. A sportsman, by definition, is one who plays fair. It also implies that ones opponent has consented to the contest. When have deer or rabbits, lynx or squirrels ever consented to be shot? How fair is hunting with party permits? Perhaps the less-used definition of sport, "mockery," is intended by these hunters?

Recently, people have admitted an awareness of the wrongness of our suicidal war against animals. At one time, hunting was a sensible use of animals, not terribly different from the exploitation of some species by carnivores and omnivores. But gradually the extent of modern hunting — and ignorance of the extent of the devastation by hunting — lead to the depression, extirpation, and sometimes extinction of animal populations. What good hunter would hunt a species below a minimum viable breeding population? What excuse is there now? We know that we have gone too far with some populations of whales and wolves. We have interfered with the health of game animals by eliminating other carnivores.

In practice, sport hunting is repugnant. It demonstrates an ignorance of human nature, of animal nature (more than just a series of tracks leading to a target), of wildlife management, and of interactions in ecosystems; it demonstrates an insensitivity to animal and human feelings and goals, and it demonstrates an abuse of power.

Animals are threatened with loss of habitat and deteriorating ecosystems, from take-over, development, human overpopulation, and pollution, as well as from hunting. Hunting puts pressure on animal populations without noticeable benefit. It causes moral difficulties as well.

Anti-hunters are as polarized as hunters in their opinions. Cleveland Amory does not consider hunting a sport. He proposes that hunters are themselves a renewable resource, a herd to be trimmed for their own good. Perhaps this is implicit in hunting laws, which have minimal punishment for "accidentally" killing other hunters through mistaken identity. Hunting other humans has been a popular fictional device; Amory and others think it would work in practice.

Hunting is a minor part of the total of human outdoor activities, numerically and economically, as well. Far more people walk outside, and take pictures, or ski, or camp. Hunting could be limited and then eliminated in six easy stages:
1. Test hunters for safety and knowledge
2. Limit the seasons
3. Limit the land areas open to hunting
4. Separate wildlife programs from hunting revenues
5. Rethink the management of nongame wildlife
6. Change the rules, making them more fair — a warmly clothed human armed with a knife or camera; for a greater challenge, hunters could "count coup" or "play tag."

Finally, let's call hunting "recreational slaughter" or "killing for fun," to distinguish it from the more traditional subsistence activity. After all, humans with brains and spears were more than a match for mammoths. And humans with super technological information systems, weapons, and vehicles have shown that they are over-matched for virtually any species, even without bringing those big brains into action. Why bother adding to our advantage? Let's grow up, though, and use those big brains for other needs and pleasures.

Author's Note: As the Editor of *Women in Natural Resources*, 1987/1988, the author, who was the only male on the staff, insisted on running this editorial to balance articles on bow-hunting and sport hunting. After initial agreement, he was later fired by the owner, officially because he had chosen a "bad color" for the cover; the owner wrote her own editorial and reprinted the entire issue of the journal, after having to buy back the computer design and data files for two week's salary to the author. The owner still has not paid a $1500.00 equipment rental to Nieman Ryan Designs.

CHAPTER 13

Aesthetic Education, Organic Dialectics, and Deep Ecology

Educational ideas and theories in this century have been dominated by the impoverished cosmology of modern industrial culture. To be effective in this age education needs to be rooted in conceptually rich philosophies, such as deep ecology. Deep ecology takes its inspiration from the science of ecology. As a science, ecology describes the interrelationships of organisms and environments, that is, the experience of living together in the biosphere. Ecology is not a reductive discipline and is not readily amenable to complete quantification. Even scientific ecology is an integrative discipline that extends beyond the boundaries of science. Ecology can be considered an amphibious discipline, with the authority of science and the force of moral knowledge. Studied through its components and relationships, ecology is a way of seeing, a perspective of the human situation in its interconnection. It is a "subversive" subject, normative and sensible, offering a "sacramental" vision of nature.

Deep Ecology

Deep ecology is a movement, formulated by Arne Naess,[1] that goes beyond a concern with pollution and resource use to consider humanity in a relational, total-field image. The movement promotes human equality, conservation, and local autonomy. In principle, it proposes a biospherical egalitarianism, the equal right of all beings to live in place. It adds a normative dimension to ecology in a framework of ecosophy, literally the wisdom of the house. As a philosophy, deep ecology investigates the normative aspects of living together, that is, ethics, and the maintenance of the affairs of communities, that is, economics and politics. As a noetic discipline, deep ecology provides information on the state of nature, recognizing that human beings are participants in nature, as well as participants in human societies.

Deep ecology emphasizes biological equality. When Charles Elton transformed the "Great Chain of Being" into a chain of eating, ecologists realized that the bottom link of the food chain, plants, was the most important. Humanity is part of the food chain, although it appropriates a large amount of the productivity of most ecosystems. The exploitative competition of humans in ecosystems is an important part of biogeochemical cycles. Humanity cannot unparticipate by choice. Deep ecology argues for diversity. In nature, variety emerges spontaneously, as the capacities of species are sorted by the environment. Variety provides flexibility in systems. The diminution of variety through human interference may debase the wholeness and stability of systems. Furthermore, aesthetic, ethical, and utilitarian reasons all support the efforts to conserve the diversity of nature. Deep ecology incorporates a broader scientific method that might be called patient practice. There are ways of dealing with the earth that are not scientific or technical; they are aesthetic or ethical. These alternatives are not incompatible with traditional science. Where the methodology of traditional science is limited and wasteful, promoting technologies that ignore or destroy values with blind quantification, the methods of deep ecology are traditional and conservative.

The Organic Dialectics of Goethe

Deep ecology considers the method of Goethe,[2] whose natural philosophy incorporates a world view of organic dialectics and whose methods are contemplative nonintervention, a passive attentiveness, and the primacy of the qualitative, where intuition and the method of analogy work towards deeper sensory participation. With contemplative nonintervention, "deep down" phenomena are only revealed to close attention, a "passive attentiveness." Goethe thought that experimentation had severed humanity from nature. His suspicion leads to questioning the effects of investigation. Perhaps as Eddington later considered, experimental equipment shows how nature can be made to behave, rather than how she acts. Heisenberg claimed that nature known is nature exposed to the human mode of inquiry, not nature as such. If this is true for physics, where the object of study is of the same scale as the means of study, then it is certainly true for the study of feeling organisms by other organisms.

In the social and animal sciences, participative involvement replaces objective neutrality. People and animals are smarter in their own context. The knower blends unobtrusively with the subject, merging and trusting. Contemplative nonintervention carries us far in understanding nature, but there are limits dictated by "respect for the unfathomable." Respect leads to patience or renunciation. Perhaps respect for mystery can go deeper than knowledge. At some depth, "the realm of the Mothers," the only knowledge gained may be tricks of manipulation. The limit may not be formal, however, only what is felt. The primacy of the qualitative deals in data — not data taken from measurements, but savored qualities: color, texture, and form. Goethe relied on an intuitive power within the senses, an "exact sensory imagination." He objected to the subordination of qualitative experience to quantitative, mathematical generalization. In fact, he banned mathematics from his natural philosophy. Essentially, he granted more reality to the sensory life than to scientific artifacts. The problem with the qualities of sensory life is that they cannot be added or subtracted; they must be evaluated, and evaluation differs with every individual. Goethe considered the validity of every-day observations, unique occurrences, and short-lived phenomena. The special instances lead to insight, in contrast to the piecework of analysis compiled in tables of samenesses. Goethe recognized that different people are sensitive to different aspects of a thing. Any investigative effort should incorporate the observations of many others.

Through organic dialectics, Goethe saw nature as unity in process. The central image of his nature philosophy was polarity and synthesis, a divine rhythm, polarity moving toward an intensified union of opposites, resulting in a higher third. Goethe's model was alchemy, the science of meanings, but purged of bookiness and confused experiments and developed into a "mystical chemistry." He wanted his disciplined study to be artistically and morally useful, a science of the whole person and unified culture. He attempted to integrate science with traditional wisdom. Goethe's theory of color is considered by many to be a foolish attack on Newton. In rejecting Newton's objectification, Goethe was concerned that the meaning of seeing, color as a direct phenomenon, would be lost in the welter of indirect artifacts. His theory of color was concerned with the evocative experience of color and light. Goethe's science discovers meaning, but, unlike Newton's, has little power. Many scholars[3] reject Goethe's science because "everything blends into everything." This is one of the main criticisms leveled against deep ecology by Richard Watson and others.

Goethe's passive attentiveness is taoistic. A taoistic approach is useful: asking rather than telling, observing rather than manipulating, being receptive and passive rather than active and forceful. Such an approach is nonintruding and noncontrolling. In dealing with living organisms, a caring perception provides kinds of knowledge not available to scientific researchers. This situation is especially true in ethological literature: Abraham Maslow, Konrad Lorenz, N. Tinbergen, George Schaller, J. Van Lowick-Goodall, and M. W. Fox have found it to be true in their research.

Deep ecology is a form of scientific animism. Nature is a feeling system. Animism is necessary for understanding the system. Animism allows investigators to behave "as if" nature were intelligent and sensitive. Deep ecology is not a single-vision science or a primitive animism; it is a scientific animism, aware of the effects of its activity. It is concerned with more than the anatomy and taxonomy of organisms; it is concerned with the mutual experience between human and nonhuman beings. Deep ecology considers the human impact on nonhuman systems and human attitudes towards ecosystems; it considers human needs for sacred spaces and wilderness; it considers territoriality, aggression, and the aesthetic reaction to the wonder and beauty of life.

The Aesthetic Education of Schiller
Modern education allows individuals to languish in an informational wasteland. A radical education, based on the aesthetic humanism of Frederich Schiller, leads them out, to a place within nature. It offers a new perspective of humanity in the total field on nature and defines balanced relationships with other species.

Schiller believed that human society could be improved by political means. But after studies on the Thirty Years War in Europe, he became skeptical of the ability of politics to create a peaceful society. He came to consider a work on art[4] historical proof that art could achieve what violence and law could not: Art educates and liberates the individuals of society in a gradual and peaceful process. In spite of the cultural forces dominant at any moment, an individual has the potential to determine a different course of action. Unlike classical humanism, which was shackled to one interpretation of the past, the aesthetic humanism of Schiller was open to the possibility of novelty.

Schiller judged society to be violent and selfish[5] and he identified the cause as an imbalance between animal and rational drives; civilization exacerbates the problem with its own imbalances, for example, fragmentation and the unnatural channeling of energy. The basis of his theory is the relation between three instinctive drives: sensuous bodily needs, form-giving reason, and the play of imagination. The third mediates and harmonizes the first two. And through this facility, human beings and their institutions gradually evolve to a higher moral plane. Improvement comes about through aesthetic [text missing] and state progressed through definite stages, in Schiller's theory. Both begin in the natural state and work toward an aesthetic state, and later, a moral state. The natural state is dominated by physical needs, but individuals can transform themselves with the play of their rational faculties. The natural object of play is the aesthetic state. Neither reason nor the state can achieve further transformation, however; that can be done only by the instrument of "Fine Art." Using a mimetic faculty, people imitate one another. Life imitates art, which is an expression of values. By means of art, people ascend to the moral state. An ecological education based on Schiller's ideas presents a whole image of humanity within nature and not a transcendent view. It confronts the past without the baggage of sentiment and the future without the paralysis of dread. The appreciation of the differences of other cultures allows human beings to enlarge their experience and identities. Art broadens the mental worlds of observers and encourages tolerance and wonder. Education in aesthetic humanism embraces three concepts: play, liberation, and community.

Play is the method of learning for most juvenile animals and a means of enjoyment for many adult animals. For humans, play is imaginative experience, entered into freely. Much human activity is play, in place in a community. Even science and philosophy are forms of play, attempts to solve the puzzles of existence. Schiller maintained that the source of both art and play is "overflowing energy." Although early ethological definitions of play emphasized its activity as an "outlet of surplus energy" (Bolwig 1963), later definitions enlarged its importance in learning and information gathering. For example, play is also defined as:

- An activity with "no immediate objective" (Hall 1968);
- An experimental dialogue with the environment (Eibl-Eibesfeldt 1970);
- Rehearsals performed in a nonfunctional context, of the serious activities of searching, hunting, fighting, mating (Wilson 1971).
- Behavior that functions to develop, practice, or maintain physical or cognitive abilities and social relationships by repeating or recombining sequences of behavior outside their primary context (Fagen 1985).

For Schiller, play is activity for its own sake, where the drives of emotion and reason are harmonized. The state of play is whole and simultaneous. It unifies permanence and transition, chaos and order, duty and selfishness. The object of the senses is life, the object of reason is form, and the object of play is living form—called beauty in the widest sense. Aesthetic play, like physical play, requires order and control. The rich flowering of human nature is possible only when the constraints of need are replaced by leisure and abundance. Liberation requires a larger perception and larger concept of rationality. For humanity, liberation means an end to prejudice or discrimination based on arbitrary characteristics. The liberation of nature and ultrahuman beings is inseparable from human liberation. Nature and ultrahuman beings are to be free from the obligation to be human (so pervasively presented from Plato to Kant) and from the status of human resource or human artifact. In Schiller's scheme beauty frees us, because we decide how an act is performed, without being restricted by blind necessity.

Humanity has taken its own opportunities, which have been codified for centuries as rights. Now, plants and animals must be allowed opportunities. The interrelatedness of species dictates the interrelatedness of rights, and these rights are necessary to the integrity of the whole planet. The extension of rights to plants and animals does not deny any traditional human rights. Community is formed by beings living together in place. The human community is only one of many. This interrelatedness makes human beings less discrete and less alone. The biological term for "living together" is symbiosis. Living together, beings do together. The Sanskrit word for "doing together" eventually became the word "ethics" in English. Ethics (and rights) are simply rules for living together. Human ethics describe only a small, self-conscious part of the rules.

Human beings gravitate into groups to live. Every culture needs its own local, sacred center, that cannot be broken if the group is to survive. Communication across the barriers of culture is necessary for a world community, but from firm cultural bases. The complete surrender of cultural identity is as dangerous as too little openness. In ecology, the saying goes, "As the community goes, so goes the organism."[6]

Radical Education
Education takes place in communities; it is the means for communities to continue. As Plotinus and Novalis recognized, education has an outward, social and civil, aspect as well as an inward, personal and self-revealing, aspect. Education has at least four ends:
1. the appreciation of the richness of nature
2. the comprehension of human existence
3. the understanding of the nature of human society
4. the training for a position in human society

Education has become more universal, but its goal, the well-rounded individual, has been distorted by its fourth aim, training for the economy. To produce wealth for the state and livelihood for the individual, education has become money obsessed. Ethics, in the second and third aims, has been neglected, since it might limit or contradict its economic obsession. In fact, the first three aims are restrictive to a growing, industrial economy. Education, as practiced by public schools, produces unprovocative individuals, adjusted to an unbalanced society. With its emphasis on play, liberation, and community, a radical education integrates all four ends.

Radical education alters and enlarges perception with the selection and presentation of relevant information and forms an ecological consciousness. The survival of human societies depends on consciousness of the global system in its complexity and connectedness. The spirit of humanity depends on the consciousness of its proper relationship with other species and the whole earth.

CHAPTER 14

Radical Education

When I first visited a college library, I was in grade school. There I had a vision of myself reading in secluded comfort, surrounded by scholarly books. I progressed towards that vision as best I could. I took courses in many subjects; each subject was no less interesting than another and led to a variety of related positions in laboratories, institutes, libraries, clinics, hospitals, observatories, restaurants, sawmills, printers, and classrooms. This long process of selection produced a few successes and many failures. But, as traditional universities failed me, I failed them. I analyzed our weaknesses and searched for a different way.

Traditional education encourages many fallacies. It assumes that its knowledge is static and adequate. It assumes that only children need to be educated. It assumes that schools are the best places to learn. Furthermore, the groves of academe are surrounded by high walls, like a maze; each discipline is self-contained and each scholar is a master of a small domain. Learning within the walls is typically a regurgitation of facts within subject and time limits for grades. The grading system seems to be a device for rationing the 'good' places in society where unemployment is used to control inflation according to an economic scheme that tries to adjust ideal theory to a limited world. The university is the final phase in the production of higher credentials to guarantee higher rewards in the economic system. The entire educational system itself seems to be political, directed at wealth for the state and a position for the individual. The system is subsidized by military and corporate interests, which depend for their livelihood on the disparity of wealth between individuals and cultures. The university is built on the same blood and suffering caused by that same disparity and nourished by continued exploitation. The dream of living in modest comfort, sheltered in a place of learning, worrying a trivial scrap of knowledge, is a dream built on the broken dreams of those who, every bit as equal, dream only of adequate warmth and food. Participation in such a dream is only at the exclusion of others. The dream of scholarly detachment insulates the conscience from the desperation of those whose accidents of color, location, or inheritance were not as fortunate. The "hardness of heart of the well-educated" was what Gandhi admitted made him most sad in life.

A true education need not be bound by the conceptual and economic limits assumed by most universities. A minimum education may train students for an economic role in society, but a good education teaches them how to enjoy living among other human beings in an ultrahuman nature and to perpetuate a good society. Poets like Wordsworth and Auden recommended that broad training in science and technology was necessary for poetic knowledge, which is part of a good education. Novalis considered that the study of the external world, through science, was only the first, half-way, step to full human consciousness. The second step was introspection, the contemplation of the self. Subject areas in traditional universities concentrate on one step or the other. Any student can achieve both steps, leading to a complete education, with time and inclination. A complete education requires intense effort, discipline, patience, and a tolerance for failure. Elite schools like Harvard or Oxford, with their richness of culture, can offer more potential than the pedestrian paper mills of Idaho and Washington, but are often limited by social fashions. Radical (in the sense of 'rooted') schools like International College, Evergreen College, or Antioch University, can offer greater benefits, with a teacher "leading" a pupil to an education (from the Latin word meaning 'to lead'). In his educational theory, Plotinus went still further and laid down a triple organization of education, requiring a social education, a personal and self-revealing education, and a synergetic one that would permit a perspective of the whole of human existence. Only schools that integrate education within a balanced society can achieve this triple objective. By encouraging students who are already working outside academic walls, radical schools can foster this necessary kind of synergetic education.

The most valuable qualities of a radical education are personal contact, which allows noncompetitive constructive criticism, and flexibility, which allows the educational process to fit an established and meaningful life-style. Radical education lets one learn how to feel and live, as well as to think. As Aristotle recognized long ago, experience is necessary for thinking. The very cloisteredness of traditional university education works against it in this respect. The university fails to teach communal responsibility, self-reliance, and physical work—those qualities most dear to Emerson. Education must embrace three concepts, according to Schiller: liberation, play, and community. Liberation is freedom from the limits of identity; play is imaginative experience; and community is the supporting matrix of life. This ideal is most closely approached in already established communities (not the artificial and involuted, temporary university dormitories) and when play and freedom are not limited by arbitrary rules and economic goals.

Radical schools provide for education within the larger community, in the larger context of work and recreation. Perhaps this attention to context accounts for the success of business training schools, such as "hamburger" university or "insurance" institute. Relevant schools stand a better chance of surviving social changes than those built on self-perpetuating administrative interests and alumni sports empires, whose vast buildings and grounds might better be turned into shelters for the poor or enclaves of employment, to produce real goods for society as part of their programs.

All human beings need a life that is protected and ordered, loving companions and contact with the wild universe. I need the amenities of a library and a small garden as well. But I cannot pretend that this little world is not part of a larger one riddled with hunger and fear—one which my actions affect. The beliefs I hold are worthwhile only if I enact them in a larger world, ethically (which comes from the Sanskrit word meaning 'doing together'). Very few schools concern themselves with the scope of ethics; that is left to the student.

Although it is pretentious to assume the responsibility for a class or culture, sincere action begins with personal responsibility and responsiveness within a smaller system. Unpopular questions, ethical expressions, and confrontations have a price. It may be only silence or deprecation; it may be expulsion, imprisonment, or death, as it does in some of the forests in South America. I trust that I have the courage to question society and express my beliefs and findings without worrying about the cost. For only through sincere, studied expression or example can anyone hope to influence the consciousness of others. And only by surrounding society with a new field of consciousness — not by attacking it — can any transformation occur. When human beings cooperate spontaneously, because they understand what it means to be human, because they understand how to treat their places on earth, only then can education be considered successful and, perhaps, lead to a more peaceful and humane world.

CHAPTER **15**

Minimum Wilderness for the Planet

Although wilderness has been defined as emptiness; and it has been defined recently as a place "where man is a visitor who does not remain" (Frome 1974), it is essentially undefined, or rather the definition is narrowly anthropocentric. The popular wilderness definition, where man is only a visitor, is utilitarian; humans visit for recreation and relaxation. There are flaws in this utilitarian approach to wilderness; wilderness is considered a resource, and it is related to human experience exclusively. Wilderness does not have to be destroyed or used or even visited to be beneficial. It is valuable as wilderness, because it is there. Humans are already using a very large percentage indirectly.

The definitions have become more complex as more about the functions and values of wilderness becomes known. There are differences between ecological definitions of wilderness (mature systems which change with time) and cultural definitions, which vary. Wilderness is more than a geographical concept; its boundaries can never be just geographical. It is an abstraction. It is hard to define wilderness exactly or objectively. Any region that can be described verbally is also a state of mind. And because it is a state of mind, it is ambiguous. Consider the biblical attitude toward wilderness, first as a howling wasteland, then (in Revelations) a sublimity greater than the world of "man," that enabled contemplative prophets to see the Divine more clearly. What originally held the disorder of humans, now holds the order of nature. Wilderness is a changing process. And, the English language becomes imprecise and mystical dealing with process thought. Benjamin Whorf pointed out the problem with verbs. A forest is not just there ("is"); it is being or becoming, treeing and animaling. Many attempts to define wilderness limit it to human experience (Smith and Watson 1979). Obviously, that cannot be done. Wilderness is still greater than or empty of most human experience.

Definitions have to consider the nonhuman nature of wilderness, as well as the pervasive human influence on all ecological systems. Definitions also have to consider traditional human reference to wild areas, as well as areas that are restored or created to become wild. Wilderness can be delineated by human use and nonhuman use, as well as by global function. Eight distinct kinds of wilderness are identified.

1. Sacred landscapes are those areas set aside out of respect, for habitats for others, where humans do not belong, and for species that cannot coexist with humanity. Examples of sacred landscapes are mountains, fisher/marten habitat, and large areas of tundra. In many archaic cultures, portions of the landscape are regarded as sacred. This may be a mountain at the center of the world, a burial ground, or the portion of a river that belongs to the fish gods. Sacred spaces are highly valued locations, with limited access and usage. Such spaces have great symbolic value to societies where they are associated with significant events (or divine manifestations). The assignation

of sacredness may result from functional necessity, to prevent flooding, limit population, or to balance the economy. In sacred landscapes no direct human impact would be permitted; mapping would be possible from satellite or airplane observation.

2. Foundation areas are zones that contain active communities where the integrity of nonhuman life is not challenged by human interference. Foundation Areas house the depth of wilderness. The depth of wilderness is what allows recolonization after disruption or human influence. Foundation areas could include common lands, that is, one of each kind of ecosystem regardless of aesthetic value. Some visitation could be allowed, but without permanent residence or machinery. Human impact would be limited to noninterventive science (perhaps based on Goethe's method of contemplative nonintervention). Examples of foundation areas would include large areas of the Amazon.

3. Reservation Areas are zones for the support of nonindustrial native cultures expressing traditional ways, that is, natural systems in which humans play a limited role—limited by a percentage of net ecosystem productivity, for example, the Miskito in Central America.

4. Preservation areas are places where the system is maintained by human activities, such as burning. Specific kinds of preservation areas could include: Boundary areas, ecotones or lands bordering agricultures, small patches influenced by human activities, or corridors. Human impact could be limited to traditional, experimental science. An example could be the central long-grass prairie in the United States.

5. Restoration areas are zones which are set in a former native pattern by human activities, but may not need further intervention. This may include the experimental restoration of ecosystems disturbed by human activities or natural events, for example, Lake Shagawa.

6. Neopoetic communities are areas that have been created through the introduction of exotic species over time, but have become naturalized. These areas have been started by human activities, but saved as self-regenerating systems and not built up or interfered with, for example the California grasslands, which have been changed with Mediterranean plants, e.g., wild oats, wild mustard, wild radishes, and wild fennel.

7. Conservation parks are areas set aside for multiple use of resources without interfering with the operation of the ecosystems. Research may be conducted to answer questions as to whether the park is big enough and shaped correctly to constitute a proper habitat for its inhabitants. Human recreation would be permitted in temporary camps and with some light machinery, for example, the Boundary Waters Canoe area in the US.

8. Wild agriculture and forests, as well as other wild systems managed for harvest of wild species, such as antelope or fish, are a final category. Although forests can be lightly harvested for wood and other products, they are predominately wild, and would be left wild, rather than being clearcut and replanted with a single species. This is an unusual category of wilderness, since we do not always consider forests and some kinds of agriculture as wild. In fact, wild species in wild systems have been shown to be more efficient in their use of resources and more resistant to other wild

species that we humans consider pests. In this category, we recognize that areas of wilderness can be used without being destroyed or domesticated.

By just wilderness, the first three areas are included indiscriminately. They all have similar values, as mirrors of existence, examples of natural, complex processes, expressions of love for nature, and wild, nonhuman places. The most important definition of wilderness treats it as a vital organ for the life of the earth, the generator of hydrological, geochemical, and atmospheric cycles. It is where nonhuman species live for their own purposes. None of these "definitions" is sufficient to identify wilderness exactly, just frame it with words.

All wild systems have had human influences for thousands of years. Even now, human influence extends into every system through pesticide drift and artificial gases. Yet, over 30 percent of the earth now has no permanent human settlements. In general, none of these wilderness areas will be inhabited (at least by industrial cultures), none of them will have permanent built roads, and management would be limited to temporary intervention.

Each of these categories would occupy different percentages of the planetary surface, depending on calculations of minima and maxima, and depending on cultural values and decisions. I recommend placing 50% of the land area in the first five divisions (eighty percent of ocean and water surfaces); then 16% in each of the last three (five for water); leaving 2% for completely artificial landscapes — industrial or city (three for water). These figures are consistent with several earlier proposals. Eugene Odum, suggests 30% forest cover world-wide, with 60% in tropical areas. C.A. Doxiadis offers 67% of the surface area of the planet left in wilderness. Paul Shepard argues that 75% of the land area should be left wild in a techno-cynegetic society.

Reserves of every kind of region need be saved. Shepard concludes that the consequences of ecological conflict in technological, humanistic civilization are either exile or sanctuary; and sanctuary is the only solution available to the humanitarian ideology. The idea of sanctuary recognizes the multiplicity of factors necessary for viable populations. But, although there are sanctuaries for frogs and ferrets, it would be impossible to establish one for every creature. Shepard discounts the humanitarian objective as considering only "worthy" species. Yet, if all species were considered, the whole planet would end up as a sanctuary. He considers sanctuary an unfeasible 19th century political solution, based on a time when space was unlimited. Shepard states that exile (extirpation or extinction) and sanctuary are allopatric choices (meaning "life not occurring together"). Allopatry (from the word for fatherland) is seen as being consistent with the tradition of personal property, domination of nature, and model of the nation state.

What Shepard has overlooked is that humanity was never sympatric. It never lived together (the meaning of sympatry) with major carnivores. Therefore, allopatry is consistent with the nonexploitation and nondomination of nature as well. Sanctuary as personal property or as nonhuman property is a moot point if habitats are saved from destruction. Domestication and enslavement are sympatric forms. A conservation program, despite Shepard's argument, cannot be based solely on sympatry.

Not all species occur together in the same place. Sympatry describes only those that do. At the habitat level, allopatry is the most intelligent use of available resources by animal communities. Large herbivores, such as elephants and rhinoceros, may choose poorer quality food and avoid competition with smaller animals and exploit an untapped food source. Many interactions between different species contribute to the mutual benefit of the members of the community, as well as to the community itself. Humanity must be allopatric with most wild species and allow them to develop independently in their own places.

Separate areas are necessary. Gary Snyder questions whether complete compartmentalization is healthy. He suggests that some land is saved like a virgin priestess, while other land is overworked like a wife, and some is brutally reshaped like a girl declared promiscuous. This argument is certainly good for human lands, but wilderness lands are being compartmentalized for the use of other species (which do not share our human treatment patterns of females).

Saving wilderness means saving large areas of land. It means placing wilderness, which is support for cultivated and industrial areas as well as natural communities, off limits to development and perhaps to any use. Every community should be allotted a place for self-development, a foundation area. Wilderness is not land that is locked up, as several U.S. senators have claimed; it is quite interactive. We are already living with it. Foundation areas are not museum pieces, as Birch seems to think. Only humanity is directly absent.

Wilderness costs are sometimes considered too tremendous to bear, in lost opportunities, resource use, administrative overhead, according to Lehmann and others, but that is without considering free services of wilderness. Their conclusion is based on the assumption that everything that can be used should be used. Wilderness areas, left by themselves, would require a minimum of management costs. Often, when it is suggested that wilderness areas be left without human management, critics conclude that humanity is not being considered as part of the environment. This is an unfortunate conclusion and untrue. Humanity does not have to convert every system to be a part of it.

A high human civilization has to limit its consumption of natural resources, its place in wilderness, to account for wilderness and future generations. Wilderness is a limiting factor in the health of human civilization. It is probably not a precise function, but we may be close to the minimum for the earth. As Gregory Bateson pointed out, it is not safe to be limited by lethal variables.

Cultures can determine the minimum, or optimum, wilderness areas to support local ecosystems; for instance, very large areas are required in tropical ecosystems or deserts, relatively little ones in grasslands and temperate forests. They can determine the natural productivity and the percentage to be used by humans, as well as artificial productivity and costs. Cultures can key their population to natural productivity for long-term sustainable existence. And, they can multiply any increase by tradeoffs, such as a reduced standard of living or exchange with another group.

CHAPTER **16**

Maximum Human Populations for the Planet

Introduction
A number of recent studies have suggested that the human population
of the earth could be much larger than it is. Several other studies have
recommended lower population numbers based on resource availability. All
of these studies are concerned with finding a *maximum* human population.
Yet, as we know, maxima are rarely stable. Human populations are limited
by the biological constraints of ecosystems, by biogeochemical cycles, by
our knowledge of these systems, and, possibly, by human psychological
and cultural limits. Therefore, an optimum human population is calculated,
using a deductive, synthetic, conceptual model based on data generated
from research on net primary (NPP) and net community (NCP) productivity.
A deductive approach is used because accurate measurements of trophic
level productivities in most ecosystems are lacking. Its synthetic character is
more appropriate to integrate quantitative and qualitative data. The model is
conceptual because of the inherent fuzziness of the systems.

Assumptions/Considerations
Ecosystems result from the interaction of all living and nonliving factors of
the environment (Tansley). These systems are profoundly affected by both
random and purposive physical and biological factors. As a result, habitats
change and organisms adapt. By modifying their habitats in the process of
living, organisms change the characteristics of the system and force further
adaptation. More important, organisms are limited by the productivity of the
system in varying degrees. Human populations (*homo sapiens sapiens*) inhabit
specific ecosystems and are parts of them. They are adapted to and limited
by the productivity of ecosystems.
  The total amount of biomass or energy produced by populations
through growth and reproduction is the productivity of the system. Gross
Primary Production (GPP) is the rate of energy storage by photosynthesis
(equal to photosynthetic efficiency) in autotrophs (plants). The maintenance
and reproduction of plants is paid for by the energy expenditure of
Respiration (R). The amount of energy stored as organic matter after
respiration is identified as Net Primary Production (NPP), which equals
plant growth efficiency. The calculation of NPP is shown by:
$$NPP = GPP - R$$
NPP accumulates through the history of a system as plant biomass expressed
as kilocalories per square meter $(Kcal/m^2)$. The kilocalorie is used as a unit
of energy flow and production; it is a useful common denominator for
these calculations. The biomass minus the decomposition in a system is the
standing crop biomass of that system. The problem of confusing production
(amounts) with productivity (rates) is avoided by considering all values per
unit area (m^2) over the entire year (m^2/yr). The energy stored in heterotrophs
(consumers) is referred to as secondary production (SP) or assimilation.

The storage of energy or organic matter not used by heterotrophs is Net Community Production (NCP). The relationship between productivities (where R_a=autotroph respiration and R_h=heterotroph respiration) is of the order shown by:

$$GPP = NPP + R_a$$
$$= R_a + R_h + NCP$$

In a balanced ecosystem, NPP equals respiration; in an accumulating system, NPP usually exceeds respiration by 1 to 10 percent. Although stable ecosystems tend to produce a maximum GPP, species, biomass, and the production to respiration ratio (P/R) continue to change long after the maximum has been achieved. In fact, as the GPP approaches an asymptote, respiration increases. In a balanced system, tropical rainforests for instance, NCP approaches zero, as adapted heterotrophs become more efficient at using production. In accumulating systems, such as grasslands or young forests, NCP can range from 20 to 70 percent. A balanced system is integrated and self-perpetuating, where production (the photosynthetic fixture of carbon) is balanced by respiration (the oxidation of carbon). As a system becomes balanced, the pressure of selection of organisms shifts; the capacity to live in crowded circumstances with limited resources is favored. Populations that depend on rapid individual turnover (r-selection) are not as successful as populations of large, long-lived individuals (K-selection). As an ecosystem ages, pressure is put on some populations by other populations. Competition and predation become more complex.

Mammals are the best regulated of highly evolved species. Their behavior is controlled and population regulated through the use of space. Most populations, furthermore, regulate their density well below the limits of the food supply, often by as much as 50-70 percent. Territoriality can be correlated inversely with trophic levels and productivity. But populations can also be limited by:

a. specificity of prey or plant source
b. size of prey or plant populations
c. predators
d. natural events or catastrophes

Human beings are mammals — omnivorous, social, bipedal, featherless, symbol-using, tool-making, game-playing, neotonous, bilateral-hemispheric, generalists. Furthermore, as Woodwell put it, humans live as "one species in a biosphere whose essential qualities are determined by other species." Mammals are bound by biological requirements that must be met if a population is to survive. These functional requirements are rather minimal for humans, however, being only food, clothing, shelter, and reproduction. Other requirements, such as respect, comfort, and self-fulfillment, depend on socio-cultural systems.

Like other mammals, humans change their habitats to suit themselves. Other mammals alter their habitats through chewing, digging, and burrowing. Rodents, such as *Ellobius* spp. or *Marmota* spp., can dislodge earth at a tremendous rate (18-120 m^3/ha/yr). In many cases these activities improve the conditions for the growth of vegetation. Mammalian grazing promotes regrowth and the movement of seeds. Bison (*Bison bison*) and prairie dogs (*Cynomys ludovicianus*) were responsible for much of the character of the American plains. Rodent caches may account for 15 percent of Ponderosa seedlings (*Pinus ponderosa*). Beavers (*Castor canadensis*) and other rodents create their own microsystems. Wide-ranging caribou (*Rangifer tarandus*) transfer energy between systems.

Humans have modified animal and plant associations in a different way, simplifying patterns of energy and chemical exchange, solidifying themselves at the end of many food chains as a dominant species. A dominant is a species with greater influence than any other in its biotic community, changing the lives of other species and the character of the habitat. By its influence of all ecosystems, humanity has become a *pandominant* species. As such, humanity reclaims, overgrazes, clears, depletes, and wastes at a level that threatens the stability and existence of many systems. One of the ecological consequences of human activity is the degradation of wild habitats for human developments (food, housing, and recreation) and the introduction of novel elements into the biosphere — elements that have not been added slowly over time as the result of natural processes. The biomass of the human species probably far exceeds the biomass of any nondomestic species, and that biomass is supplemented by the tremendous biomass of domestic animals, which is four times greater than the human biomass (Borgstrom, 1975). This biomass forms an equivalent population that consumes much of the same food as humans, such as milk, fish, and grain. The domination of humanity is related to other characteristics as well:

a. large biomass ($6 \times @10^{14}$ @ Kcal)
b. large annual increase (2%)
c. high structural organization (information and matter)
d. high energy use (globally, 13 times mammal equivalents)

This dominance has major effects on ecosystems: transient perturbations in energy relations (from oil spills, burning); chronic changes/shifts of systems (from dams, irrigation, chemical wastes); species manipulation (from the import and export of exotics); and, interference competition with wild species (as opposed to exploitative competition, which can be stabilizing). None of these effects are exclusive to humans as a species, but they are excessive, rapid, compounded, and very large-scale.

There are minimum viable populations for mammalian species (usually considered about 500 individuals). For the human species, it is unlikely that the lower limits will be approached in the foreseeable future. There are several other lower limits to keep in mind, however, as shown in Table 1. These limits are conservative estimates.

There is a maximum carrying capacity for humanity. The carrying capacity is the population sustainable on a long-term basis of renewable and nonrenewable resources or energies. For humans, this capacity must include domesticates, as human equivalents, since many domesticates compete for protein consumption. Domestic animals can extend the carrying capacity somewhat, since many of them consume agricultural wastes or use lands marginal for agriculture, but they are not as efficient as wild populations. Technology could expand the carrying capacity to some extent, with higher yield crops and resource substitution, but also it could reduce the capacity with unforeseen side effects (the use of pesticides, for example). War and social disorder reduce the ultimate capacity. Furthermore, the capacity decreases as the *per capita* use of energy and resources increases. Carrying capacity calculations often just consider food energy, but all needs—clothing, shelter, transportation, information generation, aesthetic satisfaction—must be included. Given the current political and technological situation for humanity, it is difficult to imagine how equilibrium could be reached without a significant reduction of human populations.

Previous Studies
Many theorists have estimated maximum or optimum sustainable populations for humanity. DeWit estimated the maximum human population at 1 trillion, 22 billion (1.022×10^{12}). With 750 square meters added *per capita*, for forest and recreation, this number would be trimmed down to 146 billion. With consideration for animal protein in diets, 73 billion. These calculations are based on very simple variables—the light scattering coefficient (.3), leaf area index (5), and a mean synthetic rate. Apparently, there are a number of simple assumptions, also: ideal photosynthetic conditions, a strict monoculture without any natural habitats, and human occupation of the first consumer level of all food chains, thus entirely eliminating *all* competing wild animals. An exclusively human earth would confront its sociological, technological, and aesthetic limits immediately.

Colin Clark (1967) put the population of the earth at 47 billion, at an American standard; 157 billion on a Japanese standard. Who knows how many could have been crowded in with the lowest standard? Earlier, Clark (1958) had foreseen a population of merely 28 billion, using a Dutch standard of productivity and density (365 people per square kilometer). Apparently, he neglected to account for the ghost acreage used by the Dutch. Furthermore, he assumed a very large total arable land area—in fact, 63 percent of the total surface area of the planet (by counting the tropics twice).

Burlingh et al. estimated the absolute maximum food production of the world at about 30 times the production of 1970 (and presumably an equally high human population). They arrived at this figure on a basis of quality of soil, climate, and water. It does not seem that any biological factors were considered. Weinberg and Hamilton presented a steady state model of 10 billion at 400,000 Kcal per day for the year 2050, but made no mention of how the population would stabilize at that level or what kind of economics would support it or for how long.

Most of these optimistic studies see the limits of world food supply as being determined by the physical limitation of land, water, light, and chemicals (Pirie) or by the logistics of transportation and conscience (Moore and Lappe; Gabel). They assume that areas of cultivation can be expanded into planetary biological support systems, that efficiency can be enhanced (at the risk of genetic instability), and that novel sources of food will be used (ignoring cultural limitations). These requirements may never be met, as others have realized.

H. R. Hulet assumed that plant and animal products were equally necessary. By making them equal requirements — animal and plant production at 4.2×10^{15}/Kcal each — he suggested that the earth could support an optimum population of 1.2 billion. Considering the Law of the Minimum, he offered a series of maximum populations as a function of resources (Table 2). He concluded that as technological and agricultural systems expanded, population could also. But, the rates of use for the minimum items he provided were not renewable or slow; they were based on American standards of consumption. In America, energy use is increasing at 6 percent per annum; three times faster than the population. Therefore, resources would be limited long before a maximum population was reached.

Westing estimated a global carrying capacity based on five areas of renewable resources (Table 3). His estimates range from 1.5 to 3.9 billion. These numbers are for a maximum population, however, not an optimum. It is important to remember that an optimum is almost always less, much less, than a maximum. Note also that his paper, written ten years after Hulet's, when wood production had increased significantly, gives a doubled figure for a population limited by the availability of wood. In fact, neither Hulet's nor Westing's rate is sustainable. Annual timber cutting in the United States in 1963, for example, exceeded annual growth by 50 percent. Most cutting since then has exceeded growth by various percentages. Furthermore, wooded areas are still being cleared for agriculture, housing developments, and industrial areas. Westing recognized that most usable renewable and nonrenewable resources fundamental for human life are used in direct competition with wildlife. He commented that most studies did not consider wildlife and habitats.

For a straight energy calculation of an optimum population, assuming the unavailability factor has been subtracted first, refer to Table 4. The French *Per Capita* energy use is approximately the same as the world average. Portuguese energy use is the same as agricultural trade. The main difference between this model and earlier ones is the *pattern* of utilization and the assumptions of minimal waste and *no growth*.

Whittaker and Likens have suggested that an agricultural world, where humanity lived as peasants, could support at least 2 billion, perhaps 5 or 7 billion. The current high levels of population, at a large range of standards, can only be maintained through the constant takeover of natural habitats for arable land, or through the drawdown of fossil fuels, and by economically cheating the poor and powerless. Since the quantity of wild lands and fossil fuels is quite limited, either human populations must adjust to renewable resources or technology must provide substitutes, to avoid a population crash.

Eugene Odum suggested using land area as a measure of human carrying capacity. The minimum per capita acreage requirements, with a temperate area like Georgia as a model for a quality environment, is 5 acres (2.02 ha). The percentage of areas is broken down in Table 5. The natural areas are based on minimum space needs for watersheds, as estimated by land use surveys. Food-producing land includes acreage for domestic livestock. Extrapolating this technique to the entire planet, assuming that wilderness area has been considered in the calculation of natural areas, and converting for the differences in productivity of ecosystems, the population calculation comes to 3.969 billion. This figure is very close to the 1970 world population.

Samuel Eyre, Rodin et al., and others have calculated the potential productivity of the wild vegetation of the earth at around 1.19×10^{11} metric tons per year. Eyre attempted to describe the wealth of nations in terms of NPP, with nutrition equivalents in NPP for mineral resources. He contended that one must know the productive capabilities of land in its original vegetation to compare with productivities under human management. He found, for instance, that most wild lands are more productive than most agricultural acreage. What is left out of his considerations is the amount of healthy ecosystem necessary for cycling and renewal. Furthermore, all productivity is treated as economic, to be dispensed with by the nations that occupy the land, at will. Although Eyre's NCP model is more reasonable than most models, it still puts humanity in competition with any remaining wildlife for productivity in every ecosystem.

An NCP Model
It is possible, however, to calculate a sustainable human population in balance with healthy ecosystems, using NCP instead of NPP. The population for this model would be much lower since NCP is generally lower. For instance, in tropical and temperate grasslands, NCP may approach 60 percent of NPP, although 30 percent is much more likely. Temperate forests may approach 30 percent. When NCP is used to calculate a *maximum*, the measurements are consistent with the low figures of several studies. Refer to Table 6 for calculations.

Of the NCP produced, 75% is unavailable, 5% is eaten by pests, 65% of that harvested is inedible, 80% of that edible is lost in processing, and 25% is wasted during consumption, resulting in a maximum number of *903 million* (from a preliminary figure of 82.6 billion). This figure is almost identical to a flat 1 percent rate of the NPP, subjected to the same loss percentages. Applying a fifty percent rule (Wittbecker) to this maximum figure, 903 million, results in a crude *optimum* population of *451 million.* This conservative number would insulate human populations from environmental variables and from fluctuations in productivity. Kozlovsky intuitively estimated 500 million as an equilibrium population. Lower densities of humans will always be able to harmonize more successfully with biological processes. For the long-term survival of the human species, adaptability to environmental changes is necessary. This requires a wide diversity of gene pools, which is achieved by a relatively large population divided into local, partly isolated groups in healthy regional ecosystems. The optimum size of the global human population is actually the sum of optimums for local habitats.

Before an optimum human population can be put forward as a goal, numerous questions must be addressed, including:
   1. How much land should be left in its native state? Enough to save one of each kind of ecosystem? Enough to support atmospheric cycles? Enough for exotic ecosystems? Enough for public service? Enough for

extensive recreation?

2. At what level of luxury should humans live (in Kcal/yr.)? Just basic amenities (such as clean air and water)? With the potential of having every technological device available? With an excess of flexibility?

3. What are the physical limits of resources? Should all nonrenewable resources be used, and, if so, at what rate? Should only renewable resources be used?

4. What is an optimum? Laboratory studies with rats show that, with a choice of optimum rat environments, some rats reject the optimum. Should the optimum be set for a majority (by numbers or cultures?) or for some low denominator?

How these questions, and others not asked, are answered determine an optimum human population. Knowing how many humans are alive to feed is easier than describing the basics of a quality of life. In calculating an optimum population within ecosystem restraints, few have considered minimum wilderness preservation, air and water quality, genetic minima, nonrenewable resources, appropriate technological innovation, the importance of cultural frameworks, adventure, research, beauty, uniqueness, and other intangible experiences. If human civilizations were based on balanced ecosystems, they would be more complex. The complexity encountered in trying to imagine them may serve as a Zen koan-work, to bring us to rational breakdown and to an alternative: humanity as a self-conscious, self-limiting, poetic species.

Tables

Table 1. Minimum Limits (revised from Wittbecker, 1970)

genetic minimum	5,000
fertility	50,000
ideomass	500,000
social contact minimum	1 million
evolutionary advantage	10 million

Table 2. Population Functions (after Hulet)

wood production (4×10^6 cal/yr/cap)	1 billion
energy rate	600 million
fertilizer rate	900 million
aluminum rate	500 million

Table 3. Population Estimates (after Westing)

total land	2 – 3.1 billion
cultivated land	1.5 – 3.3 billion
forest land	2 – 2.9 billion
cereals	1.7 – 3.3 billion
wood	1.9 – 3.9 billion

Table 4. *Optimum Populations (1980 Per Capita Use)*

United States	120×10^6 Kcal	189 million
France	4×10^4 Kcal	567 million
Japan/Argentina		1.13 billion
Portugal		2.84 billion
India/China		5.27 billion

Table 5. *Acreage Per Capita (after Odum)*

food-producing land	30%
fiber-producing land	20%
natural support areas	40%
artificial areas	10%

Table 6. *Simple NCP Calculation of Population*

Vegetational Unit	Area 10^6 km	NPP 10^9 mt	NPP 10^{15} Kcal	NCP 10^{15} Kcal	Population 10^9
Trop Rainforest	17.0	47.4	195.5	0.187	0.17
Trop Raingreen	7.5	13.2	55.5	0.052	0.47
Temp summer	7.0	7.0	32.2	2.24	2.04
Mediterranean	1.5	1.2	5.9	0.03	0.03
Temp Mix	5.0	5.0	23.5	6.35	2.45
Woodland	7.0	4.2	19.6	7.0	7.0
. . .					
Ocean	332.0	41.5	199.2	0.199	1.81
Shelf	26.6	9.2	43.1	2.13	3.87
Estuaries	1.4	2.5	11.3	0.07	0.06
Cropland	14.0	-	-	12.81	14.63*
Totals				90.87	82.61

* as percentage of forest/grassland in original vegetation, approximately 35% less NPP. This paper is concerned only with the values of natural productivities. Domestic lands were assigned productivities based on those of the original wild vegetation (which in fact is usually higher).

CHAPTER 17

Ecological Design and Planning

Introduction

Our local communities are proud to attract more people and larger industries, but do so thoughtlessly, without regard for the limits of population size or the rate of energy use, without sufficient consideration of the effects on the quality of our lives or on the quality of the environment. Although we make plans for people and their activities, the plans are usually reactions to growth and change. The formal development from planning results in a complex of problems, from pollution to ugliness.

We have always tried to exceed the physical and biological limits of places rather than recognize them and be guided by them. This paper suggests an approach to comprehensive planning based on the biohistory of an ecosystem, the cultural values of the people, and knowledge of the limits for sustainable development. This approach makes the limits explicit and set sustainable goals within those limits. As a synthetic framework, this approach provides for the health of the ecological system, as well as for the health of its human inhabitants.

Central Planning

At first a means of controlling people through zoning, central planning has expanded its reasons to include sanitation, economics, and aesthetics as well. Although central planning grants some consideration to support areas and aesthetic factors, cultural traditions and the natural environment are not primary concerns.

Planning in general means deciding on goals to be achieved in specific situations. For central planning, the goals are usually small and not comprehensive, such as the rate of emission of sulfur oxide, and usually end up being a compromise in cost-benefit analysis. Planning tends to neglect or dismiss the distribution of negative, uncertain, or nonmonetary effects. Furthermore, we seem to have no mechanism for developing long-range plans. Certainly, there seems to be no way to deal with long-term, slow catastrophes, from erosion to climactic change.

Most plans address *problems*, from waste water treatment to air pollution monitoring. Everything else, from employment to pests, is also considered as a problem, and not a direct effect of the cultural implementation of some technology. Most plans seem to be extremely good at compiling area data, from topographic to climactic. These plans are concerned with determining the adequacy of the infrastructure (utilities, streets, sewers) to support actual and projected population growth. Development plans (water, power) are comprehensive in the sense of seeking to meet all needs of the public, agriculture, and industry. But, they fall prey to all the assumptions of the industrial culture. They tend to be multipurpose with the aim of providing maximum net benefits through management of watersheds, fish and wildlife, and flood control. Both uses of "multipurpose" and "maximum benefits" are based on misunderstandings. Multipurpose in practice means human use, and maximum benefits have proven to be dangerous. Modern resource management strives for maximum sustainable yield, based on partial knowledge of population size and great ignorance of population flows.

Development plans also tend to call for the eventual development of *all* resources in an area; Brazil's Plan 2010, which would develop all of the Amazon with 136 high dams, is a good example. A one-world planned economy is an even greater threat. It is based on unlimited industrial production, unlimited commodity consumption, increased exploitation of nature, and the free flow of resources and labor across cultural borders. This kind of planning requires the abandonment of local controls on development, trade, or lifestyles. All countries are expected to open their markets to outside investment, eliminate tariff barriers, reduce government spending (especially to the poor), convert small-scale, self-sufficient farming to agribusiness, and open all land to resource gathering. Planning is thus characterized by a utilitarian globalism that denies value to the systems that support it.

As a result of central planning, the patterns of life have become the products of market forces and stylish transportation operating in a sterile abstract order. In America we are criticized for having a "frivolous" culture based on "savage" capitalism. Capitalism increases the pressure for uniformity, a single pattern of existence. Formal development is more concerned with an assembly-line model — simple, isolated, efficient, and easy to maintain. We become remote from, and indifferent to, the system that supports us. We acquire unrealistic images of the world and harmful values and then make bad decisions based upon them. We have not developed qualitative indicators on ecological health or quantitative measures of social health, much less an ecocentric view that would value preserves of nature for themselves.

Ecological Planning
A number of proposed plans to heal the earth and improve human communities have been presented in popular books. Many of them are too philosophical and general, suggesting that we could change values without showing how or urging us to alleviate some of the symptoms without addressing the disease. Other plans, such as the *Limits to Growth* (Meadows et al.), are too global. And still others, such as *Design with Nature* (McHarg), are less concerned with limits than with basic conservation. *Goals for Mankind* (Laszlo) offers a similar compendium of global goals that can essentially be summarized to be health and freedom for people in a healthy environment. Many of these plans offer admirable models, but little in the way of goals or paths.

A plan should consider the whole system. Communities should be designed for an optimal fit within the limits of the system. Ecological planning considers an optimum population within one ecosystem, although it is connected to others by trade for some necessities or luxuries. This kind of planning is a conscious adaptation of the benefits of technology to the traditional idea of physical, as well as cultural, limits. Direct observation and traditional knowledge yield far more "information" about the societies of animals than autopsies and mathematical models. An outline of a comprehensive plan is presented, to deal with some of the implications, as well as question them.

1. Identify our place within its natural boundaries. Most places exist in a uniquely identifiable ecosystem, with recognizable boundaries and a unique history and character.
2. Calculate the optimum amount of wilderness to preserve the natural cycles indefinitely. If the current area is less than our calculations, restore the difference and set it aside as a reserve.
3. In the remaining area, zone areas for appropriate use, including conservation, preservation, reservation, and artificial areas (with historical, cultural, and functional importance).
4. Identify the resources needed for human use, including raw materials and the productivity of the areas. This productivity can be used to calculate a base line population.
5. Apply cultural modes — in style, values, and technology — to set limits on technology and population. Preserve the cultural values. Renewable resources will sustain a population longer than energy capital like oil or gas.

As part of the formulation of a plan, we have to examine the natural and cultural histories of a place. We need to understand interactions in the ecosystem, as it was with no humans, as it was lightly settled, and as it is now, dominated by humanity.

Biohistory

The Ecosystem
Every place has a geological history. Mountains have been ocean bottoms; deserts have been marshes. Soils rest on different kinds of rock formations. Rain, rivers, volcanic activity, vegetation, and many processes including human ones have shaped the current landscape. The landscape may be flat, or hilly, or mountainous. It has a predictable range of precipitation. It contains habitats for plants and animals; there are ecotones between habitats. Animals and plants have left a history that can be traced in many cases. The interactions of plants and animals with biogeochemical cycles makes up ecosystems.

Ecosystems result from the interaction of all living and nonliving factors of the environment. These systems are profoundly affected by both random and purposive physical and biological factors. As a result, habitats change and organisms adapt. By modifying their habitats in the process of living, organisms change the characteristics of the system and force further adaptation. For example, mammals alter their habitats through chewing, digging, and burrowing. Rodents can dislodge earth at a tremendous rate ($18-120$ cubic meters/ha/yr). In many cases these activities improve the conditions for growth of vegetation. Mammalian grazing promotes regrowth and the movement of seeds. Bison and prairie dogs were responsible for much of the character of the American plains. Rodent caches may account for 15 percent of Ponderosa seedlings. Beavers and other rodents create microsystems that other animals depend on. Caribou and elk transfer energy between systems. Shrews consume major portions of larch sawfly larval populations. More important, organisms are limited by the productivity of the system in varying degrees, and the productivity is limited by light (and heat) and water.

Human populations inhabit specific ecosystems and are parts of them. They often have been adapted to and limited by the productivity of ecosystems. Like other mammals, humans change their habitats to suit themselves. Humans have modified animal and plant associations in a different way, simplifying patterns of energy and chemical exchange, solidifying themselves at the end of many food chains.

The total amount of biomass or energy produced by populations through growth and reproduction is the productivity of the system. An ecosystem has various kinds of productivity. Gross Primary Productivity (GPP) is the rate of energy storage by photosynthesis (equal to photosynthetic efficiency) in autotrophs (plants). The maintenance and reproduction of plants is paid for by the energy expenditure of Respiration (R). The amount of energy stored as organic matter after respiration is identified as Net Primary Production (NPP), which equals plant growth efficiency. The calculation of NPP is obtained by subtracting respiration from the gross productivity (NPP = GPP - R). The NPP accumulates through the history of a system as plant biomass expressed as kilocalories per square meter ($Kcal/m^2$). The biomass minus the decomposition in a system is the standing crop biomass of that system. The kilocalorie is used as a unit of energy flow and production; it is a useful common denominator for these calculations. The problem of confusing production (amounts) with productivity (rates) is avoided by considering all values per unit area (square meters or hectares) over the entire year. The energy stored in consumers, or heterotrophs, is referred to as secondary production (SP) or assimilation. The storage of energy or organic matter not used by heterotrophs is the Net Community Production (NCP).

In a Mature (balanced) ecosystem, the net primary productivity (NPP) equals respiration; in a young (accumulating) system, NPP usually exceeds respiration by 1-10 percent. Although stable ecosystems tend to produce a maximum gross primary productivity (GPP), species, biomass, and the production to respiration ratio (P/R) continue to change long after the maximum has been achieved. In fact, as the GPP approaches an asymptote, respiration increases. In a mature system, temperate rain forests for instance, net community productivity (NCP) approaches zero, as adapted heterotrophs become more efficient at using production. In accumulating systems, such as grasslands, NCP can range from 20-70 percent, although 30 percent is a good average.

A balanced system is integrated and self-perpetuating, where production (the photosynthetic fixture of carbon) is balanced by respiration (the oxidation of carbon). As a system becomes balanced, the pressure of selection of organisms shifts; the capacity to live in crowded circumstances with limited resources is favored (usually with regard to animal populations, although perhaps it applies to humans).

Human Cultures

Most places have supported different groups of people with unique cultures for recent history (at least 12,000 years), according to archaeological evidence. Prehistoric living sites have been found in caves, shelters, and camps. These archaic peoples caught insects and fish. They hunted rodents, jackrabbits, deer, antelope, and elk, and possibly mountain sheep, elephants, kangaroos, and bison. They trapped migratory birds nesting along streams, and native birds.

As the area became more densely settled, people depended more on wild plants, such as roots and berries. Settlements became more permanent, especially in winter. Groups tended to disperse in temporary camps in the summer, when getting food was easier.

Peoples developed unique identities, as Palus, Campa, Hadza, or Naga. Although their languages are different, many archaic beliefs and customs are similar. Neighboring tribes would maintain ties through trade, marriage, and sharing resources (hunting and gathering grounds).

Native people had very few large settlements. A village rarely exceeded 200 people. Although inhabitants of a village (or a band) recognized a certain amount of the land surrounding as their territory, they shared most of the hunting and gathering grounds with people from neighboring villages. Each village might contain the households of extended families, usually three generations. Clubs and societies promoted bonds between nonfamily members. Many social ceremonies reinforced the cohesion of a group. Gatherings were regular, to celebrate the first fish or last crop of berries, as well as social ties. They visited, traded, played sports, and gambled.

Archaic peoples were able to use the area without changing it fundamentally. They were able to work out appropriate solutions in harmony with their environment. Traditional forms of restraint, such as prescriptions for marriages and births or restrictions on hunts, ensured that tribes would not interfere with local animal and plant populations, much less with ecosystem cycles. Population density was controlled by the traditional approaches to resources. Cooperation and consensus, as opposed to competition and individual exaltation, permitted planning to remain informal. Cultural beliefs made planning for the indefinite future more inclusive; life was a continuity in form (from bear to human, for instance) as well as in state (from unborn to person to ancestor).

The peoples were able to convert animal and vegetable resources into all their needs for food, shelter, and clothing. When other items were available, such as horses and metal knives, they were able to trade, fish, berries, baskets, and roots for them. The population number could probably be supported indefinitely by the Net Community Productivity (NCP) — that is, using only that amount of ecosystem productivity that is not used by the plants and animals. Archaic people's contributions to agriculture, architecture, government, and planning have rarely been acknowledged.

Agricultural development was the first intensive use of many regions. As agricultural technology became more advanced and the demand for crops increased, less desirable segments of the land were converted to agriculture.

The earth has become more populated, with hundreds of thousands in urban locations. Agriculture has produced monumental yields, but only at the cost of tremendous erosion and great subsidies of fertilizers and pesticides. Dams have been built all along rivers, altering the river and fishing grounds. Changes have been made without regard to the long-term impact on the ecosystem or on its human population. Humans have simply dominated many entire ecosystems.

Humanity has become a pandominant species — a dominant is a species with greater influence than any other in its biotic community, changing the lives of other species and the character of the habitat. As pandominant, humanity reclaims, overgrazes, clears, depletes, and wastes at a level that threatens the stability and existence of many systems. One of the ecological consequences of human activity is the degradation of wild habitats for human developments (food, housing, and recreation) and the introduction of novel elements into the biosphere — elements that have not been harmoniously worked in over time. The biomass, or demomass, of the human species probably far exceeds the biomass of any nondomestic species, and that biomass is supplemented by the tremendous biomass of domestic animals, which is four times greater (Borgstrom, 1975). This biomass forms an equivalent population that consumes much of the same food, such as milk, fish, and grain. The domination of humanity is related to other characteristics as well: A large biomass (6×10^{14} Kcal), a large annual increase (about 2 percent), our high structural organization (information, matter), and our high energy use (globally, 13 times mammal equivalents).

This dominance has major effects on ecosystems: transient perturbations in energy relations (from oil spills, burning); chronic changes/shifts of systems (from dams, irrigation, chemical wastes); species manipulation (from the import and export of exotics); and, interference competition with wild species (as opposed to exploitative competition, which can be stabilizing). None of these effects are exclusive to humans as a species, but they are excessive, rapid, compounded, and large-scale. With a comprehensive, ecological, long-term plan, we can direct or anticipate changes and impacts.

A Plan for Whole Places

In place of a comprehensive plan, this thought experiment creates a deductive, synthetic, conceptual model based on data generated from research on biological productivity, the rates of resource use, and cultural valuation; it also considers minimum wilderness preservation, air and water quality, genetic minima, nonrenewable resources, appropriate technological innovation, the importance of cultural frameworks, adventure, research, beauty, uniqueness, and other intangible experiences. A deductive approach is necessary because accurate measurements of productivities in most ecosystems are lacking and exactness in values is misleading. A synthetic approach is necessary to integrate quantitative and qualitative data. In combining measures of qualitative and quantitative, it is simpler to set aside the first and then to calculate the second. The model must be conceptual because of the inherent fuzziness of the systems.

Many aspects of nature escape the precision of mathematical models. After Plato, the world has been characterized by its mathematical imperfection. The mathematics of inexact structures, or fuzzy quantities, was first described by Lofti Zadeh in the early 1960s. Fuzziness is a kind of imprecision that results from putting elements in classes that do not have sharply defined boundaries. Specifically: how much is enough? Or how many beautiful places are there? Whenever ambiguity, vagueness, and ambivalence of empirical phenomena are described in mathematical models, a binary approach is inadequate.

Classical models work well for simple and isolated natural phenomena, but are not as suitable for complex interactions, especially those with subjective dimensions. Deterministic models work well for things that obey physical laws like gravity, but not when the concepts are extended to ecological systems — systems that are adaptive — or to any system modified by human subjectivity.

Imprecision can be characterized by degrees of fuzziness. Any attempt to classify, categorize, and relate organic groupings results in fuzziness. Abstraction also loses details and is therefore a fuzzy process. Fuzzy sets provide a strict mathematical framework for studying imprecise conceptual phenomena in modeling. Fuzzy sets provide a gradual transition from the rigorous and quantitative to the vague and qualitative. The numbers in this model are fuzzy, that is, they are characterized by a possibility distribution. For example, 'the optimum population is much smaller than 2,254,543' or the 'energy maximum is approximately 1.4×10^8 Kcal per hectare.'

The central planning system is designed to take full advantage of computer applications and find a base unit of measurement. But a computerized information model is only partial; what cannot be quantified, such as feelings or relationships, is often ignored. Computers cannot very well handle personal observations, sensory impressions, or historical contexts or mythical relationships — just those things used by primary cultures to manage their resources. Although this model quantifies many things that seem nonquantifiable, it is essentially a verbal description.

This model has a small theoretical basis, but it is basically an appeal to action. The model is in harmony with strategies for sustainable ecosystems, the conservation of biological diversity, and aspects of global change. The model attempts to work out plans and policies for long-term environmental stability. The plan describes an architecture of physical and social institutions, that is, buildings as well as politics. The model uses a focus/ frame metaphor for stability and planning. It contrasts unconscious growth with conscious planning. The goal of planning is to enhance life — all life, not just human life — so it is not restricted to the human species in the present.

The human population energy of an area is related to land area, productivity, technology, and culture in one algebraic expression (Equation 1). The population (P) is a general number which is calculated by adding the total annual agricultural productivity (in Kcal) to the total annual resources (in Kcal), multiplying that sum by a technological and cultural modifier fraction, and dividing that by the annual per capita requirements (U) for food and resources. The available area (A) is the total land area minus wilderness areas, conservation areas, and other areas to be reserved. The net usable productivity (N) is calculated by subtracting the total unavailable productivity (M) from the net primary productivity (NPP); M includes percentages for below-ground productivity, various wastes, and inedibility. The energy values (E) for resources, water power or zinc, for instance are added. The sum of annual food requirements (R_f) and annual resource requirements per capita (R_r) are combined to make one figure (U). The technological modifier, (T) is based on the use of technology in extending or contracting the food or mineral productivity. The cultural modifier (C) is based on the application of cultural values in determining area and productivity. Total area, gross productivity, and energy values are known quantities, while the remaining factors must be evaluated in a mathematically fuzzy way.

$$\text{Equation 1.} \quad P = \frac{[(A \bullet N) + E] \bullet T \bullet C}{U}$$

where: $\quad A = A_0 - A_1 - A_2 - A_3 - A_4$
$E = E_1 + E_2 + ... E_n$
$N = NPP - M$
$U = F_r + R_r$
A_1 = wilderness $\quad A_2$ = conservation
A_3 = fiber $\quad A_4$ = artificial
A = area used $\qquad E_{1-n}$ = individual resources
P = population $\qquad N$ = net used productivity
NPP = net primary productivity
M = total unavailable productivity
F_r = food requirements
R_r = resource requirements
C = culture modifier
T = technology modifier

For the total figures for all essentials—food, shelter, clothing, transportation— energy is converted to Kilocalories and placed on an annual budget (and averaged over 1, 10, or 100 years). These calculations are used for the purpose of illustration; they are not conclusive or binding. The entire equation is expanded in the following discussion.

Wilderness Reserve
In spite of the uniqueness of places, there have been few successful systematic attempts to save more than patches of original vegetation. Most of the places in formal reserves are mountainous or are somehow considered less attractive or useful to people. Most systems are only partially represented.

Most conservation strategies are completely anthropocentric, from saving hunting grounds in the Middle Ages or resources this year. This plan proposes ecosystem preservation, which protects entire biotic communities:

genes, populations, species, habitats, associated traditional human cultures, and all the processes and interactions. To keep the essential services of nature, from atmospheric cleaning to soil-formation, we need large reserves. Reserves are critical elements in global element cycles; they provide a natural base line for management reference and a unique opportunity for scientific research. Large reserves would increase representation of species and save viable mammalian populations, that is, maintain the integrity of wild gene pools. Such reserves would permit natural processes to occur without human interference. We also need large reserves to derive further benefits from understanding natural processes and direct economic benefits from species. For aesthetic purposes: to see, to participate in nature (these being the basis of watching and tourism).

The desired size of the preserve is a complex function of the area's key species, quantity of suitable habitat, and minimum viable numbers of species. Large-bodied vertebrate species tend to have lower population densities, thus a reserve with self-sustaining large-bodied vertebrate populations will likely be adequate for herbivores, insectivores, and primary producers. Thus, key mammal species have to be identified. Determining the minimum number of individuals in a population to guarantee a high probability of survival results in widely varying minimum areas, depending on the key species selected. Frankel and Soule calculate that a population of 500 is needed to maintain genetic viability of each animal species. Each animal requires a minimum area; for example, each grizzly bear may requires over 100 square miles (over 16.5 million ha) for a home range. Using the grizzly bear as the key species, the minimum area for the reserve becomes 5,000 square miles (almost 8.5 trillion ha); by comparison, Yellowstone Park is only 3,458 square miles — no single wilderness areas seem to be large enough. With a home range of 250 ha, the minimum area for deer would be 200,000 hectares. With a home range of 1-5 ha, the minimum area for mice would be 2,500 hectares. Minimum habitat protection is necessary for the protection of endangered or threatened invertebrates, which are responsible for maintaining basic ecological processes through predation, recycling, and pollination.

Although there have been debates over whether a single large reserve is better than several small ones, the shape and size of some reserves is determined by habitat studies of the unique natural history and conditions. The size should be large enough so that species will not be vulnerable to "extinction vortices" caused by genetic or environmental stochasticity. In most reserves, disturbance from farming, grazing, or recreation would probably be the greatest threat.

A range of elevations across areas would minimize the effects of climactic change — and the possibility of extreme change is rarely considered in wilderness design. Soils, drainage, and land-use history and ownership would also receive similar considerations. This would allow management for diversity on different scales. The reserve would extend into the ecotones; the edge effect would benefit many species. Because local soil conditions result in a wide variety of habitat types in small areas, even small preserves would have a rich flora and fauna.

There is little information on the restoration or preservation of "near-natural" ecosystems—primarily native, not subject to major change. There are probably very few near-natural areas; most are semi-natural (pastures as a consequence of human activity) or artificial (totally humanized with asphalt and exotic species). Many a system could not be restored to its original state, since many species are extinct, but it could be rehabilitated. There are a number of methods. The inner buffer zone, being rehabilitated, would be the most expensive to create. The outermost buffer would be managed by benign neglect (despite the fact that many do not list this as a management option). Natural processes, such as fire, wind, or species explosions, would be allowed to operate freely, even if they altered the functioning of the system.

The shapes for the reserves would minimize the dangers from physical and climactic changes. The greenhouse effect could drastically alter the species distributions in reserves, with the loss of many species. Placing the reserves on heterogeneous soil types and topographies increases the chances that the temperature and moisture requirements of species would be met. Simply maximizing the size and number of reserves would enhance long-term survival.

The cost of reserves would be high. Prime grassland sells for 2,000 dollars (US) per hectare. The cost of a reserve, for restoration, has been estimated at 20-140,000 dollars per hectare, depending on the density of planting and the area. The costs of buying, rehabilitating, and managing 200,000 ha could cost 2-4 billion dollars (or $15 billion for 1 million hectares). Larger areas would reduce management costs; smaller reserves in general require more intensive management and habitat manipulation.

Resources

Perhaps the most important resource is the living land itself. The single most important system factor is climate, and this means thinking about the fluctuation of the climate over 10,000 years. We depend on water (and on aquifers) for drinking and irrigation.

For arable land the global average is only 24 percent. We tend put our buildings and roads over arable land. We dig up tons of gravel for roads. We fence off portions for range and plow other areas for crops. We depend on the plants, animals, resources, and energy we take from the land.

Plants

We have converted most arable lands to agricultural use, even though that reduces productivity. In general, according to Eyre, converting wild lands to agricultural use reduces the average above-ground annual NPP by 75 percent. Agricultural intensification also causes fallow cycles to be shortened; a shift from mixed crops to monocropping; and a shift from natural fertility to artificial fertilizers. Buying pesticides and fertilizers and the equipment for applying them, has forced farmers to operate in a condition of chronic indebtedness to financial institutions.

Much of the productivity of plants is not usable by humans. In lightly grazed short-grass prairies, over 75 percent of the production is underground

(83 percent in ungrazed). For example, in the production of forests, 65 percent of above ground production is not used by most lumbering operations (litter, leaves, bark, twigs). Of a dry weight sample of wheat, only 40-45 percent is actually grain. Of course, much of the straw and slash should be left so that the system can regenerate, and some of it can be used for paper, packaging, and press-board. Possibly the representative average of unusable material is 50 percent. Taking most of the material is not desirable anyway because mineral nutrients are locked up in plants and are needed for new growth. Complete extraction of plants would result in removal of 30 to 70 percent of the potassium in the ecosystem. Although there is plenty of potassium locked in rock particles, this is released very slowly.

Agriculture is faced with very real limits. The amount of light reaching the earth's surface is a limit that probably cannot be increased safely, despite the scientific dreams of orbiting reflectors. Most modern plant varieties have about the same photosynthetic efficiency, anyway, although some varieties have a superior leaf arrangement. Plants would be more efficient if the atmospheric composition were manipulated so that ambient carbon dioxide levels were concentrated; this might be done in greenhouses. Phosphorus is a limiting factor; modern agriculture squanders it, so that much of it ends up in rivers and then the ocean. Available water is also a limit. Many of these limits are not negotiable.

Successful agriculture depends on an artificial climax or sustained successional state. The soils have been destroyed by successful agriculture, and we may not have the knowledge to succeed at working them. The farmer should have a thorough knowledge of the physiography of the land — soils, crops, forests, pastures, mineral content, microclimate — and study the effects produced by native flora and fauna. Waste and manure need to be returned to the soil. Large-scale farming has some advantages, but even it is more efficient when done by the owners. Small-scale residential ownership is best. Smaller farms have started to fertilize by crop rotation and manure. There are successful organic farms and successful small ones that use low-energy methods. Some of these keep many kinds of plants and animals, produce less surplus, use no chemicals, grow adapted crops, and have wild lands. These farmers are closer to ecological harmony: with gardens, fields, trees, bees, fish pond, pasture, forest, and natural vegetation.

Much of industrial agriculture can choose to become more labor intensive and to multicrop, and to reduce hybridization and to use natural pest control. Sometimes farmers venture into the ecotones and even into forest areas to grow crops, which ends up costing more.

Many agricultural products, from milk to forests, are subsidized by the government and would be noncompetitive in the market without the subsidy. Eventually, consumers will have to pay the true costs, without agricultural supports. This may mean doing without or paying higher prices ($3.00 for a loaf of bread for example). There are things we can do to make agriculture sustainable and successful:

1. Diversify crops, especially adding drought resistant varieties (many places have drought conditions every summer, worse periodically). Grow wild plants to combine with food plants. Convert from intensive

cattle raising to antelope farming. Grow more varieties. Increase self-reliance on most foods, except those that cannot be grown locally, e.g., kiwi or oranges).

2. Develop new products from existing crops, e.g., oils, drugs, and fuels. Rape seed, for instance, is used in hydraulic fluids, plastic film, nylon, Lorenzo's Oil, and vegetable oil; there are new applications.
3. Use appropriate technology (solar power, field drilling, organic growing); import less energy, maybe export some. Stress low-input agriculture; low-fertilizer and low-pesticide may result in lower gross sales per area but in higher net revenue per area.
4. Process the crop (sell noodles as well as wheat or rice cakes and rice).
5. Market the crops locally; package them and advertise them, also.
6. Form cooperatives, especially for specialized market or low volume products (beets for instance).
7. Create a land trust. How do we protect farm land from pressures of development? As market value as a nonfarm increases? Zoning? Land trust to equalize taxes to farmer? Land trust funded by property transfer tax. The land trust leases the development rights on farmland.

We depend on vegetation for more than food: it makes up the content of much of our newsprint, construction, furniture, clothing, packaging. Furthermore, with shortages of minerals, many substitutes are expected to be organic. But, our monocrops are directed only to one market at a time.

Animals

The original land supported good numbers of mammals. To some extent they have been replaced by domestic species: Dairy and beef cattle, swine, sheep, and poultry. Livestock numbers over twice the human population of the earth. Over 60 percent of the cropland production is devoted to feeding them, and over 30 percent of raw materials to housing and transporting them (for example, it takes 5 to 20 calories of fuel to produce 1 calorie of meat).

Although many animals, such as cattle and sheep are raised on ranges, they often spend months in feedlots being fattened with grain for human consumption. About 95 percent of this food goes for respiration or ends up as manure. The 95 percent loss is acceptable when an animal is raised on rough ground or with native populations — antelope for example. Harvesting some wild animals may be a better alternative than agriculture; this may be cheaper than improving the pasture degraded from overuse. Wild species, especially in parts of Africa, might be more appropriate on marginal soils. Raising food animals may be acceptable using wastes and scraps that contain recoverable food, but it is not acceptable using whole grain crops or on free range. Many animal foods could come from sources unappetizing to humans, such as insects or algae. Harvesting algae directly for food is possible, but has high processing costs, poor flavor, and low popular acceptance.

Energy

Most places have sunlight, water, and wind; also small amounts of coal, gas, and oil, but not nearly enough to provide energy to residences and industries. Wind and solar energy are rarely well-developed. Power is available sometimes from nuclear power stations. The research and production of nuclear energy has resulted in hazardous waste sites. The big difference between energy resources and energy use is usually provided by the vast quantity of imported oil.

Energy use is increasing 3 times faster than the human population. Much of this energy is used in food production. For instance, an estimated 1,250 liters of gasoline (equivalents) are used to feed one person per year. (If the known reserves of fuel were spent for the earth's population at this rate, they would be exhausted within 13 years, assuming no use for any other purposes.)

Agriculture requires excessive energy. Catering to our refined tastes for flawless Platonic fruits, industrial agriculture has invested more energy in growing, packaging, shipping, and marketing many food items, apples for instance, than the energy we get out of eating them (the ratio is about 2 to 1). For some crops, such as brussels sprouts, the amount of energy used exceeds the energy yield by over 30 percent. Other crops, such as dry beans and rice, use almost as much energy as they yield. Grains like wheat still yield a net benefit (about 1 to 4 ratio). Energy for crop production is of two kinds: that to increase yields (through hybrid seeds and fertilizer) and that to reduce human labor (using tractors and drying rigs). The latter kind needs to be reconsidered. Effective use of human power in the U.S. could produce the same high yields, but using only 25 percent of the energy employed.

Technologies require energy. To produce 1 kilogram of phosphorus takes 3,200 kilocalories, including mining and processing. To produce 1 liter of fuel takes 10,000 Kcal and to run a tractor for an hour uses 90,000 Kcal. To build the tractor takes far more. The industrial revolution increased the quantity of energy, but decreased the variety of energy resources.

Energy generation itself consumes the greatest percentage (36) of energy. The main users of the remaining energy are motor traffic, manufacturing, and residential and commercial heating. Manufacturing is a heavy consumer; it includes: pulp and paper; primary metals (cars); nonmetallic products (plastics, shoes); and chemicals. Some of the energy crisis could be avoided by using less consumptive settlement patterns and natural energy utilization.

The public service function of nature provides free services to humanity that are essential to civilization. But when the free services are overloaded and breakdown, we have to pay for the repair. Further increase in flows of energy through technology will significantly reduce the capacity of the earth to support humanity, even large-scale fusion techniques. The overuse of nonrenewable resources can destroy renewable ones. For example, if acid rain from burning fossil fuels continues in the west, the primary productivity of eastern forests may be reduced by a net 10 percent due to rising soil pH. This is the equivalent loss of energy from power of 15 one thousand megawatt reactors.

An analysis of energy supply possibilities is needed to recognize the consequences of actions in terms of resources. Terrestrial energy is the stock, solar energy is the flow. There are asymmetries in energy balances, from relying too much on the stock. The future offers less stock. There is a far greater flow of solar energy than stock.

Technology could be developed to tap the flow. The stock should be kept for transitional changes, rather than being burned up by traveling or shopping (favorite human past times). In all processes in which energy is changed, some of it becomes unusable (diffused and very difficult to harness, according to the second law of thermodynamics). We should reestablish earlier energy patterns for the region, and use combined systems of wind, water, solar, organic, and fossil fuels for energy. Depending on oil for a transition is acceptable, but relatively soon there will not be any more. Singly, these sources may be inadequate, but as a mosaic they could meet the needs of decentralized communities. The energy pattern should be pieced together organically from the potentialities of a region. Consumption and production of energy must balance safely.

Energy demand has resulted in the use of high risk sources. With nuclear power, the burden of proof for safety is on the agencies themselves; its use in the absence of complete assurances of safety places unnecessary risk on future generations. Even with inefficiency, there is no need to use high risk energy generation. Buckminster Fuller claims that by using only proven energy resources, only proven technologies, and only at proven rates, within ten years all of humanity could enjoy an energy income equivalent to the United States in 1960, and nuclear and fossil fuel energy could be phased out during that time.

Passive solar heating would be adequate for individual buildings. Photovoltaic cells could provide electricity and power cars. Local energy projects using geothermal sources, winds, or the sun are preferable, since their operation does not introduce new material to local cycles. All of these sources should be characterized by a small scale.

Perfect activity leaves no track behind it. Energy production should leave as few tracks behind it as possible. Nuclear fission leaves burning, long-lasting tracks. Burning organic fuels—even in wood-burning stoves—leaves pollution and sickness. Large scale solar projects, extraterrestrial or earth-based would shift large quantities of energy around with unknown consequences; the sophisticated equipment involved would also cover large areas. Any concentrated energy use for large human populations and manufacturing may be too much. Small-scale, nonpolluting activities would leave the fewest tracks.

Minerals
Most landscapes rest on rock. Rocks vary in kind and use, not to mention ease of extraction. The same is true of metals. Many critical metals for technology have to be imported from other countries. Minerals used in advanced or strategic technologies (in alloys, catalysts, magnets, and electronic devices), are absent in large quantities, and must be imported from other regions. If no substitutes are possible, then these minerals have to be acquired through trade.

Human Population
Current Figures
The population of ecoregions ranges from hundreds to hundreds of millions. Many people now live in cities. The numbers in urban areas may not seem so large, but they are far larger than the archaic populations. Several trends are evident: A long-term trend to larger cities, and a more recent movement from cities to smaller towns and rural farming communities.

Optimum Calculations
How many is too many? How many more could there be? Is the current number above or below the maximum carrying capacity? There is a maximum biological carrying capacity for every region. The carrying capacity is the population sustainable on a long-term basis of renewable and nonrenewable resources. For humans, this capacity must include domesticates, as human equivalents, since many domesticates compete for protein consumption. Domestic animals can extend the carrying capacity somewhat, since many of them consume agricultural wastes or use lands marginal for agriculture, but are not as efficient as wild populations. Technology can expand the carrying capacity to some extent, with higher yield crops and resource substitution, but also it reduces the capacity with unforeseen effects, from the use of pesticides, for example. War and social disorder would also reduce the ultimate capacity. Furthermore, the capacity decreases as the per capita use of energy and resources increases. Carrying capacity calculations often just consider food energy, but all needs — clothing, shelter, transportation, information generation, aesthetic satisfaction — must be included.
    A number of assumptions are necessary. Calculating a population based on plant productivity is relatively simple. However, considering the need for resources and the ubiquitous Law of the Minimum, the maximum goes down rapidly. J. von Liebig's law of the minimum describes a critical minimum, under steady state conditions, of a chemical material needed for growth and reproduction. Economists have claimed that the minimum does not apply in a growing system; alas, our system has been growing through transformation and not real growth. H. R. Hulet pointed out that a population as a function of wood production would only be 80 percent of that calculated from food production — and desert areas have far less wood than the continental average. Furthermore, the population would be even less as a function of energy and fertilizer use rates. The rate of aluminum use would support only 40 percent as many. More importantly, these rates are not sustainable, being based on high American standards of consumption. A lack of some resources is not necessarily limiting, since one group could trade with other areas that need crops or products from a specific area.
    The current high levels of population, at a high range of standards, can only be maintained through the constant takeover of natural habitats for arable land, or through the drawdown of fossil fuels, and by economically cheating the poor and powerless. Since the quantity of wild lands and fossil fuels is quite limited, either human populations must adjust to renewable resources or technology must provide substitutes, to avoid an eventual

population crash.

Eugene Odum suggested using land area as a measure of human carrying capacity. The minimum per capita acreage requirements, with a temperate area like Georgia as a model for a quality environment, is just over two hectares (5 acres). The natural areas are based on minimum space needs for watersheds, as estimated by land use surveys. Food-producing land includes acreage for domestic livestock.

Probably, the entire earth could be treated according to some similar plan. The human population of the earth would be the sum of calculations of food and mineral resources for each local ecosystem. Thus, any local population depends on the limiting factors of the earth, those scarcities which could be traded between the regional populations, and each regional population having a percentage of that ultimate limiting factor (maybe phosphorus or manganese) — the percentage distribution to be determined by the regional productivity available for human consumption. Restated, a place may have enough food for over a million people, but it may not have enough wood to build them houses or enough steel to build trains; therefore, a group would have to trade food resources for mineral or timber resources, assuming that other regions are able to trade.

Samuel Eyre also devised a common denominator to consider organic and inorganic assets together. He assigned a nutrition equivalent unit to weights of metal, but this calculation depended on a dollar value for food and minerals. For example, assuming the daily standard human nutrition requirement of 3,000 kilocalories, money income from minerals can be expressed in terms of annual nutrition units. Assuming that wheat releases 4 kilocalories per gram and is 10 percent cellulose and 15 percent moisture, then 1 kilogram of wheat yields 3,000 kilocalories, conveniently equal to the daily food requirement of one human being. One metric ton of unmilled wheat is equivalent to the annual food requirement of 3 people.

Assuming that aluminum sells for $1122.00 per metric ton and wheat sells for $109.00 per metric ton, 1 ton of aluminum costs the same as 10 tons of wheat (1991 figures). If it takes 10,000 tons of aluminum to meet the needs of a current population of 500,000 (about 0.02 metric ton per person for cars, wiring, and cans), then we need to trade the monetary value of 100,000 tons of wheat to get it — enough to feed 300,000 people! And that is 300,000 fewer people than the area can support if we need to have things made from aluminum.

Population carrying capacity can be formulated using the net primary productivity (NPP) of the system. Following Odum, that only 30 percent of the area should be used for producing food, the agricultural area is set. Following Lieth, but averaging over the entire ecosystem, we estimate the productivity — either for natural or for energy-subsidized cultivated land. And, of that productivity, 75 percent is unavailable for harvest (60 percent is underground, 5 percent taken by pests, 10 percent used for respiration and reproduction), 65 percent of the harvest is inedible, 80 percent of the edible is lost in process, and 25 percent of the processed food is not consumed — leaving only 10 percent of the original productivity to nourish people. Thus, the gross productivity, in Kilocalories per year, may seem large, but only a fraction is available as food energy. Since every human being requires 3,000 Kcal per day (for adults), the maximum number of human beings, assuming that all other needs are met is easy to calculate. Of course, food is not enough. We need a large quantity of calories for trade and luxuries. Assuming that other surpluses and deficits cancel out, say excess energy from water power for cotton for clothes or basalt for lumber, and balancing only agricultural productivity and aluminum needs, we get a still smaller population.

The advantage of primary production as wealth is that the wealth is sustainable — plants are renewable and minerals can be recycled. The disadvantages are that the net community production (NCP) is not considered, which takes all of the food chain into account — the millions of other species; furthermore, dollars are used instead of human work units — using human work units, the number of hours of labor to produce a standard measure, would make wheat and aluminum much closer in price; and the technological production (greenhouses or algae farms) of food is not considered.

It is possible to calculate a sustainable population using NCP instead of NPP, however. For temperate grasslands, NCP may approach 60 percent of NPP, although 30 percent is much more likely. The population calculation for NCP results a yet smaller number. These population figures have been maximum for the productivities. Because the climate is variable, the ecosystem is ever-changing, and humans have unforeseen effects, we should strive for an optimum number, which we could arrive at through an arbitrary multiplier like .5 (Wittbecker, 1981). Assuming the multiplier is 50 percent, the optimum population becomes only half, obviously less than many current populations. This multiplier is considered as part of the cultural factor.

Each way of calculating a population has become more comprehensive and cautious and has resulted in a smaller population. The target population for planning depends on how cautious we are. Should we gamble and go for more people? If we calculate the maximum wrong, we break the system, and we may not know how to repair it. On the other hand, what if we go below some minimum? The likelihood of this possibility is low. The archaic peoples did not approach it at very low numbers. The minimum number for genetic health, for the entire species, could be 5,000 individuals. For a guarantee of fertility, 50,000, and for a minimum for social contact, again 50,000.

Many people fear that a reduction in human natality would cause human development to stagnate. Natality is the birth of new ideas, also, producing an ideomass. J. B. Calhoun's idea of ideomass is a multiple of population and individual potentiality. The brain and culture give human expression almost endless possibilities. Possibly, a minimum population for ideomass is 50,000. The concept of ideomass could support a stable or declining human population without a reduction in the quality of life.

Resource Use by Population

Just as important as any calculation of the population in an area is the rate of resource use by that population. Grasslands were the basis of early civilizations, and few regions have been altered as thoroughly or as devastated by human occupation, by overcultivation and overgrazing. Using too many resources, such as water and timber, too fast can make deserts.

We are digging up metric tons of minerals per hectare. We are draining aquifers for irrigation. We are mining the soil for everything it has and letting the rest erode. Hundreds of billions of tons of top soil are lost annually and millions of hectares of agricultural land are degraded through erosion annually.

We are using 10 to 30 times as many resources as most people did 200 years earlier. Are we 10 to 30 times as happy? Indigenous peoples did not use resources at an accelerating rate. They were limited by their technology, but more importantly by their wants, as taught in myths and stories. Our high rates are clearly unsustainable. The difference may be in our technology and values.

The Impact of Technology

Technology can be used to expand or contract resources. Technologies have the capability to minimize the use of resources. Buckminster Fuller noted that one quarter-ton satellite made transatlantic cables (requiring thousands of tons of copper) obsolete. In another instance, breeding, fertilizer, pesticides, and modern equipment have certainly increased agricultural production, but the negative impacts of genetic loss, soil degradation, erosion, and pollution decrease both the actual and potential productivity. When all factors are combined, and total energy cost is compared with energy production, the result is disappointing and does not compare well with traditional methods, using draft animals and human labor (Wittbecker, 1976).

In another instance, technology has greatly increased the kind and quality of materials used for buildings and machines, especially aluminum and other light metals and silicon constructs. Yet, the scale of technology produces pollution that reduces the productivity of natural and agricultural systems. Unbridled, unconscious technology has given us benefits, but only at the cost of irreplaceable stocks of energy and environmental degradation. Instead of expecting technology to triple or quadruple our wealth, it is more likely that it has barely had a positive effect (Daly and Cobb, 1989). Making technology appropriate, responsive, and conscious may go a ways to increasing its positive impact. Two technological processes are especially important, substitution and recycling, for minimizing waste.

Substitution. Substitution is one way to avoid shortages of resources. For instance, petroleum products have substituted for rubber from trees for many uses, including tires; aluminum, relatively abundant, is substituted for copper, even though its electrical properties are less desirable. Plastics, also from petroleum, are used to substitute for wood, leather, and metal in many industries. Some substitutions may temporarily ease pressures on biological systems, but they depend on the supplies of petroleum, a practically nonrenewable resource. Therefore, in considering substitution, we should concentrate on materials that are plentiful and local rather than scarce or remote. There are not substitutes for everything, so there must be a minimum of irreplaceable stocks.

Recycling. When a resource, especially a metal, is in generous supply, it is used for many nonessential and trivial purposes. When the resource becomes scarce, it becomes more valuable; its uses are curtailed and its presence is tracked — few people throw away gold or platinum. Recycling could lower some of the costs. For example, the asbestos fiber content of tailings produced and discarded in the 1930s has been found to exceed the grade of new ore now mined. Some urban refuse is often richer in metal content than natural ores currently mined. Carbon and petroleum products can go through several levels of recycling, even being burned for energy at the end of the process. Recycling also uses resources and energy and creates pollution. But it does reuse resources and it is wiser use.

Production and Waste. Every process produces some material that is not immediately useful or some materials that cannot efficiently be brought into a repeating cycle of use. We judge that natural processes also produce waste, although we have no way of knowing how the long term processes, deposition of carbon dioxide in tundra for instance, may be useful in maintaining limited levels of some elements, such as oxygen or carbon. Unfortunately, human waste has some differences in type and scale: many materials have no processes or organisms to break them down and recycle them — no solutions have evolved to break down plastic; sometimes the quantity of an artificial toxin is immeasurably greater than natural toxins.

A redefinition of waste is needed. We need to understand the capacity of the environment to absorb waste over time. There are no sinks on the earth where waste vanishes. Things are only moved around; eventually they return. Nonefficiency is not a synonym for waste, however; cathedrals are not wasteful, except as office buildings; gourmet dinners are not waste, except as minimum subsistence; the richness of nature is not a waste, because it is so intricately involved in the process of life. Many things considered wasteful, such as wild, open space, are required as aesthetic needs.

Some of the products of our industry are not wanted or considered valuable. Waste is sometimes visible as heat or pollution. Pollution is a symptom of imbalance and improper resource utilization. Advanced communities have to adapt biologically sound processes. A serious problem is our lack of understanding of the extensive, long-term effects of pollution on the atmosphere. Once it is determined that materials and wastes, especially nuclear or chemical are dangerous and long-lived, they should be minimized. There are ways to reduce generation of wastes by acting on the flow: reduce number of products; reduce the quantity of waste in each product; increase the durability of each product; and make sure that the cycle is a spiral. The best results might come from a mix of these strategies.

New technology has been a primary force for change for decades; but some technologies, like computing or genetic engineering, may lead to the opposite effects from the intended ones. The use of such tools may have unexpected effects. For instance, mass-produced computers may lead to individual autonomy and the invasion of privacy. New energy technologies could have the same effect. Small-scale water power plants could lead to decentralized living; or energy-saving devices could result in more energy being used. Assigning a multiplier to the effect of technology is difficult due to the contradictory impacts of technology and due to the values of different cultures, but, odd as it may seem, the multiplier is usually not more than 1.

Choosing Technologies

We assume that improving human lives requires new materials, machines, and techniques. But, we do not have measures to judge whether technologies are beneficial or harmful. We could start questioning technology: who it benefits, what the negative side or cost is, how it fits in a network of technology, and how can it be reversed if necessary. We could assume that technological changes are always somewhat harmful and study them for a long time before implementing them. Technology is not neutral or value free; there are inherent social consequences.

We debate the virtue of a technology through public discussion of its strengths and weaknesses. We could reject technologies that damage the social fabric. Society has the option of rejecting technologies like television, but perhaps not the will. The complexity of the network of technologies make that difficult. Television is linked to computers, satellites, power stations. Computers are basic now to most innovations.

Computers look like friendly appropriate technology, useful at home for empowering people in democratic organizations. But he wonders if the computer benefits the centralized corporation more than individuals; in his opinion, computers set the movement back.

Certainly computers can be useful for certain aspects of education. But, we must not forget what function the computer is assisting or let computers displace the skills themselves. Education should include a core of mathematics as classical education always has. Poetry or narratives should be memorized still, as well as sometimes typed on a keyboard. We cannot let the skills degenerate or disappear; for instance, we cannot teach computers to recognize the medicinal qualities of plants if we have forgotten what they are. The ultimate importance of education is how to live harmoniously in place with other living beings. Computers can be an important part of this education, but not necessarily the primary part or even a critical one. Computers can also be an important part of politics. Bryan and McClaughry think teledemocracy is great; Winner and Mander think it would not work.

Personal transportation is also problematical, especially cars. Only 8 percent of humans own cars, but, these 8 contribute disproportionately to air pollution, acid rain, and gases that contribute to ozone depletion and greenhouse effects. By allowing changes in distances, by forcing dispersed needs, the car becomes indispensable. Nor have we considered what changes in social mores and even physical health result from our passionate embrace of the car. And, now trucks and planes have become romantic as well. Perhaps we should consider restoring and expanding rail transportation, bicycle, and walking paths.

McLuhan suggested thinking of technology in environmental terms, since it envelops us and becomes difficult to perceive. Technology has become our *umwelt*, the psychological bubble that colors our perceptions. We are inside a manufactured environment and we are coevolving with it. It is the framework of awareness. Some technology, television, works to homogenize cultures, by imposing the same framework for experience.

Being inside the technology, it is more difficult to notice the changes that we undergo to accommodate it. Television is fast. The natural world is usually slow. Television attunes us to fast rhythms. To understand nature we need to be slow. How do we decide the appropriateness of technology?

A favorite Indian pastime was the telling and retelling of stories. Television also has repetition, but not the same depth of involvement. Radio stimulates images in the way books do, because neither impose images, although neither of these forms may be as involving as real story-telling. We have assumed improvements in technology are improvements in human existence, but we must examine them constantly, if we are not to lose some of the characteristics, such as empathy, that we value most highly. Some communities may wish to examine cars, computers, and television, and eliminate some of them or modify their use.

Limits to Nature and Technology

A frame of reference (in physics) has a uniform motion at a constant velocity in a constant direction. Each frame is considered a locality and obeys the principle of locality, that is, what happens in one frame does not depend upon variables subject to control in another. The rules that apply to part of the universe from a limited perspective are not universal. A global system is the sum of localities and may have unique characteristics of its own, that is, the universe has characteristics that local frames of reference do not. Frames of reference are easier to define in physics than in ecological or social systems, with their extended complexities. Nevertheless the principle of locality can be used to explain many difficulties in cosmology, evolution, and ethics.

The limits of the universe, like the speed of light or quantum of a field, put limits on freedom. Events are limited to localities; size limits function; history limits development. The earth is suitable for life because: solar radiation has stayed within certain limits for four billion years; the biogeochemical cycles of oxygen, carbon, nitrogen, phosphorus, sulfur, water have stayed within certain limits; and the environment has been constant enough for organic evolution, but variable enough for natural selection to be challenged.

Life involves a vast number of interacting structures. Living consists of complex behaviors whose limits are defined by rules of order that can be empirically described. Biological order is built on physical and chemical orders. That is why life is limited to such a narrow range of conditions. And that is why the most complex orders are vulnerable to changes in their substrates; energetic radiation can alter and destroy an individual, or a small change in climate can destroy crops and civilizations. Complex orders always depend on simple orders.

The order of human life in one locale is a cosmology. The very circumstance that makes each cosmology unique — being in a unique place — ensures that it fits a place. Particularly in agricultural societies, cosmologies are gauged closely to seasons. The cosmology of a culture makes the world manageable by limiting it. A local cosmology is also tuned to the limits of the local ecology, within the knowledge of interactions (the long-range ecological consequences of drainage, irrigation or overexploitation contribute to the deaths of cultures). Many archaic cosmologies are a form of fitness and limitation. Most groups try for adaptation before domination (refer to Reichel-Dolmatoff). Leopold describes ethics in terms of limits, when he states that the extension of ethics is "actually a process in ecological evolution. Its sequences may be described in ecological as well as in philosophical terms. An ethic, ecologically, is a limitation on freedom of action in the struggle for existence."

Technological innovation also is subject to diminishing returns, so its potential for resolving economic weakness is also reduced. Thus, there are real limits to long-term productivity growth. Joseph Tainter recommends that the best key for continued socioeconomic growth and for avoiding declines in marginal productivity is to obtain a new energy subsidy. While this strategy may have worked in the past, it assumed that there was new energy available and that its use would not overburden ecological systems — those assumptions are no longer automatic or safe.

The earth is finite and its resources are finite. It is very difficult to quantify the limits exactly for human purposes. But, there is a limit of productivity that can be taken over by humanity, and there is a limit of energy use before the system is disrupted, and there is a limit of interference in an ecosystem before its collapse — humanity has exceeded those limits in different places and times. The trend is still towards the increased disturbance of natural systems.

The ancient world had limited energy sources, but many peoples were able to expand to avoid declining marginal productivity — the expansion was only temporary, however. The use of fossil fuels has generated great expansion , but humanity shows no inclination to learn the lesson of limits. Every case involving growth has been only temporary. Nevertheless, humanity has invested incredible amounts of energy on housing and transportation — not only building units but operating them. The costs in Kilocalories is tremendous. Yet, these items too must be related to the limits of the regional system. These items too have an optimum number.

Within unknown limits, we need to establish a pattern of use that is sustainable and leaves other possibilities open, that is flexible. Mineral resources, like aluminum and copper, although finite and nonrenewable, may have indefinite cycles of use. Self sufficiency and autonomy are reduced as specialization increases. Resources are used more efficiently, but then there are fewer reserves or flexibility in allocation of resources.

Technological multipliers are efficiencies that allow a limited increase in the carrying capacity of the entire ecological system. The load on the system can be expressed as the resource demand multiplied by number of people. Trade diminishes some resource or food capital, but can be used to avoid some minimums. The Critical Minimum is a factor limiting carrying capacity. We do not know what it is.

Cultural Values

Cicero stated that human culture is a second nature. Lewis Mumford wrote that the great gift of civilization is a cultural tradition to live within and to provide a vision of renewal to lead us forward. Identity and meaning are not possible without orientation to the past and future.

There have been two basic cultural patterns. An archaic one that offered tribal egalitarianism and limited impact and a technological one based on central government and industrial consumption. The first could not compete against the second, but it has not been completely eliminated either. The combination of the best of archaic traditions, with appropriate technology and knowledge, could guide our future.

Athelstan Spilhaus suggests that a good index of the quality of life is the number of choices or alternatives a society provides its individuals (similar to McHarg's subjective notions of negentropy as free choice). Western culture has valued the pioneer life of the individual — independent, self-sufficient, free to choose different life styles — but with more people, the choices are being subtracted. The more we value open spaces, the fewer people the area can support and still have the open spaces.

It may be that the archaic people require 10 times the space as rural settlers; and, it may be that these settlers require 10 times the space as their urban counterparts. Or, it may be that people would prefer to have more discretionary time and less work time, as was possible in archaic societies.

There are numerous advantages of the archaic way of life: fewer working hours (about 4 per day); more leisure to talk, sleep, engage in rituals; a diverse and healthy diet; deliberate underproduction, below the maximum levels; deliberate control of population growth below maximum levels; deliberate under use of resources, resulting in small ratio of people to resources. Subsistence economics means simply that surpluses are not accumulated. This might make them more vulnerable to food shortages, although the ratio reduces the possibility; furthermore, industrial cultures have far higher incidences of starvation, now. We are subsistence now; we cannot leave after we have cut trees or eroded the soil. Perhaps we will be less tempted to exploit land for short-term profit, now, since we have to remain after the profit leaves.

The greatest happiness for the greatest number could be achieved by devising a measure of quality of life from the accumulated wisdom of nutritionists, medics, psychologists, and sociologists. It could yield an optimum quantity and quality of the commodities of life: how much space, how much food, how much air and water. But would that be enough? Indicators of the quality of life cannot all be expressed numerically; they are elusive. Quality depends more on the spirit of the society than a count of material possessions. Even orthodox ecological criteria are not adequate to evaluate the quality of a particular environment for human life. Nevertheless, we can try to assign a cultural multiplier to our aesthetic and support space needs.

Because both archaic and modern societies are in disarray, it may be necessary to treat values in a formal ecological economics, rather than depend on cultural traditions or theft or luck to provide them. This economics would make explicit the conclusions of common sense and good science. The plan requires changes in economic and political institutions to be effective.

Summary for the Model

Ultimately, our cities depend on agriculture, and agriculture depends on wilderness — for recycling, pest control, genetic diversity, soil-making, and water purification, among other things. Our cities also depend on cultures for their vitality, and cultures depend on wilderness for their context and imagery. Therefore, we must preserve wilderness areas and cultural knowledge first.

Most places have adequate resources to feed and care for their current populations — at a reasonable level. The rates of use of energy and some minerals are far too high. Our demands and values are not related closely enough to the particular beauties and limits of our home ecosystem. We will have to impose limits on ourselves.

Implementing the Plan

Economic Things

Because human and natural systems are interlocked, there is a common framework for ecology and economics. Economic decisions are based on human reference and not nature, but human reference is not large enough. Economics is the study of both material and energy budgets; it is a rheology — a study of things. Ecology is the study of natural budgets, material and energy; it is the study of beings in a living context.

Economic and ecologic systems interact. Economics must recognize that ecological health is vital to its own continuation; it cannot make large mistakes. Individual economies could be linked with climax vegetation and not successional vegetation, where change is costly. Production needs to be stabilized in a mature economy, like a climax system, where processes and cycles are constant. Human economics grows from positive feedback due to feeding on the wealth of the past and the future; positive feedback creates instability. Negative feedback is a characteristic of stable systems. A mature economy must be based on natural laws and ethical principles. Natural laws include thermodynamics and ecological theories. Rules of economics, laws of nature, and ethical principles must be related. An ecological economy is survival-oriented, not profit-oriented.

Economic Goals

What are our economic goals? Market leadership or development for all of humanity? Subsidizing corporations does not seem to be a good way to achieve the latter (especially when corporate profit grows even when standards of living drop). Regarding consumption levels, how much is enough? When does consumption stop adding to satisfaction? How necessary are Chilean grapes or New Zealand kiwi to healthy consumption. By matching consumption to ecological limits, the way is open for more psychological measures of wealth.

An economics based on ecological understanding would have many different assumptions. For instance, the capital of an ecosystem would be its physical environment and its gross primary productivity; interest would be the net ecosystem productivity. The production percentage would be the amount necessary to keep the ecosystem healthy. Cultural capital would be the wealth of human knowledge about environments, and cultural interest would be experimentation.

Expanding Capital. Traditional economics has a three capital model of human wealth. Land, labor, and manufactures. Clearly, this is not adequate; each kind of capital should be expanded. Land is the entire ecological system, complete with other species and biogeochemical cycles, preserves, as well as agricultural areas, resources, and artificial modifications like dams. Labor depends on the traditional capital of a culture, the beliefs and myths and rules for behaving, the institutions. Manufactures depend on culture, and land (resources), and on technology as well.

Modern economies, embracing the metaphor "nature is capital," draw on the accumulated "capital" of ecosystems for production. By ignoring the real cost of the capital, as well as the costs of natural services, such as nutrient recycling, soil building, and atmospheric renewal, these economics create a temporary wealth. Decisions regarding resources are made on short-term economic grounds and lead to material shortages and environmental degradation.

Economics must internalize all costs for product cycles, from agriculture to manufacturing. This means finding the real costs first, starting with environmental degradation, lost employment, increased health care, tax credits, defense. The real costs of energy usually far exceed what users directly pay, for nuclear, coal, oil, or gas. When workman's compensation premiums are paid at a rate determined by a company's accident experience, then the cost of an unsafe operation is already a recognized direct cost of doing business. These external costs can be internalized directly with taxes, especially one on carbon-containing fuels. Another possible tax could be an anticipatory tax for degradation. This would encourage more efficient use of sources and development of more cost-effective technologies. For agriculture, this means reducing waste and pollution, and emphasizing diversity and disease-resistance instead of gross yield.

Diversifying Institutions. The simplest economic transactions occurred between individuals who gathered food or made tools and then traded with other individuals. With the increase in specialization and complexity, came individual traders, then guilds, and finally corporations.

Corporations were formed with the historical promise of providing social services to the national states that chartered them. In the name of secure trade they often explored territories and guarded resources with private armies. Unfortunately, as private good became identified with public good, corporations became less concerned with social service and more concerned with greater profit through greater technology.

As legal entities, corporations have formal expectations that have become imperatives: Profit, for instance. According to Milton Friedman, the only social responsibility of a corporation is to make money, by striving after profit as an efficient agent of production — although Friedman admits that the corporation should conform to the rules and norms of society. Other imperatives, equally important if less touted, are constant growth, aggressive competition, objective, i.e., amoral, decision making, efficient exploitation of resources, and the quantification of values. Unfortunately, as an entity with limited responsibilities, these imperatives tend to dehumanize workers and consumers, homogenize people and cultures as markets, foster no commitment to place, and interfere in natural processes. Corporate forms, with their characteristics of simplification, naiveté, homogeneity, and incompleteness, turn wild landscapes into flatscapes, where variety disappears and significance is ignored for the comfortable standards of meaningless continuity.

Corporations, as economic institutions , do have many responsibilities: ecological, social, community, political, and individual. The first responsibility is to discharge its specific function. Then, it has a responsibility for its impacts; this is one of the oldest principles of law. The institution has a duty, and self-interest, to discharge its function with a minimum of negative impacts. Profit making is a necessary part of business, but not the sole reason for business. The best business serves public goods as well as private interests. Environmental and social problems should get as much attention as profitability, because they are as much a part of the process as sales, finance, and production.

Corporate law holds that management must act in the economic interests of shareholders. Communities could change that law to start to control corporations. The corporation is a noncorporeal entity. But because it is a fictitious person, speech in the form of advertising is protected under the First Amendment. So, corporate speech has little in the way of regulation. Therefore, it may be necessary to regulate corporate speech under a new amendment. Corporations sponsor wonderful shows on nature as a resource. So we become attuned to corporate interpretations and objectives. We need to be more attentive to their real objectives.

We need different kinds of regulations for corporations, ones that limit destructive activities to environment and workers, ones that encourage responsibility to nature and community. This may require changes in economic organization and legal institutionalization. Corporations should be anchored in the community. Although they must not act beyond their competence, a larger community responsibility is merely expanded self-interest. They need to build concern and responsibility for the common good into their vision, values, and behavior. Common good does not necessarily emerge automatically from conflicting interests.

The corporation is defined as a collective citizen; we need to enforce the duties of one. How to make the corporation accountable to the community? The problems of corporations are structural, inherent in the forms and rules by which the entities operate. Corporations increase their power because of our failure to grasp their nature. We could restructure the market to favor long-term investment over speculative profit. To give back to the community, we could require a percentage of new stock offerings go to government for the public. The state grants the charter to a corporation; the state could revoke the charter to protect the interests of the state (as citizens). Corporations have many human rights but few human responsibilities. Even when their actions cause death, the corporation cannot be jailed or executed. Perhaps we should change that law also and be willing to disband it.

Stressing the Development of Wealth Over Growth. Powerful images can influence cultures over centuries. The principle of plenitude, restated in Christian terms, presents that an intelligible creator gave an earth of unlimited bounty to humanity for its use. This principle seemed to be confirmed in the Renaissance with the discovery of the richness of heaven, microscopic life, and unexplored continents. Many modern political ideologies and economic systems have been shaped by the principle of endless wealth. Adam Smith calculated that the real price of anything was just the toil acquiring it. This image is dangerous because it ignores the evidences of the limits of the earth.

Economics has always been concerned with measuring wealth. Wealth once meant tangible things, land, ships, houses; then labor and production; it has come to mean negotiable symbols such as cash and stocks or unlimited information. Yet, no single description is adequate.

It is said we are in an information age, that information is the ultimate resource, that land and resources are less. But information without form is nothing (in-ation?). Information only lets us use resources and land more efficiently. The basis of wealth for a long-time will be land. Even more so because we do not know its complete value (as many native peoples have found out when coal or pharmaceutical plants were discovered on their lands — as a source of information).

The narrow definition of wealth means that it can be increased only by producing a bigger supply of goods or reducing the demand for goods. This assumes that wealth is defined as supply divided as demand. If supply is limited then wealth can still be increased two ways: reduce the expectations of individuals or reduce the number of individuals. Supply may be mostly material things — but not status, for instance — while demand has the more psychological dimension. Therefore, wealth will always have a psychological dimension. This dimension is not limited by strict logic. Wealth can therefore be expanded without being limited to supply or demand for materials.

Economics is distorted when reduced to quantity and technique; there is always a psychological and ethical dimension to be accounted for — motives, values, needs, aspirations. Economics needs to be restructured to take this consciousness into account. The assessment of personal or cultural wealth, for instance, is mostly psychological; wealth may be measured by how many valuables one has, which may be physical, like feathers or salmon or gold or land, or by how by much status, which may be behavioral, as enjoying deference or a good reputation.

Pitirim Sorokin indicated that the wealth of an area was a function of its physical attributes and its culture. In fact, the attributes are only possibilities until appropriate cultural perceptions and technologies exist.

The redefinition of wealth in an ecological framework would increase human enrichment and natural preservation.

Economics tends to devalue many things. Our economy is crippled by the unspoken principle of immediate interest maximization. We allow economics to discount the future value of benefits. That also means that our children's and grandchildren's lives are worth less than ours; are they? A common strategy in Rome was to defer the true costs of government by debasing the currency. The cost is shifted to the indefinite future. We have been doing the same, but the future is becoming more definite.

Our recognition of risks, from smoking for example, is irrational. Environmental risk assessment does not seem to require valuing the entire ecosystem or its parts. We could charge, however crudely at first, for environmental services, such as soil building, carbon fixing, flood control, erosion control, detoxifying poisons, and sequestering heavy metals.

Ecological and economic processes and values are often the same. Benefits can be assessed in a common metric. The GNP is an inadequate way to measure well being. It measures only increases in spending. The International Standard of Economic Welfare (ISEW), developed by Daly and Cobb expands consideration to literacy and longevity as well. More is needed, however, to measure wealth and happiness in addition to general welfare. The measure needs to be expanded to include more of the human growth needs recognized by the psychologist Abraham Maslow. It also needs to recognize an expanded definition of the self, to include the ultrahuman beings who support us.

We make tradeoffs in social systems without assigning dollar values. We could do that with ecosystems. The valuation could be scientific or be based on human labor. New values of natural resources are being recognized now by economists, such as option value, that is, reserving a resource for future use or existence value or paying for something to exist that will not be used.

One thing business can do is put a price on nature. But, let us make it a real price, reflecting the real cost of replacement. Let us base the cost on human labor and technology, so that 1 gallon of oil is worth a million dollars, as Buckminster Fuller once calculated. Let us make all those prices be high, too high rather than too low.

Nothing is value-free, not technology, not education, not economics; we just do not always see the values. Values are time dependent; ecological time is much longer than social time. Our social values depend on ecological values.

The goal of economics should be mature development, not growth. Growth has been a substitute for equality; in that sense it has been necessary to forestall revolt. There is no necessary association between development and growth. Development means the introduction of an innovation. Economic development will still require technology. Ecologically sound technologies could minimize stress to the environment.

Limits of Economics

Both ecology and economics attempt to understand and predict the behavior of complex, interconnected systems where individual behavior and flows of energy and material are important. There are many other common or similar processes: resource allocation, optimal behavior, and adaptation. For each ecosystem, and at each level of technology and kind of social structure, there is an optimum size of population that offers a high quality standard of living. The optimum in this sense is a working one based on our knowledge of all of the factors — and we cannot know everything. No one has complete information about the current environment or the results of their actions. Complete knowledge is not necessary, however — only the knowledge of a law of the minimum. For humans, trade can ease the law but not repeal it.

Rachel Carson was one of the first to show that we lived in a world of limits. The limits may not be absolute, in the sense that history shows that advances in technology expand the availability of resources (less is needed to produce more). But, the same history also shows that human populations expand to the new limit of misery allowed by the new technology. New advances tend to increase the size of humanity, not its happiness or wealth. Society is growing at the same time as needs are growing, resulting in greater demands on natural systems.

The transition to a sustainable state does not mean returning to an all-natural, that is nonhuman, condition for ecosystems. Human activities have always had some impact (as do the activities of all species) on ecosystems. Sustainable development requires recognition of the large number of limits.

The popular definition of sustainable development is inadequate; the Bruntland Commission defines it as "meeting the needs of the present while not compromising the ability of the future to meet its own needs." But, as Herman Daly has pointed out, this definition is contradictory in practice, where it really means "expanding the needs of a growing population without inflation." Daly offers 3 rules of sustainability to make the concept consistent and meaningful: (1) harvest renewable resources only below or at their rate of regeneration, (2) limit wastes to the assimilative capacity of the local ecosystem, and (3) require part of every profit for investment in renewable resources.

The ecological approach to development makes it irrelevant to discuss global limits to growth. Local limits are far more significant to a majority of people. Regardless of how much food exists, people will starve unless they can get it (as is happening so often, now). Every community is forced to accept some upper limit, beyond which it cannot grow any further. Further growth results in destruction or disruption of the community itself and the natural communities on which it depends.

Complex societies depend on production of resources. Increased complexity requires more information processing and more integration of disparate parts. The costs of communication increase. Complex societies need control and specialization. Yet, investment in complexity yields declining marginal returns because of the increasing size of bureaucracies, increasing taxation, costs of internal control.

At some point society is investing heavily in a course that is less and less productive; increased costs just to maintain status quo. In a mature ecosystem, a larger percentage of energy is used for maintenance of the system, until net community production approaches zero. The system also becomes more efficient, supporting a larger biomass with the same amount of energy in weblike food chains. If society were to parallel this development it would probably be very stable. Societies can fail (and disappear) when they become inefficient and spend more energy than exists in the system flow to maintain the system.

The economy is not an automatic mechanism for good. We cannot predict which transactions will have good effects. We have to understand it and direct it. We expected faithfully that the market would promote the general welfare. But, people work to maximize their own good, as Hardin has pointed out, and self-interest makes it difficult for us to acknowledge our dependence on nature and on others. Economics started as a branch of moral philosophy. In a large sense, morality is a set of rules for living together, and economics is still a branch of moral philosophy.

The theory of complexity shows that complex systems do not allow predictions; they are influenced by factors that are not statistically significant. Yet, the climate is roughly predictable and stable, where the weather is unstable and unpredictable. The best economic policy is probably one that tries to balance long-term productivity and competitiveness with short-term benefits (tradeoffs).

Political Things

Archaic nations governed their areas independently. Their political principles were similar: all land is communally owned by the tribe, although household goods may be personally held; all decisions were made by consensus in which everyone participated; chiefs were not coercive rulers, but teachers and leaders with specific duties limited to their realm — medicine, war, or ceremonies for example.

When the Europeans or Chinese settled many areas, they brought their centralized governments. The original goal of the American republic, according to Jefferson, was to make each person a participant in the everyday affairs of government. But the government (state or federal) has become gigantic, managing the area from remote locations of power, and participation has dwindled. Despite a recent emphasis on personal responsibility and international cooperation, our political institutions have not responded.

Central politics overwhelms local politics. It dominates the process of decision-making. Politics deals with words, which are arbitrary symbols for events or things. The wrong relationship of things and symbols can result in misguided politics and violence. Decisions are made on narrow political grounds. Citizenship in industrial cultures is the abandonment of responsibility on the assumption that others know how to manage things; government is the assumption of responsibility, without knowledge, that leads to immense and interrelated problems.

Besides size and power, there are other things that make governing difficult: division of labor (are there professional citizens?), centralization, and technology. The interrelations of these things necessitates discussing them at the same time. Political institutions are not givens or timeless. Taking from the strengths of earlier forms, it might be possible to modify government to be more effective. The first step would be to form an independent government for each ecosystem.

An ecosystem is a good candidate to be an independent political unit. It is a governable size. It has clearly defined boundaries, as an ecosystem, that is the ecological community including humans, and not a state or county whose boundaries have been determined by rectangular grids at human whim. The kind of government, with more clearly defined, might look different.

Goals of Politics

The government of a community is a framework to maintain the lives of people. For the original archaic peoples, tribal teachers were adequate. In our representational republic, representatives are less adequate and less capable of institutional change.

Expressing the Purposes of Government. Central government has lost sight of its own purpose, which is not the sum of special interests or its own perpetuation. Government has always had other reasons for existing. Some reasons are to:

1. Hold a vision of the common good, where 'common' means common to all beings in the ecological community as well. Make goals conscious, with some flexibility to enhance the vision over time. Balance public and private interests.

2. Coordinate the means to satisfy the long-term needs of the community. Balance freedom and regulation. Tie rates of consumption to the limits of the system — this means controlling resources and land use, in essence determining the physical shape of the community.

3. Regulate the community. Tie it to cultural values. Determine the closure and openness of the community; rates of increase or decrease, through births, deaths, or immigration. Encourage or discourage some forms of technology or trade. Provide work opportunities to members.

4. Protect the community from internal and external threats: natural disasters, criminal elements, and other communities. Most of these threats are unavoidable. Some are long-term and rare; others are constant and of low intensity. Some are part of the human condition; others the result of historical balances that cannot be restored easily or quickly. Be aware of them and minimize the damage.

Increasing Participation. John Dewey believed that personal face-to-face communities were necessary to democracy. The local communities need not be isolated as they have been in the past; they are more open and active, connected to other larger communities. Democracy requires trust and goodwill; these arise more easily in communities of acquaintances.

Citizenship is too complex for television or even electronic global villages; it must be in the community, in person, in place, where individuals can learn about each other in context. Government by local meeting assumes the common sense and wisdom of the common person in an open exchange of belief and need. It requires trust and esteem.

Often this kind of involvement takes more time than just voting annually or having one person decide. The effect of presenting a problem before an American Indian council was to slow down response by passing it to the entire constituency and getting a consensus. This ensured due consideration.

To encourage the participation necessary for effective democracy (or communism or socialism for that matter), government should solicit public opinion. Land-use agencies do so. Offer real power to people — power should devolve to lowest level — by changing the local political institutions to start. Montana or Vermont offer examples of how to change local participation.

Taking and Yielding Powers. Central government must shrink that local government can expand. Some things must be done at a national level, such as the protection of watersheds, rivers, and the atmosphere, to make sure of minimal or median protection. Some protection must be enforced at the international level, also.

Following the federal model, delegated powers go to the highest level, and reserved powers to the community. So, a new national government would coordinate internal and external defense and security. Maintain law and order. Set ground rules on economic exchange to ensure fairness. The most important responsibility of government is to set standards for itself and its institutions. The constitution would instruct the courts to interpret clauses as narrowly as possible.

The new nation would have an Administrative department to handle taxation, budgeting, and purchasing. It may coin local money (perhaps on the model of the Local Employment and Trade System — LETS — on Vancouver Island in Canada, which records credits and debits on a computer, which can then only be used locally). There may be departments to protect the civil rights and liberties of the people and a department to protect the environment. Environmental disputes could be resolved by mediation (as was developed in Seattle in 1980s). The nation would also conduct foreign policy, provide technical services to communities, and maintain regulatory offices.

To avoid insularity, being set against the rest of the world, we could create an office of global communication, which could set up connections similar to sister cities program. It might be beneficial to join confederations of other similar areas (especially those that could offer complementary crops) or a larger union, such as the United States, for preferred trading.

Spending on education, roads, welfare would be done at local level. There is some risk, especially with education or wilderness restoration, but the breakdowns and errors will not be devastating as with centralized planning. Citizens will need to do the work of government as well as make the decisions. They should have total control over some things. The judgments of the people are more important than the efficiency of those judgments. It may not be necessary to have many separate authorities or committees; it might be better to integrate policy making bodies so they are not too specialized (Bryan notes that highways, schools, and waste are related).

With centralization of functions, money has become the primary source of security for most people. Welfare as giving money to those who do not have it may reduce homelessness or disease, but it cannot restore family. Family needs a supporting community context—institutions. Decentralization, and the power it would return to local communities, may also help the family as a source of security. Money is an enormous simplifier, but many things cannot be simplified. Decentralizing would make government and economics enough of a small scale to be understandable.

Decentralized communities fitted to their ecological location are more suitable and livable than urban spreads. Some cities may still be fairly large and dense like those envisioned by Paolo Soleri. Some may be smaller and rural like those suggested by Murray Bookchin. Their relation to support areas would be more explicit and include large amounts of natural and domestic (crops) vegetation. As much as possible, the cycles of materials would be closed.

As cities become more sustainable, their forms may change. They may become more compact, with more multiple-use streets, as a focus for human activity (less for cars); buildings use solar power, efficient heating, perhaps integrate roof-top crops; integrate older buildings into new groupings to integrate services, play, and work, with living; local public spaces and services; recycle waste into cycle; regenerating derelict land, either as green area or new construction.

Preconditions to a sustainable, steady state, economy include pollution control and the redistribute resources more equally. The redistribution of resources and improvement of environmental quality are more important than increased production by sophisticated technology. This strategy calls for social and educational organization more than technological style. Styles of technology must be determined by culture and context. Such development requires a local authority working with suitable economic and ecological conditions. No authority can be effective without the participation of the people.

Limits of Government
When a place has a reputation for being small and livable, it attracts more people, until it is no longer small and livable. But, imposing the limits and keeping from growing are problematic. Government could impose limits on birth through licenses (perhaps risking rebellion), through limits on housing and public services (possibly causing shanty towns), or through peer pressure (contributing to social disorder) Nature is self-organizing, and, society is self-organizing, but we need to recognize some limits and define others, and take responsibility for keeping to those limits. Limits are fundamental to understanding nature and life.

Prescribing an Optimum Size. To restore participation, we have to consider the limits of participation. The current large populations of many places may seem to be too large for direct democracy, as does any of the projected optimal populations.

The number 25,000 is large for direct democracy. Many of the cities and towns in an area are approaching this size; thousands have already exceeded it. The size has to be small enough for people to meet and "exercise government" (in James Madison's words). Kirkpatrick Sale concludes that the optimum size for direct meeting is 500. Probably participation becomes more difficult as the size increases. Bryan and McClaughry suggest 2,500 as a maximum, since in larger groups people cannot all know one another and the assemblies become a debating forum for a few.

Putting these figures together, we could design neighborhoods to be 500-2,500; these neighborhoods would make up communities of 5,000-25,000; and the communities would bring a regional (or national) population to 400,000. Differences in size seem to be a unit of 10. Each Community legislature would be 40-60 people. Perhaps these sizes are close to the optimum (these numbers are similar to C. Alexander's *A Pattern Language*) About 3,000-4,000 people are needed to support an elementary school. The Swiss are a good model for government levels — with a national government equivalent to Swiss national, communities equivalent to Swiss cantons, and neighborhoods to Swiss communes.

There is a point in critical mass reaction where the mood of the mass becomes indifferent. As long as the size is small enough for recognized identity, people will behave with concern. At a larger size the ideology, which is capable of anything through indifference, takes over (refer to Leopold Kohr).

Small communities are essential to the democratic ideal. The uniqueness of place gives belonging and identity; the whole community gives meaning and richness to life. The population density of some places may cause some difficulty, however. People will not be within walking distance, but may have to travel 20-30 miles to meetings or communicate remotely through telephone, computer, paper, or friends.

Protecting Ways of Life. A government could impose a limit on the intensity of development of the entire area, with transferable development rights (TDRs) assigned to each unit of area. Any project would have a TDR value (25 for an apartment building, for example, or 1 for an individual home); TDRs would have to be bought or traded from other landowners. All land held in communal ownership and leased to farms and businesses, except preservation and conservation lands.

Compartmentalization avoids the need to compromise every ecosystem for human use. Multiple use systems should only be part of the picture. First, government could ensure a protected environment of mature ecosystems, then productive systems, and multiple use, and urban. This could be done through function (not activity) zoning. The landscape needs to be zoned (compartmentalized) to provide a safe balance between protected ecosystems and used ones. Restrictions on land and water are one means of avoiding overpopulation or overexploitation.

Long before the limits of food or space are reached, or the ecological balance is lost, or a vital minimum is exhausted (phosphorus, for example), the quality of life will sink lower. Regardless of how much protein or energy can be provided to support human life, human happiness will be problematic in large, insecure populations. The question is not how many people agriculture and technology can support in one place at once, but what kind of life is possible for those who have no choice but to live in that place. The limiting factor is that condition in the environment that approaches the limits of tolerance of individuals. The population density may be the limiting factor. It may be living space. It may be wilderness. It may be beauty — aesthetic space (Wittbecker, 1970).

At a limit, the cost of change accelerates. We seem to understand technological limits, to sailing ships and computer chips for instance, but not to individuals or groups, not environmental or ecological. Calculating these kind of limits is difficult — too much data, too much uncertainty.

Drucker points out that economists from Adam Smith to the conservative F. A. Hayek argued that it is impossible for governments to control or manage the economy, especially in an information age. Recognizing, on the basis of mathematical models of complexity, that detailed management of the biosphere is beyond human capacity, a government should minimize its management. The biosphere is dominated by natural communities of which we are largely still ignorant. Detailed planning of complex open systems is not necessary. Planners are not in a position to attempt detailed models of future situations because many relevant parameters remain unidentified, and many of those known cannot be quantified. Plans can be made within the limits of variables, although it is not safe to be limited by lethal variables, as Gregory Bateson recognized; closeness to limits reduces flexibility, that is, uncommitted potential for change. Vagueness and lack of detail are acceptable in planning, because people will fill in the details. Furthermore, it is almost impossible to plan every detail of a dynamic chaotic system. That does not mean stagnation, that a rice field must always remain a rice field and a town a town. What the government should preserve is the pattern, not the details, limits not directions. The limits are to be applied to scale not development.

Therefore, we must limit human intrusion. Government should zone some segments to be free from human activity. and tailor human-made systems to approximate the form of the natural systems replaced. Interference is a broad term for the negative side of human activities. There are numerous forms of human interference: overexploitation, introduction of exotic species, pollution of air and water, subsumption of habitats (in shape and size). Interference is caused by large human population growth (and its requirements of poverty), inadequate metaphors and images (too anthropocentric or short-term), uncontrolled change or transformations (colonialization or revolutions), and political or economic failure (wars or market internationalization).

Paying Costs and Leveling Extremes. Relative to European communities, we have less funding for public services, such as parks and public transportation. We traded public support for higher levels of private affluence, which has not made us any happier. In fact, we are more insecure; we can be richer or far poorer (and then second-class and neglected).

Many cultures should, as Sweden and England and Japan have tried to do, weaken the connection of material reward for achievement. Income distribution is too unequal. Full internationalization would bring only greater extremes, which we can least afford.

The communities could levy taxes on property. But there is a discrepancy in the wealth of communities. The nation could collect income taxes, and communities could claim a percentage of taxes collected. The community and nation could both tax the same bases: income, sales, meals, property, fuel. The nation would set a ceiling on each; the community rate could be zero on some. Or the communities could do all taxes and give the national government a percentage, although differences in wealth may be maintained; then the state could return a percentage to make up equality in education, environmental protection. The important thing is that taxes are used to direct development and reflect the true costs of the society that people want.

Government could change taxation procedures to reduce growth instead of stimulating it. Parkinson suggests that taxation beyond a certain point yields declining marginal returns. Government could use a single tax rate flat at some percentage (from 10 to 25); and then pay everyone a fixed amount of income for basic needs (from 3,000 to 10,000 dollars).

Similarly, property taxes could be appropriately scaled to use. One way to keep farms as farms is to tax land by use. The more important the use, farming for instance, the lower the taxes. Buying farmland for shopping centers would result in discouragingly high taxes.

It is difficult to persuade people to pay more in taxes, to vote to keep less. But, through education or understanding, a culture could expand the understanding of the self and expand self-interest in that way. Some catastrophe might work towards equality, but that might have other high costs to society.

Convening a Constitutional Meeting
Starting is simple. A constitutional convention could be called in 1992 to work out a new national government, possibly more radical than the 1972 Montana constitution, which required a local government review process; it mandated real change in the form of government, although few changes took place. The convention would suggest the forms of areas, according to watersheds or other boundaries. But local people would decide the areas. An example can be found in the Washington State statute of 1967.

Once the constitution sets up representation in a legislature, then the legislature has authority to make policy. No limitation on taxing and spending, except maybe common sense; people don not respond well to tax rates over 30 percent or so. The legislature would work to protect the uniqueness of a place and its history.

Because of the differences in size, and the need for even the smallest community to be enfranchised, it may be useful to adopt a floterial district system as in Idaho. A representative could then represent 2 or 3 communities.

The convention could draw on the cultural depth of the region. Goals that have been only thoughts or hopes could become explicit expressions. The nation could apply for recognition as an independent nation through the United Nations, and could give recognition to other nations, such as Scotland or the Karen.

The Unrepresented Nations and Peoples Organization (UNPO) promotes the respect of human rights of all peoples through nonviolent tolerance and self-determination. UNPO offers an international forum for nations and peoples whose causes are not addressed by the UN or existing bodies. UNPO provides assistance to 26 groups, representing 50 million people, including the peoples of Mari and Tibet; the Lakota Nation, Mohawk nation, and All Indian Pueblo are observers.

International Context
Differences in Histories
Rapid European expansion occurred at rates rarely exceeding a growth of 1 percent per year, and with unparalleled opportunities for expansion into sparsely settled areas (North America, Australia, South America, South Africa). American expansion took place over a lightly inhabited continent. The rest of the world does not have these opportunities; violent exploitation and population growth have restricted or wrecked their possibility for equal material development. Growth itself requires greater material support for houses, schools, medicine, and jobs. Because of historical imbalances in the exploitation of some nations by others, those other nations that have amassed more advantages must be willing to engage in some form of equalization (which means a lower population or standard of living for them). More people suffer than ever before. Good intentions are not enough — inappropriate benevolence can wreak the same havoc that greed and indifference have.

Differences in Wealth
In the 18th century trade was mostly complementary; countries exchanged products, for example, England and Portugal exchanged wool and wine (trade characterized by partnership is similar to mutualism as an ecological interaction). In the 19th century, trade became competitive; the US and Germany each sold chemicals and machinery to each other to create customers (similar to biological competition). Now, in the 20th century trade is adversarial, aimed at dominating an industry; the Japanese intend to control electronic products (similar to interference competition). Adversarial trade challenges traditional assumptions. Protectionism and free trade would both be disastrous, according to Peter Drucker. Drucker recommends forming economic regions that transcend protectionism and free trade and work through reciprocity. Reciprocity is a one for one trade between countries. Cobb and Daly point out that international trade should be governed by community interest, between two communities, to restore comparative advantage. Perhaps the most intelligent response is consciousness of how trading has changed Drucker also suggests a legal

system for international investment. The rules would protect community interest, even though the trading would be done by individuals.

The international economy is now transnational, controlling national economies. The nations, regions, and corporations are only partially dependent. The symbolic economy of money dominates this transnational economy. In some respects, the transnational economy is stable. due to the diversity and separateness of ecological zones. Zonal diversity can be used to develop different resource production systems in a local area. This increases the likelihood that production will fluctuate nonsynchronously so that support for human populations is constant. Of course, keeping flexibility and reserves could have the same effect. By forming a complex society, the scale of production is raised from a local group in limited territory to a regional population in diverse territories. Even hunter/gatherers engage in similar "energy averaging" systems.

Because of the great contrast in productivity between hot and cold regions and wet and dry regions, many nations, such as French Guiana, are wealthier in terms of ecological productivity than traditionally wealthy nations, such as Germany or Italy. Because of great disparities in the location of mineral resources, many nations, such as Russia or Saudi Arabia, have far more trade units than others, such as Japan or Morocco. Many nations, such as Switzerland or Japan, or the Campa or Mbuti, have succeeded in providing for their people, regardless of the distribution of arable land and accessible minerals. In the future, it might be prudent to divide nations into categories of successful and unsuccessful, rather than developed or undeveloped.

Nations must be as self-sufficient as possible as soon as possible, and this implies fewer people. Lower densities of humans will always be able to harmonize more successfully with biological processes. For the long-term survival of the human species, adaptability to environmental changes is necessary. This requires a wide diversity of gene pools, which is achieved by a relatively large population divided into local, partly isolated groups. And this requires healthy regional ecosystems. The optimum size of the global human population is thus the sum of optima for local habitats. And the optimum, remember, is partly determined by a percentage of the global availability of critical resources. Furthermore, many recently externalized resources, such as clean air, water, and wildlife, ignore human boundaries and circulate in their own patterns. International cooperation is required to protect these common goods. The concept of total global carrying capacity must include such obvious things as tourism, which also reduces the population of a local area.

Commons and the United Nations

As the economy is transnational, so there is a transnational ecology. The atmosphere and the seas, for instance, require transnational policies and laws. The UN could create large numbers of common areas that function according to international arrangements. The UN should have authority for surveillance capacity to detect illegal activities in common areas, such as the oceans, then enforce them. The UN needs to be the international system of justice that has been lacking, whose lack has permitted great injustice and exploitation.

The UN must work with traditional cultures and realistic planning, to minimize untested conclusions. Rational planning can catch up as it develops. The UN framework is an open, flexible, and partially-planned global relation, instead of a finished, closed, completely-planned society, as imagined in utopias. It accepts the imperfect nature of humans and the changing ambiguity of nature. It detoxifies cultural rivalries. Racism, sexism, ageism, and speciesism lose their importance in a cooperative society of advanced communication, equality, humane scale, and meaningful preservation.

The global system may be too difficult to manage and democratize, however. It might be better to try manageable communities first and let the whole change as a result.

Summary

Cultural and Personal Changes
Everything changes, the earth, ecosystems, patterns of habitation, and cultures. Those cultures that survive the changes usually have an openness to new values and ideas that allow continuation in a broader context. The basis of belief is in tradition. But, cultures must avoid narrowness, with a minimum of human and nonhuman rights.
Conscious change requires the recognition that something is wrong. The problems are easy to list. And they are all interconnected (partly as the result of an inappropriate world image). But, we have to power to choose a new future using a healthier image.

Human ills cannot be cured by a return to idyllic hunting and gathering groups or to small agricultural societies or to classic civilizations. There is no possibility of complete return. Much of industrial nations are urban; agricultural countries pack their surplus peoples in cities. Many traditional cultures no longer exist; others are disintegrating under pressure from industrial cultures. Nor can there be a jump to a complete technological future, where technology transforms hydrogen into wealth for everyone.

The establishment of an ecosystem, or an area of traditional culture, as an independent nation would not be unrealistic and utopian. For some people, now, even freedom from hunger and sickness is utopian. For other people in the industrial system, the choice of a fulfilling profession is utopian. Grinding poverty, economic dislocation, homelessness, are more painful than a transformation to a sustainable, independent community. Already much of our environment has been transformed by cash crops, mining, tourists, highways, high-rise housing, and condominiums. Physical disruption has been more extensive than the transition to sustainability could cause. Industrial culture has replaced older patterns with great suddenness. This cannot seem more sudden than the loss of a home or place. Industrial cultures have reduced people's control over the means of production and power. This does not offer less control. Whole communities are being destroyed by industrial scale. Our social structures are already changing rapidly and impractically. Let us just make the changes conscious and more practical. This ecological plan offers movement towards common, achievable goals.

There will be questions regarding the wisdom of independence. Some will want to decide boundaries by ecosystem; others through culture, watershed, or political power. These questions can be answered in meeting. The quest for ecological balance means that some ecosystems must be maintained by systems managers, who often overmanage. The larger the human impact, the more control is necessary. This plan seeks to improve people's circumstances by enlisting them to save their own environments and their own ways of life.

People cannot be given material equality instantly. Providing work for everyone is one way to narrow income differences. The government, communities, and families must provide it. Worthwhile work requires imagination. The large work force employed by military contracts in industrial countries will be dislocated at first, but that employment is supported by taxes, which could be reallocated for construction and deconstruction.

Crime and civic unrest will not disappear. Dangerous weapons, from automatic guns to tanks, and dangerous products, including nuclear reactors and biocides, would be strictly regulated. People will still choose badly sometimes. But, if a form of government is bad or ineffective, they can alter it. They can learn from mistakes or unintended effects. The scale is small, so the catastrophe is small. There will always be some injustice, inadequacy, and unpredictability. Large political and economic institutions have only made it worse. People may object to giving up too much or not gaining enough.

Goals and Images
Our problems reflect an unbalanced and immature image of the earth. The collective image that people make is a world, derived from the German word meaning "man-image." The image is constructed metaphorically, but considered "as if" it were true. Root metaphors are comprehensive and dominate our attitudes towards things. If the image is incomplete or does not fit environmental conditions, it may cause the whole culture to fail. This has happened with many cultures over thousands of years, from the first Mesopotamian Empire (2150 B.C.) to Mississippian complex centered at Cahokia (1250 A.D.). Modern industrial cultures also have defective images.

An image of a culture can be stated as a series of principles, such as, in industrial cultures: "the universe is mechanical, humanity is master, and all persons are equal." Metaphors limit cultural possibilities. The same kind of principles, restated by a Yakima, might be: the "universe is personal and orderly, events are primary, and the tribe is first." Perhaps, we could combine them and expand them: "the universe is self-organizing, being human means participating in it, and persons should have equal opportunity within their groups."

The realization of hard realities does not mean a descent into chaos or poverty. We will have more flexibility if we choose our way and salvage much of the industrial revolution, at the same time preserving a good part of our ecosystem and lowering our population and rates of resource use. Natural systems are characterized by resiliency and flexibility, and often high productivity. We need efficiency, but a higher efficiency — an efficiency in life, not one department of life. We need to keep ourselves below the limits that ecology and technology can extend. There must be a positive flexibility; potentials not used must be preserved for future use to accommodate dead ends, mistakes, or change. As Gregory Bateson noted, we have not allowed a sufficient margin of safety as a buffer against climactic fluctuations, social or environmental changes.

We can cultivate our place. We must give deep attention to the land, understand its ecological relationships. Attend to the whole culture in place. As technology needs our constant attention and involvement, so does democracy. We need to make the interdependency comprehensive so that we can take responsible action. Many things can distract us: politics, television, temporary rewards. We can change the institutions by altering their legal status.

Civilization can follow a circuitous route, practicing rigorous self-discipline and economy (more than if we had started earlier). There must be fewer people than now. There can be a selective reduction of industries. Large industry might still be needed for certain things (televisions or computers), but the technology must be appropriate for many different cultures. There should be a proper mix of handicraft labor, intermediate technology, and heavy industry. The root problem is how to live with technology in a mature manner. We need an ecological awareness at all levels; a humane, existential ecology, where humans are part of the system and aware of it. But that may not be enough; we may have to legislate limits or induce adherence with economic incentives, if awareness and reasoning are not enough.

The goal of planning is community success and personal happiness based on self-reliance in food and shelter, self-sufficiency in agriculture, and self-limitation in size and desires. If human patterns were based on mature ecosystems, civilization would be far more complex; human values would allow for the welfare of humans, animals, plants, and land. We have to be wise enough to be disciplined, to leave wilderness for other beings, and yet to make good places for ourselves.

CHAPTER 18

Design Proposal for Moscow Vision 2020

Thank you for asking for responses concerning the future of Moscow (Idaho). As any city is supported by a much larger rural area, I have included Moscow in its ecological context, the Palouse grasslands and submontane coniferous forests. I have enclosed a draft of a paper written last year (parts of which have been published) and I have written a few paragraphs about trends and goals. I would be pleased to contribute to a public presentation.

What Moscow may look like if current trends continue: From demographic trends, which will probably accelerate over the next 25 years, we can expect Moscow to have in excess of 30,000 people from increased births and immigration from denser metropolitan areas. The city will have more malls and strips, possibly one continuous one from Moscow to Pullman, and more and cheaper apartments and houses. We can expect people's actual income to decrease as jobs change from production to service, and there will probably be higher unemployment, more homelessness, declining job satisfaction, and much fewer employee benefits; disparity between high and low incomes will probably continue to increase. Although there will be less water, less hard energy (petroleum products), fewer tress, smaller crops, and a shrinking sense of community, there will be more traffic, more pollution, more laws, more crime, more population movement, and more unrestricted trade (resulting in even lower incomes). As our options run out, as we manage for maximum yields and complete control, there will be even less flexibility for change, creating the possibility of cultural decline and collapse.

What Moscow may look like if people recognize ecosystem and cultural limits and transform their values into goals: The population of Moscow would be adjusted to the net ecosystem productivity of the area (considering trade for necessary resources for high technology, e.g., copper or tungsten) — perhaps only 12,000 or less, easily a good size for direct democracy. Businesses and housing will be more centralized, possibly in an arcology. A wilderness area of a good size would be restored to guarantee ecosystem processes, such as produce clean water and air, and optimum populations of wild animals. Agriculture, which still depends on wilderness for soil, recycling etc., would become more diversified and regenerative; new crops and uses for crops would be generated. Soft energy paths would be developed to ensure a constant supply of energy for human needs. People would decide on what kinds of technology to allow and how fast to replace it; many older technologies would still be useful. Railways and buses might increase dramatically. Recycling would continue to increase and waste generation to decrease. Through taxes and tariffs, residents could protect the results of their labor. Economic and cultural institutions could be diversified. A vision of the common good would make our goals and a way of life conscious. A democratic government, with increased participation, could regulate the community and protect it from internal and external threats. We can plan to create a successful, sustainable community based on self-reliance and self-limitation, but it requires conscious participation in a political framework.

CHAPTER 19

Transportation and the Pullman Comprehensive Plan

The Pullman (Washington) Comprehensive Plan is being made even more comprehensive. The City Planners, the Planning Commission, and the public are engaged in a total revision of the plan — every aspect is subject to review. New sections on the environment and on economic development are being written.

One of the traditional elements of the plan has been "circulation." This element is being expanded to include pedestrian and bicycle circulation, as well as the movement of vehicles, mass transit, rail, and air.

The goal of the circulation part of the plan is to reduce our dependency on motor vehicles. One way to bring this about, according to Pullman City Planner Pete Dickinson, is to reduce the need to drive by providing dispersed services, such as neighborhood food stores, that people can reach on foot or by bicycle. This could be accomplished through rezoning and redesigning. The downtown, meanwhile, would be reinforced as the cultural center of the city.

Transportation in general would benefit from being prioritized for efficiency of moving persons and products, rather than just vehicles, according to Wiley Hollingsworth. Emergency vehicles, pedestrians, bicycles, car pools, and buses could be favored over low-occupancy vehicles through combinations of exclusive lanes and intelligent stop-lights.

For the time being, the pedestrian/bicycle plan is separate, but it will be included in the overall plan. This plan was moved to a higher priority as a result of a public survey (75% out of 1100 people sampled) showing that people wanted improved bicycle paths and more opportunities to walk. The highest priority of the plan is to build more sidewalks, widen the pavement of some streets, and establish marked bicycle lanes. Bicycle trails around Pullman will be marked. City Planner Dickinson emphasizes that walking and biking are going to be promoted in the new version of the plan.

Changes for vehicular traffic have already begun, with the widening of Main Street out to Forest Way. Rather than build a by-pass around Pullman, which is considered unfeasible, the plan calls for a new road (Coliseum Road), an extension of Terre View Drive, and improvements to Airport Road — thus establishing a "quasi-by-pass." Mass transit, or bus service, will not be decided until after land use has been determined.

Widening streets is the traditional unsuccessful approach to accommodating and controlling traffic. In the past, more car lanes have edged out bicycle lanes, as happened with Bishop Boulevard. The history of road-widening has followed the failed patterns in big cities; until we stop widening streets for cars and start changing people's movement habits, we will keep making traffic conditions worse.

The Pullman/Moscow corridor, the path between two cities and to universities, is not addressed directly in the plan, although Dickinson mentioned that the city has a "lead role" with Whitman County (through the

Whitman County Comprehensive Plan) and Moscow in planning for it. Yet, the city may annex property up to Washington Water Power to build a new sewer treatment facility and create a satellite village in the area.

Most traditional plans, and the Pullman plan seems to be no exception, are reactions to problems. The negative effects of planning and nonmonetary effects are usually missing from such plans. Bioregional context is also missing. A sense of limitation is missing. The plan considers public facilities to accommodate growth, without addressing whether more growth is needed or wanted. Many people in Pullman want a small town, with good cultural amenities, but there is no organized support for staying small — the plan could address people's vision for a small town.

The current plan has several major inconsistencies with people's desire for a stable, appropriately-scaled community. The approval of low-density, single family homes works against sustainable community and ignores important developments in co-housing (*Co-Housing Journal*) or eco-villages (Sim Van der Ryn and Peter Calthorpe), not to mention arcologies (Paolo Soleri). Many of these developments are being realized in Europe and other parts of the US (New York, Massachusetts, California, and Arizona). Pullman could attract ecologically-responsible developers by articulating a vision of high-density housing with fewer and smaller roads.

Dickinson mentioned that Washington State University (WSU) is the driving force for growth in Pullman. But, the university has no zoning, and the city has no control over WSU. The state legislature has more control over the destiny of Pullman, than the city itself, by allowing or directing the growth of the university. WSU expects 8000 more students over the next 15 years, as the bubble of baby boomer babies mature, most of whom will require city services — but what happens when the bubble bursts, leaving a calcified shell of unused services? The goals of WSU seem be related more to maximizing its growth and income than to academic excellence at a proper scale.

A truly comprehensive plan would design the community for optimum size within ecological limits. Water limits, at least, have been considered through a voluntary agreement with Moscow. The sewage treatment capacity of Pullman is an immediate limiting factor to growth. Traffic congestion is an obvious symptom of the uncontrolled growth of people, use, and vehicles without recognition of limits. Population limits are never mentioned; nor are incentives or disincentives to limit population. The plan does not relate the shape of Pullman to long-term values of its residents or to a healthy population-technology-consumption equation that drives our impact on the local environment.

CHAPTER 20

An Ecological Impact Study of Rainforest Lotion

1. Ingredients

Water. The water is filtered and deionized. Is it sterile or softened? Where does it come from? Springs? Glaciers? Olympic Rain Forest?

Aloe vera gel. Aloe vera is an anthraquinone compound from the plant leaf from a South African lily-like plant (anthraquinone is a coal tar color — like D&C Violet 2 — from phthalic anhydride and benzene). Where does ours come from? It is used as an emollient (usually an oil or wax to make skin smooth). It is 99.5 percent water, and the rest is 20 amino acids and carbohydrates. The AMA says there is no scientific evidence for any benefits in cosmetics, but there is no known toxicity when applied to skin. It crossreacts with benzoin. However, if taken internally, it can cause severe intestinal cramps and sometimes kidney damage. It can cause contact dermatitis and cancer in rats from oral doses. It acts as a purgative — it was once rumored to induce abortion if swallowed. Nevertheless, Aloe vera has a good reputation that would contribute to the product.

Soluble collagen is a protein substance derived from the connective tissue (skin) of young animals. Where do we get the animal tissue? Cats and dogs have been known to lick wherever it is applied on humans (usually in shampoos). For a long time it was difficult to stabilize; if ours is soluble then quite likely it is not so under natural conditions and it has to be made so by the action of a detergent (sodium sulfonate as agent?). The AMA says it cannot affect the skin's own collagen by topical application. It has no known toxicity, but allergic reactions can occur. Many people do not want animal products in cosmetics. We might consider replacing it with legumin (vegetable casein, protein).

Saccharide isomerate is a mixture of sugars derived from using an alkali and water on a mixture of glucose and lactose (milk sugar from the milk of mammals — which can cause tumors in mice when injected under skin). It is used as a general base in cosmetic compounds. (Saccharides come as mono, di, and poly. They are carbohydrates. Isomerism means same number of atoms but in a different structure.)

Propylene glycol ($CH_3CHOHCH_2OH$, 1,2-Propanediol) is a clear, colorless, viscous, bitter tasting liquid. It is a common moisture-carrying vehicle that has good permeation through skin (better than glycerin and less expensive). Basically, it keeps the cosmetic from drying out. It is a hygroscopic liquid, soluble in ether and miscible in water. In cosmetics it often replaces glycerin as a humectant (substance to preserve moisture content). It has properties as a fungicide. There is a slight toxicity, and it has been linked to sensitivity reactions. The dermatological action is almost nil; it is not an irritant or a sensitizer. There are safer glycols that can be used — butylene and polyethylene glycol for instance — although these have slight toxicity also. Sorbitol might be even better (distilled from berries of mountain ash, berries, cherries, apples, and seaweed); it has no known toxicity.

Glycerine ($CH_2OHCHOHCH_2OH$) is a by-product of soap manufacture, a sweet, warm-tasting, oily fluid obtained by adding alkalies to fats and fixed oils. It is a solvent, humectant, and emollient that absorbs moisture from the air. It is a hygroscopic liquid, miscible in water. It has negligble toxicity, although a gross overdose can cause systemic effects. Generally, it is nonirritating and nonallergenic. The dermatological action withdraws water from the epidermis; it also aids in healing wounds.

Carbomers (Carbopol, Carboxypolymethylene). What number will you use? 934, 940, 941, 950? A carbomer is a white powder, slightly acidic, that reacts with fat particles to form a thick stable emulsion of oils in water. It is used as a thickening, dispersing, and emulsifying agent. There is no known toxicity.

Dimethicone copolyol $\{CH_3[Si(CH_3)_2O]_nSi(CH_3)_3\}$ is a protective film (a silicone oil that clings to the skin, making it smooth, protecting it from heat or dirt—silica, from rocks, is dried and heated in a vacuum to hard, transparent, porous granules); it is white and viscous. Dimethicones are water-repellent liquids, with low surface tension, stable to heat, and resistive to many chemical substances, except for strong acids. It is used in industrial barrier cream and in hospitals to prevent bedsores. Very low toxicity.

Orchid extract, from orchids. Where does it come from? Sustainable plantations, greenhouses, collection from wild plants?

Sodium hydroxide (caustic soda, soda lye) is an emulsifier alkali; it can burn the skin and mucous membranes. Its white flakes readily absorb water. Orally, it can cause gastric upset (vomiting) and prostration. In the lungs, it can scar the tissue. FDA limits it to 10% in liquid drain cleaners. Can we do without this? Sorbitan palmitate, derived from sorbitol, might be a better emulsifier, certainly safer and much pleasanter sounding.

Disodium edta (ethylenediamine tetraacetic acid) is used as a sequestrant (a substance that combines with metals in solution to render incapable of effecting other ingredients; a preservative that prevents physical or chemical changes). It is often listed as a coloring for shampoos. It can irritate the skin and mucous membranes; and, it can cause asthma and allergic dermatitis. For preservatives, parabens and quaternary ammonium compounds are much better and safer.

Fragrance. What kind of fragrance? From where? This is just to satisfy the FDA labeling requirement, right? It should be from a natural source.

Glutaral, or glutaraldehyde, is an amino acid that occurs in green sugar beets. It is used as a demulcent (a soothing, creamy substance used to relieve pain in inflamed mucous surfaces). It has a faint, agreeable odor. There is no known toxicity.

D&C color. What color number? red no. 6? Dye can cause allergic reaction. Grape juice, or elderberry juice, can do the same coloring, and it's much safer and has a better reputation; and, it looks better on labels. Many people do not like to see artificial colors in their cosmetics.

Suggestions. Olive oil is superior to mineral oils in penetrating power. Polysorbates and poloxamers seem to be safer emulsifiers than the sodium and disodium ones. Proteins and amino acids are thought to help water penetrate the skin; so instead of collagen, use some of the 8 essential (not manufactured in the body) aminos.

The most effective skin sealers are petrolatum and lanolin. Dryness is due to insufficient water in the skin layer stratus corneum; although an emollient retards evaporation of water, oils do not usually penetrate. Skin moisture is replaced internally, as a result of consuming water.

2. Packaging

Materials. Minimize packaging by eliminating paper or cardboard outer materials. As long as the packaging meets the minimum purposes (safety, hygiene, transportation, storage, identification, and communication), it does not need to be more (large, stylistic, luxurious). Simple rules: 1. Packaging should never cost more than the product; 2. Do not use more than is necessary. Simple packaging may be more attractive to consumers.

In order to minimize the total impact of the product and packaging, we need to understand the complete life cycle of the package, from collection and production through manufacture, distribution, end use, and disposal.

The outer delivery container could be recycled cardboard or a recycled foam box. If cardboard, then biopacks (flexible, absorbent, dust-free, paper pouches filled with straw) can be used to cushion the item packages; they are easy to reuse or store, but if discarded, they decompose—it's a good alternative to polystyrene. Recycled materials can play a significant part in reducing overall resource consumption. Virgin materials should not be necessary for store delivery containers or the item container.

If the item container is artistic or more expensive or more durable, we might consider making it part of a refillable system, where the container is returned or refiller packs are offered for sale (some laundry soaps are offered this way, now). Another alternative is refilling the bottles in refill shops (The Body Shop offers this option for its goods). The lotion would be kept in stores and dispensed from large bottles into the item packages.

Glass is a good material, easy to recycle, but fragile and heavy (this would increase shipping costs). Plastic can be recycled and is clear. It is lighter, safer, and almost as robust and durable. Polyethylene terephthalate (PET) is a good recycled plastic (commonly used for soft drinks). It is clear and the clarity promises a product with pure, natural ingredients. Furthermore, it is fully recyclable and an infrastructure already exists to collect and reuse it.

3. Human Use

Since the product has been used extensively in hospitals, it might be helpful to refer to that as a selling point.

The label states "skin lotion," but perhaps it should say hand lotion; at least one tester tried it on her face and found it unpleasant. The only other real considerations for use are: ingesting it accidentally—this is a risk, but it is a risk for even the most natural products; applying it to sensitive areas, such as eyes or genitals—again, a risk, but not too much of one; and mixing it with other cosmetics or medicines—not much has been investigated on interactions. Most of the benefits have been mentioned. Are there others?

A number of things, mostly dealing with the larger aspects of impact analysis, were not completed. Several of these considerations have to do with the manufacturing site: (a.) Where is the manufacturing site (Sandpoint?) and how was it selected? What interfaces are there between the manufacturing and people? (b.) What are the ecological effects of the manufacturing process? Is the process sound? How efficient is the use of energy and material? What are the main effects, products, effluents, and wastes? (c.) What is the potential for significant habitat alteration? How many acres are removed from natural or agricultural production? Other considerations have to do with community relations: (a.) What does the company offer to the community besides jobs and taxes? (b.) Are recycling systems in place, or does the design include reuse and redirection?

4. Recommendations

Although the product has been tested with and found satisfactory by hospital and medical personnel, we recommend slight changes in the ingredients in two phases:

(1) remove the caustic, e.g., sodium hydroxide, and unnecessary ingredients, e.g., D&C red; these are easily replaced, and
(2) replace some of the ingredients with safer versions or substitutions, e.g., propylene glycol with sorbitol.

It is not necessary to make the product purely organic, although if you did you could probably increase the price over 300 percent with only a small increase in cost.

For packaging, a standard size recycled plastic (PET) would probably work. You also need to address what pollution is produced in the manufacturing process and what happens to unused chemicals.
We would be pleased to have the opportunity to examine your production site and make more complete recommendations.

5. Package Copy

Why save the Olympic Rainforest?
The Olympic rainforest is the only temperate rainforest in the United States. It is not only unique, home to many unique and wonderful species, but it affects much of the climate of Washington and the Northwest. The disappearance of this wonderful place would be an unfortunate ecological disaster.

How will buying this lotion help the Olympic Rainforest?
Many of the ingredients of this lotion are made from plant products that grow in Western Washington. Using such products makes the forest too valuable to just cut down for a short-term profit. By buying this, you will also promote a growing awareness of ecological concerns in your local business and residential communities. And, also, we contribute a portion (over 5%) of the net profit from your purchases to rainforest preservation.

Where can you get more information?
The Rainforest Action Network is a nonprofit organization that cooperates

with other environmental and human rights organizations on major campaigns to protect the rainforests of the world. Write to ...

Cultural Survival is an nonprofit organization that works to help tribal people defend their lands and ways of life. They provide educational information about products that promote rainforest preservation. Write to ...

For information on our product, write to:

Olympic Rainforest
1129 Second Avenue
Seattle, WA 91111

Side panels:

About our packaging. This container is made of PET, a 100% post-consumer, recycled plastic. The label is printed with soy-based inks on recycled paper, which are more environmentally friendly than petroleum inks and virgin paper. Opposite Side panel: Ingredients ...

Discussion.

Numerous companies, including Rainforest Moisture, Ben & Jerry's, and Rainforest Products (Mill Valley), refer to RAN and/or CS on their packages. We should make contact and get a list of ingredients that they sell in order to help them and to use them.

Package Design.

The rain forest is highly stratified; trees generally form three layers: very tall emergent trees that project above a canopy layer (a continuous evergreen carpet over 80 feet tall), and understory that increases in density near breaks in the canopy. This stratification might be good for design ideas for packaging.

For example, you could use three sizes of bottle that fit together in a tree-like pattern for each set. Or, the bottles could be simply different heights.

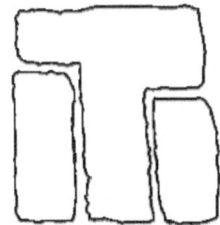

Figure 3. Sample Packaging Sketch

Author's Note: The Nieman Ryan Design team was fired by Olympic Rainforest for pointing out the toxicity of some of the ingredients. The company still owes us $1125.00 for expenses.

Chapter 21

A Proposal to the Potlatch Corporation

We have read about and applaud Potlatch's intent to become a green company. We also have a long-term interest in ecological research. The staff of Nieman Ryan Ecological Community Design (NRECD), besides acquiring vital experience in community design, has availed itself of the products of many companies, including Potlatch. We would like to contribute our expertise back to our providers and partners. We would like to help Potlatch increase its long-term wealth through comprehensive ecological design.

Corporate solutions to environmental problems tend to be addressed through relatively rigid departmental structures. Furthermore, the corporation tends to sit outside the community that contains it, importing staff , food, and equipment from distant places, for reasons ranging from prestige to economy of scale.

To make the corporation more fluid, to overcome its economic and intellectual isolation from the community, and to stimulate new discussions of community order, we have several general suggestions:

1. Establish an ongoing self-study group that would investigate how the corporation has been, is, and will be related to the rest of the community. This group would evaluate the corporation's contributions and responsibilities, from departmentalization to waste management. This group would incorporate gender, ethnic, and outsider perspectives; it would criticize and discuss values. Its goal would be to increase self-understanding of the corporation and to encourage reform.

2. Create a unit whose responsibility would be to use local resources for the operation of the corporation. This unit would gather information, raise questions, make recommendations, which would strengthen the economic capacity of the local community. Buying locally makes more sense in the long run; money circulates in the local economy, generating even more money in other businesses. All food, energy, materials, water, air, and waste, that flows through the corporation would be examined for its social and environmental costs. The unit would involve the staff in the operation of the corporation.

3. Establish an interdisciplinary center to address issues of culture, emergent problems, regional impacts, such as those related to global problems of deforestation and CFC production, and even comprehensive anticipatory design (such as we practice at Nieman Ryan).

The actions we recommend would not impinge on the corporation's charter or violate its standards of behavior. They would tend to heighten awareness of the historical and technological character of the corporation and the unspoken assumptions upon which this character is based.

We think that a few structural changes would optimize the

corporation's energies by promoting more organic and holistic ways of behavior. A few changes would make people's work more relevant to the problems they experience outside the corporation.

We would help strengthen and renew the corporation's traditions so that it contributes more to the larger community and increases the community's ability to be self-sustaining. Please let us know if we can make a formal proposal.

Figure 4. Nieman Ryan Trademark 1976

Author's Note: Potlatch claimed that they had sufficient expertise and would not require any advice or assistance.

CHAPTER 22

Ecological Design with Nieman Ryan

N i e m a n   R y a n
Ecological Community Designs

Why?

Nature is self-making and self-designing, but we humans now influence every natural system, taking what we need from some ecosystems, misusing others, and interfering with the rest. We need designs to restore balance between human needs and natural processes.

At NR, our ecological designs focus on whole communities that work in the same self-sustaining and self-limiting ways as nature. By consciously creating meaningful order, NR develops ways of producing widespread community wealth while positioning the community for a long, sustainable future in a healthy environment.

Who are we?

We are a core design group, generalists with a working knowledge of the arts and sciences, plus a talent for integration, who devote our efforts to creating and implementing ecological designs. We assemble teams of consultants — designers, architects, artists, ecologists, and engineers — for specific projects to work with corporate officers, government officials, military units, citizens groups, and planners. We are associated with the Marsh Institute, a nonprofit educational and research corporation; we donate 10% of all profits to conservation organizations.

What do we do?

We design communities. For corporations and neighborhoods, we recommend how to:
- revegetate yards and plant sites
- reforest and replant fields
- reclaim damaged and paved areas
- integrate product and material flows into biogeochemical cycles; shift to renewable energy sources; recognize links and dependencies
- adapt sites to functioning ecosystems to increase diversity and stability
- mediate technology and community; redesign and reuse buildings
- simplify and minimize maintenance
- make alternate paths of education, communication, and financing

For towns and cities, we show how to:
- mitigate the problems of urban development, such as heat increase, waste, and water runoff
- remake patterns of convivial activity to restore contact with nature and amenities in urban life

- redo private and commercial developments
- restore wilderness, conservation, and park areas
- set an optimum population within resource limits and carrying capacity
- internalize and manage costs and resources; minimize external subsidies
- revitalize the sense of place and direct changes
- promote diverse economies to become self-reliant and self-sustaining; coordinate and cooperate with other regions.

We design places as organic wholes to promote the well-being of individuals and the common good. We strive to employ all the characteristics of good design: right scale, place specificity, simplicity, efficiency, fitness, resilience, durability, and redundancy.
How do we do it?

We participate with you in a complete design process, guiding your involvement and commitment to the art of living together as a community.
- First we review the situation, observing patterns of movement, population change, land use, building and development, boundaries, limits, and life. We conduct ecological and functional analyses.
- Then we record all of the resources, from physical resources to cultural resources. We survey the area and create base maps, from geological to zoological maps.
- Next we evaluate the interactions in terms of impacts, needs, goals, and limits. We assess the whole system and create a series of plans, from the site plans to value plans.
- We start to design, which is a community process requiring the participation of all people (including the elderly, handicapped, and poor, as well those ultrahuman beings who cannot voice their concerns). We synthesize simulations and models (conceptual, capability, and suitability). We make another series of plans, from landscape plans to policy plans, within a master design.
- Finally we implement the design together and start to maintain it. We use appropriate measures and techniques, emphasizing native species over an adequate time period to ensure the stable processes of transformation. We provide services for continuity and management.

Our designs are open, respectful, and continuing, derived from an understanding of natural processes, stability, and change, as well as from the range of human needs. We stress suitability, fitness, health, flexibility, frugality, maturity, and diversity.

How are we unique?
We address all levels of design, from the conceptual to the political, and are involved in all stages of the process.
- We relate a project to its total context (a fourth level of design); we are concerned as much with cultural survival, justice, and wilderness preservation as with efficiency and aesthetics.
- Our perspective is ecocentric; our vision is of the whole community in which we dwell. We apply ecological concepts, such as networks and

carrying capacity.
- Our designs are anticipatory, flexible, pluralistic, polyvalent, and polytechnic. We try to make open guidelines for long-term decisions.
- Essentially, we work backwards from values and goals, and from the bottom up and inside out, drawing designs from the genius of place.
- We participate in place, care for all inhabitants, and assume responsibility for the designs.

Our immediate goals are to reverse degradation and reclaim places for communities, but we also work to increase public awareness of the interdependence of communities, to create environmental quality, and to transform public values by generating new metaphors for living.

Where do we work?

We are specialists in temperate ecosystems: coniferous forests, great basin grasslands, and nontidal wetlands. Although we are centered in the Pacific Northwest, we are able to consider any location on the planet.

How much will it cost?

We charge a flat hourly fee and a flat cost rate for equipment (and for recovery of external expenses, such as tests and special equipment). For example, grassland restoration costs about $1500 per acre per year, based on the first two years. Design for a corporate plant site costs about $500 per acre; a home site can cost as little as $400. We estimate that the percentage of expenditure on our design is the rough equivalent of the costs of waste and pollution. We like to learn, so we absorb any costs of our learning. Reducing your costs is a part of the design.

How long will it take?

The restoration and operation of a community is a dynamic process; there is no fixed term. The study may take less than a month or as long as a year, depending on the size of the project; we put a three-year cap on the design itself. We offer five-year maintenance plans for remediation or restoration. For longer projects, we provide strategies and monitoring support.

We provide an intelligent direction for uncertain circumstances in an indefinite future. Our design is a quest for fitness of form and purpose combined with beauty into an organic whole.

The NR logo is based on the Egyptian hieroglyph for water, which also has the sound value of "n."

For information on our projects or for free consultation, please contact us.

Author's Note: Nieman Ryan became Rian Ecological Design in 1997, then in 2003 merged with a Florida design group to become Rian Garcia Calusa.

Chapter 23

Another Earth Day (Another Holiday)

The earth day celebration is over. What were we celebrating? That we *are going* to save the earth or maybe just still thinking about it? Perhaps we were celebrating our intention to go on a material diet or an opportunity to spend money on t-shirts and buttons. Perhaps earth day is a new spring-time variation of a new year's resolution—a temporary awareness, a limited intent, and a reason to party before business as usual. Or, perhaps it is a modern penance that allows us to buy a place in heaven by promising to save the earth with tokens.

The token changes and vows do help, but are they enough? Will a little conservation avoid a great human disaster? Aren't these easy remedies reminiscent of medical cures for diseases that could be avoided by simple denial (smoking, overeating, or stress)? The implication is that a few small things, such as using less water or recycling bottles, will save the earth—that ozone depletion, rainforest destruction, population growth, and the polarities of wealth will somehow be corrected, as governments and industries continue as before, adjusting their labels green.

We were told that saving the earth starts in the home. No wonder corporations gave their blessings to this event—most pollution and waste is industrial and agricultural! Which issues have higher priority? Deadly local ones (toxic waste dumps, topsoil loss) or deadly global ones (greenhouse gases, chemical runoff)? Is alarm justified, or is caution enough? Should we listen to the ecological Cassandras or the economic Neros?

Are we too lazy to follow through with the effort that we already celebrated? Are we too cheap to deduct a required percentage from profits to pay the real environmental costs? Are we too crazy to stop our growth? Where is the will and the vision to really make radical changes? We have, in fact, taken the easy way at every branch. We have assumed that corporations will choose the proper path of production and regulate their pollution. Yet, we know that they put profit first and only prevent pollution when forced to do so.

We have wasted twenty years attacking the symptoms and not their technological or social origins. We must acknowledge the failure of our remedial efforts, our failure even to address the flaws of our ideologies.

I don't want to participate in the wrong games. I don't want to drive fewer miles in a high-powered gas guzzler—I want to travel by train; I want my radio shipped by train and not truck; I don't want my vegetables shipped at all, but grown locally. I don't want farmers to do slightly less aerial spraying of fertilizers and biocides, I want organic produce. I don't want to recycle aluminum and plastic, I want returnable glass containers. I don't want safer coal-burning centralized power plants, I want local solar power. I don't want to give more money to the homeless, I want my tax money to help them build and keep homes. If industries can't help me with what I want, then I want my government to encourage them and channel them, tax them and regulate them.

We need to propose and execute national policies to steer technology. I want my representatives to ban CFCs, to ban burning, to tax nonrecyclables, and to tax "bad things," like pollution—and if they don't, then I'll run for office myself.

The earth does not need to be saved or healed, as if we could do either. The ways of life that we remember and prefer, the places that depend on other species and natural processes—these can be saved. Our own divided minds, that let the poor be enslaved by the wealthy, that let "good" animals be domesticated and "bad" animals be eradicated, can be healed. The sacrifices will have to be great; the changes will have to be radical. But, the celebrations will be meaningful only then.

CHAPTER 24

Meaningful Jobs

Recently, I noticed a book entitled *Future Jobs*. As I browsed through its list of service and managerial position descriptions, I became uneasy. Most positions seemed boring or meaningless. Nobody seems to notice that we have a longage (Garrett Hardin's correct term) of human labor and a shortage of responsible positions. Some kinds of jobs were not even mentioned; others used to be functions performed by people in stable communities, but have disappeared. I thought of a dozen right away.

1. Mr./Ms. Know-it-all. For grocery stores (five per store or so), she would explain where foods come from, what value they have, what sprays or biocides are used, what effects they have on environment and local economies, and what substitutes may be used. For instance, she could tell shoppers to use soap and bicarbonate for washing clothes instead of harsh detergents or expensive organics.

2. Recycling squad member. He could go house to house and collect things, or scavenge roadsides or dumps for reusable items. He could be self-employed or part of a city force. 'Pardon me, you weren't going to throw that away, were you?'

3. Neighborhood watcher. She would be a monitor; do nothing but observe and offer to anyone who'll listen what was observed. She could contact other agencies about suspicious activities, such as cutting trees or breaking into houses. 'What's that in the drum your dumping?'

4. Corporate clowns. One would be assigned to every President or major officer of every corporation. His/her purpose is to make fun of, and call into question, every pronouncement and euphemism offered to dodge real questions and responsibilities. Remind them that someone thinks everything they say is funny. What do biz-speak statements mean? 'As trade barriers come down, many of our customers face global challenges.' This kind of seriousness can be deadly. This position is parallel with political clowns, who would accompany every politician above county commissioner.

5. Small farmer. He grow crops in a sustainable, organic way, sells locally with minimal packaging, and charges the true cost of growing the item. Small farmers are more flexible and responsive to consumer needs, such as growing organically. He could rehabilitate the parking lots of many malls, making them bloom again.

6. Gadgeteer (Mr./Ms. Fix-it). She would rebuild lawn mowers, fans, tillers, or maybe combine some of these machines into new patterns (a tiller with a fan to cool the operator at the same time); or refit machines, say refrigerators without freon. She could survive on the junk of civilization.

7. Neighborhood artist and poet to tell stories about communities. He would (re)invent modern myths for cultural survival and development. He would read stories for pay. Stories thus presented may be enhanced

by each retelling. Artists can transform the effluvia of modern civilization into creative visions of what the civilization is really like.

8. Tree planter. Like a doctor, he will have to know when to cut and remove exotics as well as when to plant appropriately. The rush everywhere to plant something green sometimes commits trees to a lingering death and sometimes results in weed trees taking over the habitat and changing it, as it has in Tucson, among other places. 'What kind of trees used to grow here?'

9. Ecological designer. She would design communities within habitats; she would understand what shapes and relations of buildings are effective and exhilarating as well as the suitable context in the surroundings. She would decide appropriate materials to minimize energy use and local impact. 'Why are most buildings empty most of the time?'

10. Roving minibus driver. To fit between taxis and buses, in neighborhoods or between neighborhood and city, she would pick up people who could not make schedules or time their emergencies. There would be no definite times or destinations; these vehicles would move randomly down streets searching for passengers.

11. Rememberer. She would remember everything of local importance: High school track scores; the first bus; rebuilding the rails. Collect important symbols and pieces of the local variation of civilization — all important things that happened in this place from the Pleistocene on. Manage small-scale neighborhood museum for locally important artifacts.

12. Environmental officer. He would protect the human environment from improper dumping of out-of-place substances. Remind us of our responsibilities to live lightly and improve our places. Protect nonhuman places reserved for other beings and the natural cycles that provide services for human civilization.

I do not think that I would mind doing any of these things myself, and I would certainly like to have them around when I go shopping or looking. What do you think?

CHAPTER 25

Behind Glass

Glass has advanced our civilizations by permitting the easy separation of fluids and reactions. But, it leads to an objective attitude towards living beings and nature. Being "behind glass" has become the metaphor, first for a scientific approach to knowledge, then for a utilitarian ethics and for a teleological ethics. Only recently, in theoretical physics and in ecology, has it been realized that there can be no perfect detachment or objectivity; there can be no perfect insulation from the object of study. The new paradigms in physics and ecology reveal an inextricable participation in nature, in the objects of study, that becomes the basis for an ecological ethics.

The Beauty and History of Glass
When most liquids are cooled, they relax into the state of lowest free energy; they reach a temperature at which their structure changes to a crystal. By contrast, glass becomes rigid without giving up its liquid structure; it never crystallizes because it is blocked from reaching a condition of lower energy by its rapid cooling and by the 'confusion' resulting from its configuration. In a liquid, atoms are joined together in a random structure, without a regular three-dimensional pattern. With glass, cooling produces no discontinuous changes; it simply gets stiffer until it is 'solid' with the internal structure of a liquid. Glass has similar properties to other fluids, but the time scale is greater. It is a 'slow-motion' liquid. It would take millions of years for glass to flow at room temperature (68° F).

Glass is a solution, a homogeneous mixture of substances having different molecular structures. It can be formed with a variety of compositions. The largest constituent of common glass is silicon dioxide, from sand. There are several properties in common to almost all glass. Glass is impermeable. It surface does not allow penetration by other gases, fluids, and solids. It holds materials at high or low temperatures. It holds a vacuum. It separates gases, fluids, and solids.

Glass is transparent. Structurally, glass is a large molecule that contains no internal surfaces or discontinuities that approach the dimensions of the wavelength of visible light. Thus, light passes through glass unhindered. Glass is opaque at infrared and ultraviolet wavelengths, however, since these waves resonate with molecular vibrations in the glass.

Glass is brittle. It shatters suddenly when subjected to stress. Although glass can be five times as strong as steel if formed with a perfect surface, surface defects make it less strong in practice. Flaws from corrosion or abrasion concentrate any stress on a few interatomic bonds, which break and then spread rapidly, since there are no grain boundaries to stop the spread.

Glass was first used as a glaze for ritual objects in northern Mesopotamia in the fifth millennium B.C.. Glass vessels, formed on a removable core, did not appear until 1500 B.C. and spread from Mesopotamia to Egypt. The Phoenicians spread the technology to Cyprus,

Rhodes, and Italy by the fifth century B.C. Using the new technique of glass-blowing, which appeared on the Syrian coast around 50 B.C., the Romans began using glass for plates and windows. Glass became so common that it was a sign of lack of affluence (the wealthy switched to gold). Rome became a city of glass. Glassmakers were confined to an industrial suburbs because of pollution from furnaces. It has been estimated that 40-80 million acres of forest were cut over several hundred years for glass manufacture, firewood, and metal smelting, resulting in the treeless, eroded landscapes of the Mediterranean so much admired by natives and tourists, as well as in the necessity to import food and wood.

Science and Knowledge
From medieval times, scientists found that glass could separate materials and distill liquids. Astronomers and microscopists found that glass could be shaped to focus light waves to reveal the very small and the very distant. Portions of the universe were placed behind glass in a laboratory world. Glass was very useful. Scientists came to rely on its advantages, but they were unconsciously imprisoned by its limits.

As science cut the connection to direct observation, it became as blind as mathematics to the 'outer' world. The formality of science made statements about the outer world tautological. This proved to be a problem with quanta, species fitness, and psychological needs. Scientific hypotheses form filters like glasses. They cannot be shed entirely, but their effects can be understood.

William I. Thompson saw that industrial civilization intensified human culture and miniaturized nature. The forest was miniaturized in clumps of trees; animals in domestic images; plants in a garden; and nature under the glass roof of the Crystal Palace. We have created a hard glass between the mind we are projecting and the object that is receding. Sometimes we doubt if there is anything on the other side that can be seen. Glass forms a window for consciousness, which swamps the mind with the 'error of the eye,' in Marshall McCluhan's phrase. Vision, whose mode is successive and not simultaneous, is emphasized and split away from the total sensorium. After human consciousness places the glass, human needs shape and tint the glass. Utilitarian economic thought casts a thick, convex glass, to focus on the individual. Romantics use a tinted, concave glass, the better to diffuse to see the whole.

Neither art nor science can let us understand life or nature itself. Scientists have experimented with biological processes behind glass (in vitro) in the laboratory. The primary commandment of Jacque Monod's ethic of objectivity is 'thou shalt not participate' in the workings of the world. But, detachment from nature is detachment from the basis of knowledge. We distance ourselves behind glass. This detachment is the greatest threat to the welfare of nature. It permits the vivisection of the "voices of existence," as Neil Evernden warns.

We use glass to protect ourselves from the ambiguity and messiness of nature. We have made an experiment of ourselves; our mentality has evolved behind glass. We have isolated ourselves by technology. We have seen more on television, but are moved to do less. C. P. Snow has commented that watching megadeaths by starvation in Africa on television screens could mark the end of any moral community of humanity. We are behind glass and fear it will break. Augustine remarked to the Romans: "What glory is there in the largeness of empire, bright and brittle like glass, and forever in fear of breaking." Reason alone cannot cure what it caused. Glass cannot divide humanity and nature; nor can we humanize the planet without dehumanizing ourselves. We need to break the glass. We need a sanctified vision of life from a deep participation.

Participation
The physicist John Wheeler believes that since law, field, and substance exist after a theoretical big bang, the universe owes its existence to trillions of acts of registration. The phenomenon comes through an elementary act of an observer — participation. Wheeler questions whether the universe might not be brought into being by the participation of its individuals. Quantum mechanics strikes down the neutral observer; participation is vital. Wheeler notes that "To observe the electron even, the experimenter must shatter the glass — must reach in with instruments." The quantum principle destroys the ideal of the observer behind glass. The universe is not the same after measurement; the observer becomes a participator. The act of measuring changes the measurer and the measured.

Earlier, Albert Einstein said that no event can be postulated without the presence of an observer. But no observer can see the whole system, and any thing or being can be an observer or participant. A world without participants is impossible. Nor are there any lone observers. The observer is part of a natural or social community. Humanity participates in the natural world, so nature is part human history as well. All beings participate in the relationships that make up their worlds.

Human participation in community is celebrated by ritual. Through participation, consciousness is released to an 'other' so that it thinks in the participator. Full participation removes barriers; and 'otherness' pours through. The difference between the self and the other disappears. The fundamental aim of ritual is the harmony and well-being of the community, its coordination with the harmony and nature of the cosmos, of which it is part, and the integration of the individual. The function of myth and ritual is to engage the individual in local organization, by creating intensely shared experiences. Participation becomes the impossibility of *not* being a part of a society.

Other modes of knowledge surround science. Theodore Roszak defines true knowledge as 'gnosis,' of which scientific rationality is only a small part. It is gnosis that is needed to perfect the universe and soul, mutually, with the spirit of love. Gnosis is the whole spectrum: the hard, bright lines of science, the hues of art, the dark voids of religion. Gnosis is augmentative knowledge, in contrast to the reductivity of science. Paul Tillich calls gnosis "knowledge by participation." And that is what humans need to do — recognize that they automatically participate in everything, that they know by participating, and that they cannot unparticipate by choice.

We can put things and beings behind glass, but we lose them. Relationships are so strange and complex that they cannot be understood behind glass. The glass creates an illusion of objectivity. We need a wild universe to live fully, to be fully conscious. When we understand our roles in nature, we will not be stewards or managers, but participants in experience. Then, we are, and will be, in life, not behind glass.

Figure 5. Marsh Institute Trademark, 1970
(based on a dandelion to represent global and local ecosystems as well as political and cultural forms)

CHAPTER 26

Better Living Through Chemistry

This interview was conducted in Seattle, Washington, at a conference in 1989, by Alan Wittbecker, a Research Associate in Wildlife Ecology and Editor of *Pan Ecology*. Before his retirement, Doctor X was the Research Director for a major U.S. chemical corporation; the author of numerous professional articles, he holds several U.S. patents. He serves now on the Board of Directors for a nonprofit educational corporation.

Ed: What is chemistry?

Dr. X: At one time, everything was chemistry. Alchemists believed everything could be transformed into something else. Now, chemistry is a scientific discipline that deals with the relationships of molecules. And, of course, you cannot ignore that it is an industry, major contributor, major polluter, and temporal discipline.

Ed: What good is it? What has it done for us lately?

DR. X: The synthesis of DDT saved millions of lives; people who would have died from insect-borne diseases. The collection of penicillin saved millions of lives. The invention of nylon saved millions of lives.

Ed: Nylon? How?

DR. X: Its use in parachutes, tires, fittings, clothing, bottles, fittings, upholstery, carpets — this carpet, it looks like.

Ed: What about the negative side of chemistry? The production of poisons, the production of pollution? DDT did save many lives, but it also is responsible for millions of animal deaths — billions of insect deaths.

DR. X: Most people consider human lives more valuable than mosquito lives. As a scale effect DDT is poisonous, but in this sense so is oxygen; no one gives pilots pure oxygen when they're flying. Many of the negative aspects are just side-effects.

Ed: Of course, you know of the great difference between oxygen, an important ingredient in the functioning of our body, and DDT. DDT is artificial and concentrates through the food chain; something oxygen does not do. Furthermore, there are no such things as side-effects, just effects (we call them side-effects because we didn't intend or anticipate them, and calling them side-effects lets us pretend that they are less important or accidental — remember Garrett Hardin's first law, that you cannot do just one thing).

Dr. X: Many elements in nature are poisonous in concentration. Any wolf who eats a bear's liver is going to die from the vitamin A. Furthermore, since humans are part of nature, human products are also part of nature, made from naturally-occurring compounds.

Ed: True, but their sudden introduction into a system that has evolved slowly for millions of years poses a threat that natural toxins do not. As for wolves, they rarely have the opportunity to eat bear livers, but pesticides, or biocides, concentrate down the food chain and are unavoidable.

Dr. X: Nature is more sudden than you think. Many natural processes such as volcanic eruptions and floods concentrate toxins at incredibly high levels, killing millions of living beings.

Ed: Being aware of the violence of natural processes, I wonder if we are adding unnecessarily to them with pollution from the production of silly and trivial things. What has chemistry contributed to global problems: the losses of species, the greenhouse effect?

Dr. X: There are many natural greenhouse gases: carbon dioxide and methane, for instance.

Ed: Shouldn't we be concerned that we are upsetting some kind of systemic balance? Other gases can have greenhouse effects in parts per trillion presence. CFCs for example. Shouldn't we ban CFCs, with what we know about them?

Dr. X: What do we know about them? CFCs are artificial compounds, chloroflourocarbons. Thomas Midgley Jr. invented the two most important CFCs, dichlorodiflouromethane (CFC-12) and thrichloroflouromethane (CFC-11) in 1930. Dupont manufactures them under the name Freon. CFCs are extremely effective refrigerants. Without this kind of refrigeration, people could not live in desert climates, could not ship foods thousands of miles, or could not store foods as effectively.

Ed: Should people live in the deserts? Navajo and Taureg live in deserts without refrigerants, because they have developed adaptive architecture and cultural traits. As we know now that we have experience, shipping fruits thousands of miles is inefficient, wasteful, and of questionable taste (they still cannot be shipped when ripe).

Dr. X: Doubtless our architecture could be greatly improved; this would improve the efficiency of CFC use.

Ed: And the dangers of CFCs?

Dr. X: Refrigeration units are closed systems. There's no reason why they cannot be kept closed. CFCs last about 75-100 years. They're also used in spray-can propellants and as foam blowing agents. They're extremely stable and nontoxic and nonflammable. The refrigerants CFCs replaced, sulfur dioxide and ammonia, were really dangerous.

Ed: Immediately dangerous, yes. CFCs pose another danger, a long-term, broad-spectrum, invisible danger. They accumulate in the stratosphere, where they are broken apart by sunlight and react with ozone. Worse, the CFCs absorb wavelengths that carbon dioxide does not. Far worse, a single molecule of CFC-12 traps 20,000-times more heat than one molecule of carbon dioxide. The marvelous science of chemistry cannot seem to be dissociated from the dangerous of its applications. Is there any way to salvage CFCs from the tonnage of used and discarded refrigerators and electrical equipment?

Dr. X: There should be ... Excuse me while I think for a moment. Yes, they could be, but the cost would be high, initially.

Ed: Have man-made fibers given people better lives, on the balance?

Dr. X: Yes, of course. Much better lives. Long-lasting clothing, much more choice, and many more clothes, actually. Heart valves, veins, artificial skin.

Ed: Uglier clothing, buildings stuffed with plastic junk, addictive drugs. How large is our artificial chemical environment?

Dr. X: You already admitted that man is natural, so human products are natural. If you mean artificial in the sense of artistic, most human cultures make as much of the environment as artificial as possible, decorating, changing, and developing everything they can.

Ed: Yes, it is the result of artifice, but I meant in the sense of being exotic and alien — not developed in an evolutionary time scale, not developed in context. How can nylon breathe like cotton, decompose like wool after it has been discarded?

Dr. X: In the evolutionary scale, nylon and other fabrics may indeed break down. Their ingredients are all natural — coal and water — only their arrangement is new.

Ed: Let's look at some magnitudes. There are:
120,000 people exposed to the toxic chemicals used in electronics,
2,000,000 workers exposed to benzene,
9,000,000 Michiganites exposed to polybrominated biphenyls (PBBs) for flame retardant,
135,000,000 exposed to unsafe levels of CO,
240,000,000 exposed to the 2,6000,000 pounds of pesticide in use.
Why are these chemicals dangerous? Benzene, for instance — what does it do to people? Can it be neutralized? Why are these 'artificial' things so dangerous?

Dr. X: Well, substances used to be drawn from nature and combined. As we learned more about them, we learned how to fragment substances and alter compounds for specific purposes. Nylon, for instance ...

Ed: What about accidents, explosions, chemical spills? Is danger inherent in the nature of large-scale production?

Dr. X: With the scale it is. Scale always magnifies risk, but of course, it is the nature of life to be risky. We may have eliminated too much of that already.

Ed: Can the chemical industry operate as it has and does now without any kind of moral responsibility toward human and ultrahuman lives, towards biogeochemical cycles and life on earth? What about the Union Carbide spill in Bopal, India? Was that avoidable? Should Union Carbide be held responsible?

Dr. X: The last really huge chemical explosion before Bopal was in the early 1900s, in New Jersey (?) almost 80 years before Bopal. Considering the sheer quantity of manufacturing, chemical processes are very safe. And, usually, large accidents result from human inattention, so in a sense they are avoidable, but given human history, they are certain to happen once in a while. Yes, Union Carbide should be held responsible, through insurance. Accidents are the reason for having insurance, which is just a way of spreading the risk among many — and after all insurance companies are profiting enormously from their risks.

Ed: Should Union Carbide pay?

Dr. X: Yes, they should admit culpability. They should pay what they were insured for. Perhaps more, to improve public relations. Unfortunately, some individuals and the host country have responsibilities, also. I wouldn't mind seeing an international body help with regulations and responsibilities, perhaps through the UN.

Ed: Do companies like Union Carbide or Dupont conspire to conceal information about the danger of some products or the processes to produce them?

Dr. X: Carbide and Dupont are profit-making ventures in a progressive society. Within the moral restraints of that society (which aren't that many when it comes to the free goods of nature or the use of nature for disposal), Dupont is probably better than many companies. Yet, some mistakes have been made.

Ed: Who should bear the costs of safe disposal of hazardous chemicals? The companies, the government, the users, the taxpayers?

Dr. X: Obviously the best answer would be the company, if all companies were faced with the same restrictions and costs. Many things may not get to market if the company had to pay disposal costs.

Ed: Are there too many chemicals on the market? Do we rely on chemicals too much, for pain, fertilizer, pesticides, fabrics? How many things are unnecessary?

Dr. X: A free market will always produce more than is necessary. Do we rely on them too much? Maybe. Air conditioning, for instance, has become such a necessity that architects take shortcuts on designs, more people move to more peripheral areas, the devices themselves are not made well.

Ed: Are chemical companies, like Dupont, willing, like auto companies in the early 1900s, to buy and suppress the competition (auto manufacturers bought out trolleys and trains, then scrapped them)?

Dr. X: For example?

Ed: Here is a chronology of events:

 1900 — hemp is used for linen, ropes — many things;
 1916 — hemp pulp technology is invented (for making paper);
 1930s — W. R. Hearst mounts news campaign against marijuana (a product of hemp) and against Mexicans, in retaliation for some injury;
 1935 — nylon is patented by Dupont;
 1937 — Dupont patents a sulfuric acid process to pulp wood into paper;
 1937 — the Dupont annual report urges stockholders to invest in radical changes — the government is to force 'acceptance of new ideas' of social reorganization;
 1937 — U.S. government passes the 'marijuana tax law' outlawing this product of hemp;
 1938 — Popular Mechanics introduces a new device to harvest hemp stalks.

Coincidence or could Dupont been involved in a conspiracy? Could nylon have competed with hemp in the 30s? Hemp, remember, is strong and rot-resistant; its pulp can be turned into paper and particle board; its fibers into linen, canvas, rope, and twine; hemp hurds (the particles remaining after the fiber is removed) could be used to produce methanol to fuel engines. Furthermore, an acre of hemp produces the equivalent of 4 acres of wood pulp; it needs no chemicals or fertilizers since it has no weed or insect enemies. Lately, French scientists developed a type of hemp with no THC (the active constituent of marijuana). Would you favor more research on hemp?

Dr. X: There is no reason why both fibers could not coexist. I was never aware of any conspiracy, but then I was just a college student then. I'm sure that many companies devote energy to eliminating or beating the competition. When the auto companies bought out there competitors there was no indication that automobiles would cause such physical and social problems. Fortunately, we can go back and rebuild trolleys and railways; we can tax cars and trucks to encourage more efficient transportation. People wanted cars instead of trolleys—how else could GM buy whole companies? Sometimes people want things that are not good for them. Chemistry should not be a scapegoat for bad social decisions.

Ed: I have been reading in the alternate presses about the sheer criminality of many companies—still polluting, even now against the law, still dumping banned products on foreign markets, still plotting to avoid responsibility. On the other hand, in industry journals and other publications, companies like Dupont and Union Carbide are receiving environmental awards for reducing damages from their processes or for setting aside acreage for wetlands and forests. What is the real face of such companies? What do you see as the future for chemical companies? Will they be able to survive regulations and public consciousness?

Dr. X: I think companies are schizoid, two-faced. Many of them learned to play with one set of rules, and now the rules have changed. They are under attack by regulations and protesters. I think they will incorporate the new rules, or they will not survive. They are human accomplishments—and like most human accomplishments, have unforeseen effects or costs. I only worry about the scale. With more and more people to clothe, shelter, and feed, chemicals will be used more indiscriminately, to keep up with demand. Only global regulations can help them.

Ed: Can chemistry repair damage? Or would it be better to let natural processes adjust over time?

Dr. X: To some extent, we can use our knowledge to repair our damage. For instance, lime can be dumped into lakes to combat the effects of acid rain. Perhaps some process can be invented to neutralize the production of acids. Humans have shown great inventiveness when they needed to.

Ed: Still, many lives and entire species are being lost, and it is not likely we can restore them. Is it the form of knowledge itself that is causing so many problems? The Baconian/Cartesian paradigm with its emphasis on reduction and analysis?

Dr. X: Reduction and analysis are extremely important for making experiments and understanding the results. Physics and chemistry benefited immensely from the methods. Of course, these methods are not as effective when applied to animal and human groups. And, maybe they shouldn't be.

Both chemistry and medicine will stand to benefit from the classification, understanding, and preservation of new species from the rain forests and other biomes (and many others can offer new substances and derivatives). Be sure not confuse the science of chemistry with its application by industries for profits. Any discipline, that can do good, can be subverted for personal or institutional gain.

I think that science, especially chemistry, is the solution to the problems caused by technological applications. Developing chemical and biological techniques may let us improve the environmental degradation and raise the standard of living in the process. Much of the problem with waste and suffering and disease is political and logistical, not scientific. Regardless of what science provides, it will be useless without the social and political discipline to apply it. I have faith in the rational scientific process, more than in the mystical and impulsive vagueness of trendsetters or followers.

Ed: Thank you for your time. By the way, I noticed that your clothes are all nylon. Mine are all cotton, of course.

Dr. X: From Qiana, actually. One of a kind. Shall we have a drink?

Author's Note: "We did have a drink, followed by several more. During our conversation, Dr. X asked that his identity be protected so that his pension would not be modified by his former employer. Knowing that company's policies and actions, I agreed."

CHAPTER 27

Revolutionary Ecology

As awareness of ecology has grown, so has awareness of its limits as a science and as a movement. Nathan Hare characterizes ecology as an elite, "white" science. William Tucker echoes this, writing that environmentalists are elitists who preserve rich resources for their exclusive recreation. Both critics level all groups to the lowest denominator. But, these groups also can be discerned as part of a large ecological movement, with a unique philosophy, science, economics, and politics, with a higher denominator.

Hare laments that blacks and their environmental interests have been ignored by the ecology movement; indeed, he states that the two stand in contradiction. He states that ecology ignores the needs of the poor. Perhaps this is so, but racial justice ignores the basis of all wealth, ecosystems with human members. Both groups are ignorant at great risk. Yet, both groups are largely ignored or underfunded by government. The wars on pollution and on poverty are as ineffective as the wars on people. In one sense, he is right; ecology and blacks often end up competing for federal crumbs from war machinery bread loaf.

Ecology seems to be the faddish successor to middle-class concerns for conservation, a distraction from poverty and war. But, ecological action and wilderness protection, like earlier forms of social action, such as labor laws, women's rights, and minority rights, have been the result of leisure-class elitism, the projects of those people with enough to eat and time to reflect. Its origin does not make it less important, however.

People in the "culture of poverty" (Oscar Lewis's term) have less concern for the future. Those who worry about being assaulted in their homes usually do not display as much concern for threatened species. Those who worry about their debts, jobs, or health are less likely to be concerned with acid rain or Amazonian deforestation.

Worse, many responses to the environmental crisis have been superficial, dumping garbage into the poor neighborhoods or sending pollution out of state. Many environmental groups have been concerned with the health and recreation of their members or the maintenance of their own hierarchies of power. Reactions appear regularly in minorities as an anti-ecology sentiment.

Limits
That sentiment is unfortunate. If ecology is unpopular it is because, at a time when advertisers are expanding our desires for things and pleasures, ecology is describing the limits of nature and the limits of humanity in nature. Industrial consumer propaganda is infecting everyone with the unfulfillable desire for temporary goods and flashy mobility. The false hopes of industrial culture do more damage than the uncertain warnings of ecologists. The problems of pollution, overpopulation, environmental degradation, and inequities are the result of disregarding limits while increasing our

domination and control of nature. Our successes in medicine, agriculture, and technology are unsustainable.

Only in a dangerously unbalanced economy is well-being for the poor tied to more wealth for everyone (especially the wealthy, who get most of the increase), or is adequate employment tied to expansive growth. The purpose of capitalism is acquisition, at any cost, even at the destruction of a rich and unique heritage, and not equal goods for minorities. Industrial growth threatens the fundamental structure of the environment that supports humanity.

Hare states that blacks suffer when colonizers use the resources and labor of the colonized to develop and improve their own habitat, while leaving that of the colonized undeveloped. This is certainly true, but the discrimination occurs at a cultural class level all over the earth. The underlying institutional structure is the enemy, that structure common to the war on poverty, the war on weeds, and the war on other cultures. The structure of industrial institutions threatens everything with an expediency that has been used to justify slavery as well as wilderness exploitation. The prisoners of addiction (rich) and the prisoners of envy (poor) are both slaves (Ivan Illich's term) in a consumer society.

Racial or class groups that seek a share of the unbalanced economy perpetuate the inequality and hierarchy of the industrial state and legitimize the institutions that make the pollution and thrive on slavery.

Economy cannot be divorced from ecology; the myths of limitless growth and free goods cannot continue. Ecology seeks to conserve goods, and the greatest of those goods is a self-renewing and self-balancing nature. Ecology is more than a pollution warning for restricted beaches or special devices for internal combustion engines. It is a warning about the imbalance of ecosystems and global cycles.

Many of the things that humanity has to face, in terms of limits and discipline (population control and austere consumption, for instance), will label the environmental movement "anti-poor" or "anti-people". But, that does not negate the importance of the movement or mean that it does not offer the best chance to avoid greater misery and greater catastrophe in the immediate future.

The effects of this movement are more immediate to the poor, since they have less insulation against pollution and catastrophe. It is true, also, that many solutions to improve economic quality will have more adverse effects on the poor classes than on the rich. Ecological economics cannot be considered without some notion of distributive justice.

Justice

The warning to alter the trends of consumption and usurpation cannot be used as a rationalization for continuing the present social inequities between blacks and whites or between classes or between hemispheres. Hare and others have complained that the environmental movement is a "cop-out" in the struggle for justice. Hare is right in stating that social justice must be established before any solution to environmental degradation can be found. The human victims of greed, violence, and oppression, need justice. But, long lasting justice is impossible without the proper ecological relationships and responsibilities. Without ecology, social changes are doomed to be short-lived and painful. Nor can ecology ignore social injustice. Without social justice, ecology is doomed to impotent theory. There are two houses for humanity, the urban system and the life support system of nature. Ecology attempts to recouple both into a harmonious whole. A rat-infested tenement is part of urban ecology, as much as spotted owls are part of old-growth timber.

Ecological justice cannot be created without economic changes, in pricing and profits, as well as in the counting and discounting of real resources. Furthermore, reforms in justice advanced to manage the crisis, rather than eliminating it, are cosmetic correctives to an irrational, unbalanced society.

Ecology can be a revolutionary process, based on the need to reconstruct industrial societies within ecological limits. It is not landscape decoration, as Hare dismisses it; the ecology movement is trans-class, embracing all economies and all ideologies.

The concern of ecology is not elitism as Tucker charges, but the growth of institutions, energy use, population, consumption, and waste, resulting in the thefts of tradition, uniqueness, choice, a convivial environment, and wildness. Growth, greed, and apathy are social and ecological issues, not simply class or race issues. Economic inequities and class problems are the more visible symptoms of social and ecological imbalances. Freedom, empathy, and balance with nature are not specific to a class or race, or gender or species — these are universal interests shared by all beings, human as well as ultrahuman.

More than any other movement, the ecological movement needs to mobilize popular involvement for environmental health and balance. Humanity will not die without gadgets, jets, or televisions. Humanity may not die without clean air and water. Humanity *will* die without meaningful work or play, or without meaningful relationships with others and with wild nature. *That* is what the egalitarian science and subversive movement of ecology can provide.

CHAPTER 28

Clear Thinking about Animals and Nature

In his fashionable but severely flawed article in *Audubon* magazine, "Fuzzy-Wuzzy Thinking About Animal Rights," Richard Conniff touches on several of his own misconceptions about animals, rights, and nature. He mixes fallacious arguments with an ignorance of nature in his indulgent attack on the animal rights movement. We applaud *Audubon* for presenting such a divergent perspective, but ask for more intelligence and balance the next time.

Conniff's Fallacious Arguments

Conniff tries to present his arguments logically, but the arguments contain many semantic and pragmatic fallacies. He depends on these fallacies to make his points, or rather, general sweeps of irritation (while accusing others of "defying" logical consistency). It is tempting to annotate every sentence in his article to display the kinds of logical fallacies, from name-calling to misplaced concreteness, but a few examples should suffice.

Conniff misunderstands the context of animal rights; it is not as he says, "Muffin the cat" or the "eviscerated elk calf" — it is all of nature. His first fallacy is in mistaking a part for the whole. Then, there is the fallacy of accident, where it is argued that a specific case be subsumed under a general principle; the form of this fallacy is: 'A calf is killed by a mother bear, therefore every living being is killed by another.' The appeal to authority (*argumentum ad vericundiam*, even a suspect authority, as in this case) is a pragmatic fallacy: 'Suzy Chaffee says nature is violent, so it is.' The fallacy of complexity appears throughout. Here arguments have multiple assumptions, and attacking one part has the appearance of attacking others, which may be worthwhile. Thus, Conniff considers an argument against unnecessary cruelty to fur-bearers to be against every kind of trapping (including that for relocation).

The fallacy of ignorance of purpose (*ignoratio elenchi*) uses an argument to support a conclusion that is not the proper conclusion of the original argument. Conniff sets up and thrashes the idea of uniqueness by substituting the idea of superiority, without really defining either. Conniff states that the idea that "humans have no moral superiority" contradicts the idea that "we have a ... moral obligation to treat animals more humanely." In fact, there is no contradiction, since egalitarianism is quite compatible with our obligations as a successful species. Elsewhere, he misunderstands human complexity and uniqueness to mean superior; in his terminology, where humanity is "most noble," human complexity is used interchangeably with species superiority. His contrast of superiority with distinctness and racism is incoherent.

Conniff notices that the modern world thoroughly accepts the worthlessness of human beings while attaching sentimental importance to individual animals. What he does not notice is that human worth has become degraded exactly because animal worth has been degraded over centuries. Despite of the ethological studies of Lorenz, Fox, Schaller, and others, the modern Cartesian interpretation of animals as soulless automatons dominates scientific epistemology. Human worthlessness is the logical continuation of this flawed thinking.

The two erroneous preconceptions he lists (that "furry critters are cuddly" and trappers are "Neanderthals") are his, not the animal rights movement's. We know that critters are living beings who bite and scratch, and we know that trappers are business people (although the business is often wasteful, crude, and government subsidized). He equates trapping with profit, then with animal control. The "animal wrongs" movement has gone to great lengths to show that trapping is not the best or only or last way of animal control. In some situations, for instance, guard dogs are far more effective than traps (as prevention is always better than cure).

In some places, Conniff does not offer arguments at all; he just trivializes the animal rights movement with jarring word associations compiled from overheard conversations in ski resorts: "Hitler," "lowlife scum," "moral fascists." Ironically, Hitler justified the extermination of Polish intellectuals by analogy with the same laws of nature (nature as "slaughterhouse") that Conniff espouses—and if Conniff is aware of that he is strangely silent. He seems obsessed with Nazi themes and links them constantly with animal rights and environmental protection.

Conniff misses the point of many of the quotes he uses. In comparing six billion chicken deaths with six million human deaths, Ingrid Newkirk is comparing institutional scales of suffering, not equating chicken lives with human lives. Similarly, the needs of a laboratory mouse are not equated to the needs of a human child. What is noted is the value of millions of mouse lives wasted every year in misdirected or frivolous research—we are genocidal, in our destruction of monkey and mouse populations.

Conniff actually seems resentful that actors are often the first to campaign against inequality, homelessness, or discrimination. Perhaps because the scientists, business people, and politicians are afraid to speak or to take a stand. He seems resentful of a few conscious rich or elite (yet not of unconscious politicians or oil company executives) and not of the injustice or the institutions that promote it. It is not animal rights that denies support for homeless humans, it is politicians like the 'business' president or the 'environmental' president, with whom Conniff eagerly identifies.

Conniff reduces animal suffering to public relations 'visuals' and the animal rights movement to a public relations war. His pseudo-identification with activists is an argumentative device. His words are cold. Identifying himself as a misanthrope, he then equates the recognition of human abundance with misanthropy. He characterizes the essence of the animal rights movement with an extreme position, then dismisses its target as "easy." Animal rights is no more "easy" than human equality—look how long that is taking. Perhaps he means that it is as apparent as human equality of opportunity.

Conniff's Ignorance of Nature
Conniff is coyly candid about his ignorance of wildlife, of gray and red fox, but not about his ignorance of natural processes, which he uses to support an untenable position with sophistic arguments. Conniff mentions his own divorce from the source of his food, and even though he discovers the source of bacon, he continues to see the hog as a food product. Does Conniff still eat bacon? Does he even now have any concept of a billion beakless, caged, overmedicated chickens being processed into 'breasts' and 'wings'? Self-loathing is not the answer, but neither is self-delusion.

Ignorant of an ecological image of nature, Conniff buys the old, flawed Tennysonian image that nature is red with "tooth and claw." He quotes Suzy Chaffee that animals in the wild seldom make it "to the old-folks home" (which is itself a symptom of modern human problems related to worth — in most nonindustrial societies, there are no such age-ghettos, and the elderly have active positions of respect). He states that animals "suffer cruel deaths from starvation and disease." But, nature is not the charnel house he and Suzy imagine. Few animals die from predation — even in the Serengeti, biologist George Schaller estimates that less than 10 percent are killed for food. In some species, flamingos and elephants for instance, natural (i.e., nonhuman) predation approaches zero. Most animals live their entire lives with some forms of parasites, bacteria, and viruses, but in healthy animals these do not cause constant suffering. Many animals, especially those in human-disturbed habitats, do suffer from hunger occasionally, but most species in wild habitats do not. Animals do not have the same kinds of cultures that humans have, of course, so there are no old-folks homes, although some animals, such as wolves and apes, do feed and care for elderly or crippled individuals.

Conniff worries that the suburban "pantheism" lacks any sense of the first law of nature, which he presents as "Eat and be eaten" (a naive suburban idea — not the law of any Ibo, Inuit, or Crow; furthermore, pantheism is a weak argument for animals, since pantheists have little reason to protect individuals when everything is alive. Not many animal protectors seem to be pantheists). Conniff gets more extreme: "Nature is a slaughterhouse — vast, brutal, gory, and efficient." Perhaps we treat nature as a slaughterhouse, a human institution for processing living beings into packaged nutrient slices, but it is not. Perhaps it seems so because we fear wilderness and death, without understanding either, insulated in our human flatscapes. Conniff draws an inappropriate conclusion to his own misunderstanding, by saying that if we followed the example of nature, we would kill "whatever we wanted, whenever we wanted" — which it seems we already do, from domestic food animals to 'pests' like porpoises, rhinoceros, elephants, coyotes, and alligators.

Conniff claims that it is an illusion that nature is essentially benign. It is not an illusion — nature is essentially benign, peaceful, creative, and very stable. Humanity poses more of a threat to many species than other species or natural catastrophes, like hurricanes or earthquakes. Conniff claims that it is an illusion that death is the exception rather than the rule. Death is not an illusion, but a very natural regulator of life. Without death, we would probably all still be prokaryotic cells in the ocean. Death allows change and development.

Animal Rights without Conniff
There may be real flaws in the animal rights movements, but Conniff does not mention them, much less argue against them. Like most proponents of a position, people in the animal rights movement exaggerate some facts and ignore others. Extreme statements have been made by many people for and against animal rights, or large human populations, or diseases, or exploitation—some of these are tactics to get attention or to counter the shouts of profit. There are sometimes contradictions, but these do not invalidate or detract from the movement. Activists may tend to protect individuals rather than species and habitat or to reduce immediate suffering rather than the conditions that promote it, but that is a human choice. Some of the activists are too sentimental or too idealistic, but this extreme is preferable to the utilitarian, ignorant attitude exemplified by Conniff. Interestingly, those who tend to be too abstract, such as Singer, are ridiculed wrongly by Conniff as ignorant.

From a human perspective, sad things do happen in nature: flies do blind frogs or suffocate reindeer; owls and mink do overkill at times, but these occurrences are not typical or the rule. It is not possible to end all suffering in nature by intervening in every life. Killing and eating animals is natural for predators, humans included, but institutional interference in natural cycles is suicidal, and the scale of human intervention is nature is devastating to entire species. Using all the parts of a prey (including fur) is also natural behavior, but humans do have alternatives to fur, and human scale has made it imperative that we change our ways or our scale. What was appropriate behavior in native American Indian societies is absurd in industrial societies full of remote, naive, dependent specialists.

Most people are not against the use of fur in traditional archaic societies, just against it for fashion and status using rare and endangered species. Most people are not against individual trappers, fishermen, or loggers, just against indifferent cruelty or the institutional destruction of old-growth habitats and whole animal populations for economic purposes. The economic scale is what makes some trappers and fishermen resent loggers for ruining their livelihood. The unnecessary suffering and use of animals to extirpation or extinction is what makes ethical people protest against trappers and fishermen. The scale of suffering on factory farms is what makes vegetarians try to convert meat-eaters.

We do have the need, as Conniff implies, to exploit (as a biological term) other species. But, we do not have the right to interfere so completely in their habitats and natural cycles that the species perish. We do not need to save animals from natural death, as some activists allege, but from unnecessary suffering for silly human purposes (from cosmetics to overeating). We do not need to interfere in natural ecosystems, but we do need to intervene in restored or preserved areas for a while, just to repair our own devastation.

Conniff's tract argues speciously against important ethical positions for animal rights, seemingly not realizing that animal protectors are more concerned with stopping institutional cruelty, and not, like Conniff himself, with pinning labels on the actors. Conniff does not realize that saving animals from unnatural and wasteful deaths is part of saving much of the earth from ourselves.

Author's Note: *Audubon* magazine refused to print this article as a rebuttal or as an abridged Letter to the Editor. I cancelled my subscription.

CHAPTER 29

Runaway Forces

In his recent book, *G-Forces: Reinventing the World: The 35 Global Forces Restructuring Our Future*,[1] Frank Feather presents an ambitious outline of steps necessary to achieve global order. Amidst the tonnage of free-floating assertions are a few grams of common sense, sometimes contradictory, but welcome nevertheless. For example, Feather proposes a decentralized United Nations, a global Equalization Tax system, and a global bank (which we already have in the World Bank — what we actually need is a Global Trust Account). A decentralized UN would result in the super powers, such as the US and China, being closer to the periphery of world affairs, having surrendered much of their authority and sovereignty to the United Nations.

Despite these few good ideas, which other authors have previously suggested and incorporated in their own works, Mr. Feather's book does not hold together — it is unsatisfying. Feather covers severe social and environmental problems with several coats of thick ignorance. Perhaps these problems seem small to him because of his remoteness — just why is it that optimistic books like this never come out of Calcutta or Manila, just Los Angeles or Toronto?

Just what are these forces that give themselves to the title of this book? Sex and population. Hunger and food. Energy and environment. Power and government. Lifestyle, values, spirituality, health, wealth, employment, and shelter. Except that many of these are not forces at all, in any meaning of that word. Some are needs, others are biological drives (that may have some effect on forces); some are patterns, and others are social structures. The environment, in fact, is merely our complete surroundings, that contains forces, such as winds, tides, gravity, or electromagnetic lines.

Feather combines a custom hierarchy of needs, based loosely on Abraham Maslow's theory of motivation, with Alvin Toffler's concept of cultural waves, as described in his book, *The Third Wave*. Feather expands the waves to six and plots them against ten levels of needs, by adding some of his own "forces," such as wealth and employment, and subtracting some of Maslow's needs, such as esteem and self-actualization — in fact, he concentrates almost exclusively on physiological needs, ignoring Maslow's growth needs, such as wholeness and self-sufficiency. This chart is truly wacky (and I use this as a technical term here). The sixth wave, called "Outer Space," boasts global freedom as a lifestyle, cosmic sexuality as a sex drive (does he mean a lust for asteroids or for every other human being or what? — whatever, he seems to repeat the Freudian mistake of deriving love from sex), a commonwealth for wealth (does this mean that the entire earth is for all human use?), and global freelance for employment (does this mean that everyone has to move to where the info-work is, or just have a computer?). The needs and aspirations for the world in a fourth-wave society of information (1990-2045), as far as shelter alone goes, is for a "Big House/Condo, Stylish Clothes." Oh, my.

The third wave need for food is to be satisfied by "Well-nourished Clean Air." The fifth-wave need for sex is to be satisfied by "Recreational Super-sex." The fifth-wave need for spirituality is to be satisfied by "Monism." The fifth-wave need for lifestyle is to be satisfied by the "Leisure Ethic." The sixth-wave need for shelter is to be satisfied by "Super Cities" and "Space Wear" (perhaps because we will have no atmosphere left). Feather is confident that the world has a remarkable potential to achieve a leisure society — perhaps like hunter/gatherers have had for the past ten thousand years, although we are exterminating them or trying to convert them into consumers of "Super Sex" or Space Wear." He devotes many pages to barren discussions of population, environment, and government.

Population
Feather states, without qualification, calculation, or reflection, that the carrying capacity of the earth is "at least 30 billion." He also states that the world does not lack resources, and that meeting the needs of a large population would generate vast economic activities. Indeed, he states that the unprecedented increase in population presents the "greatest economic opportunity ever afforded to humanity." Really? If so, then today's population should present at least a great opportunity, but it does not. The poor are too poor to buy the "Full wardrobe" and "Well-nourished Clean Air" needs in the Feather's current third-wave society. Furthermore, our present population of five and a half billion has decimated rain forests, depleted aquifers, polluted all the oceans, driven countless species to extinction, and interfered in biogeochemical cycles — what is left for a projected eleven billion to mismanage? He repeats that the population explosion is not a problem, without understanding the direct relation of population size to poverty and hunger — population is a critical factor in any total impact equation.

Feather dismisses the problem of food, praising the "Successful" Green Revolution — we have known for over a decade that it really failed, ruining soils and farmers — and stating that "agri-food technology" is keeping pace with food requirements — alas, in spite of Feather's word, food is not a global problem at all, it is a series of local problems exacerbated by global market pressures, and supply has not kept pace with requirements. Furthermore, Feather ignores the significant ecological costs of the green revolution and global agri-industry methods, which are horrendous; soil erosion alone is reducing crop potentials below the needs of current populations. India, for example, is not self-sufficient in food, as feather implies; it has chronically worsened and become more dependent on aid. China seems to be self-sufficient, but it is unlikely that hunger has been eradicated there as Feather hopefully crows.

Feather recognizes some of the immediate crises facing humanity, but population is a slow-changing, long-term crisis, easily dismissed; other crises, such as starcation, are also easily dismissed. After all, poor people usually succumb to disease before starving to death, and disease is a medical category. Feather does not understand that, in the absence of personal and cultural controls, population increase will wipe out any economic gains and interfere with the natural wealth we need to survive, mush less need to increase industrial output.

Environment
The planet's ecological problems are dismissed by Feather as "mythical." The chaos in the "conduct of our husbandry of the planet is simply due to our lack of long-term vision, long-term planning and adequate geo-strategic management processes." Obviously, Feather does not consider how often the management processes are, in fact, the problem itself. Our management of wildlife populations for maximum sustained yield, for instance, has destroyed many local populations and species. Nevertheless, Feather is confident that our problems, "including those of the planet," will continue to yield to solutions as knowledge develops. Frank, we have the understanding now, but not the wisdom to use it; we have the tools, now, but not the will or understanding to use them.

Feather recognizes that deforestation is a serious threat and that forests have traditionally been mined for trees, but he is again confident that the ecological deficits are temporary and being corrected voluntarily by industry. He states that the timber industry (why is everything an industry now?) is replanting adequately and that the US Forest Service is the "best in the world at forest management." What world is that, you ask? The same one where old-growth stands are given away, complete with new expensive roads, for wasteland elsewhere. Feather is ignorant, or bought. Planned forests are not sustainable, as Weyerhaeuser and others have found out. So, the USFS and private companies are mining faster than ever, ignoring the few laws that exist to protect forests.

Then, Feather admits that CFCs "will continue destroying ozone for a few decades" but the process will slow down and stop, and ozone will be reestablished by the planet's activities. Meanwhile, just use sunscreen 9000, and you'll be safe. Hey, it probably won't interfere with trees and crops will it?

Feather's solution for planetary wellness is to provide appropriate levels "of diet and nutrition" (that is, money and technology) to both the overwhelmed and the undernourished parts of the system. Truly, this is a recipe for disaster. As the response to gout is not more sweet liqueurs, the response to starvation is not more electric woks. It is the richness of the toxics that is poisoning the planetary systems; it is the circulation accelerated that is destroying ecological systems. Feather's proposed solutions would make things fall apart faster.

Global Government
Feather traces the faltering of the United States to John Kennedy's assassination; the country was "shot" then and governed by incompetents ever since then, according to Feather. But Feather is sentimental, without discerning between Kennedy's foibles and successes and every other president's. Kennedy's policies probably would have done less for civil rights than Johnson's and less for alternative energies than Carter's (not that any of these policies are perfect models of success) Others have traced the

US decline from the Civil War or from the Spanish-American War. Feather should start with De Tocqueville and rethink the history of politics.

Feather does not recognize that his country has borrowed so heavily against the future that it is virtually bankrupt. But, his solution is to join the US with Mexico and Canada, to make an even more unwieldy and ungovernable state. In this he runs counter current to modern trends, which favor smaller states. Furthermore, he is willing to jettison the past to ensure a "better future," and he urges us to break the "boundaries" of the past. The past, alas, is where people find the strength to face the future; without a past that extends into the future, we will have a short, shallow life as a species.

Feather believes in complete globalization. He judges that a mature civilization will only be achieved when we evolve a "true system of global governance for the single global tribe." Humanity will then govern "itself—and its world—in a geo-strategic and opportunistic way." Feather is concerned that skeptics will dismiss his vision as a dream, but it is more likely to be rejected as a deeply flawed nightmare.

Feather recognizes that the nation state system stifles creativity, but he does not admit that a global state would stifle it far more. Local systems with local spheres of excellence (with many people who are good and a few who are best at things) would foster far more creativity than one large system with one pool of excellence and one best person. He concludes that it is the "teeming populations of the planet" with their largely untapped potentialities who will achieve global social development. The teeming populations have yet to respond.

Feather claims that antipathies among people are slowly fading, mentioning occasional friction with Sikhs, Tamils, and Shiites. Without equalization schemes and smaller arenas of politics, the antipathies will keep growing, not fading. Feather also claims that, in a global government as a global actor and full participant in governing the planet, the nation state will gain "more" authority and responsibility, not less. This has not been the historical pattern. Globalization has resulted in less authority and responsibility so far.

Summary
This book contains many errors of content and style. For instance, Abraham Maslow is erroneously identified as a psychiatrist; he was, of course, a renowned psychologist, but he did not have a medical degree. Missing commas after series and clauses make reading more difficult. The home-made neologisms also add to the confusion.

Feather claims that it takes an enormous leap of faith to believe that progress is possible from a "bowl of rice at home" to a "sprig of parsley on a plate in a nice restaurant" for all of earth's four billion underdeveloped people by the year 2050—except that by then of course the population will be in excess of ten billion and the poor will probably number nine billion of those—unless we make an enormous leap of denial. His unconscious reference to a statement by Robert Malthus, 200 years ago, that no country should have "more people than could enjoy daily a glass of wine and piece of beef for dinner" indicates that he did not understand Malthus' thought: That there are limits to the capacity of the planet. The book depends on tautologies and misdirections to make many of its points. Feather solves many problems by redefining problems as opportunities. Thus, the population problem becomes an economic opportunity. Waste is simply unused energy, Feather states, without understanding waste, energy, or entropic change. For example, he cites the twenty-fold increase in incinerator capacity in Japan as successful recycling, without addressing the sheer increase in imports or the sudden increase in acid-rain thousands of miles down-wind.

The book is full of neologisms and clichés. What, for instance, is geo-strategic planning? Is it really different from strategic planning, which also has severe and unrecognized limitations, or from ecological planning, which is simply unrecognized? What is the process of info-globalization? Does this mean the spread of information? We are told to "dare to dream," that the "torch has passed to a new generation," that we are "party to the biggest revolution in history" (and that revolutions only go forward, an interesting and unhistorical concept, with wheels or cultures). He continues: "All we need to do is marshal our collective resources," "take the necessary steps," and "be headstrong," yet "humbly confident" that together we can "reinvent the world."

Feather concludes that people must think globally, and learn to "think community," cultivating values of "humility, self-sufficiency, benevolence, inovation [sic] and responsibility to the planet." Unfortunately, the majority of his suggestions work against his own conclusion. Mr. Feather misses the important lesson of "think globally, act locally" so aptly said and demonstrated by Rene Dubos, Hazel Henderson, Garrett Hardin, and others. Everything is globalized inappropriately by Mr. Feather. Essentially, he recognizes that the planet is ungovernable, and suggests that global politics is the answer, that nation states must find a way to stop the anarchy of disjointed, conflicting national policies. In fact, we need just the opposite of Mr. Feather's proscriptions; we need local sovereignty with even more anarchy of policies, each appropriate to its culture and location. And, we do need a revitalized United Nations for the truly global problems, but this book never gets past the unsupported claims and slogans.

CHAPTER 30

Speed, Global Power, Local Wisdom?

In his book, *Powershift: Knowledge, Wealth, and Violence at the Edge of the 21st Century*, Alvin Toffler foretells a dramatic redistribution of power, from slow countries to fast. Speed is the critical factor for Toffler; he states that, historically, power has shifted "from the slow to the fast," whether speaking of "species or nations." Certainly, being faster to the industrial market has advantages for many international corporations. But, this kind of speed is not applicable to species. Slow species have survived as well as fast (either adaptively or neurally); many fast dinosaurs perished before their slower mammalian contemporaries, for instance. For species, size and flexibility seem to be more critical for survival than speed.

Speed

Toffler notes that the industrial revolution stepped up the metabolism of economies, but does not seem to make any distinction between good or bad metabolism (fever as well as excitement speeds up a metabolism). Truly, we are speeding up our use of resources without knowing where they are coming from or going to. Modern economies, embracing the idea that "nature is capital," draw on the accumulated "capital" of ecosystems for production. By ignoring the real cost of the capital, as well as the costs of natural services, such as nutrient recycling, soil building, and atmospheric renewal, these economics create a temporary wealth (similar to the healthy flush of a fever, perhaps) and a long-term imbalance. When an economy falls out of balance with its local environment, massive disruption often results; industrial economies have only avoided disruption by trading advantageously with other economies, by using fossil fuels, and by promoting institutional inequality.

    Continuing his paean to speed, Toffler states that fast economies generate wealth and power faster than slow ones. But, what kind of wealth? Financial or cultural, agricultural or symbolic? And, what kind of power? Mechanical or organic, political or personal? Industrial economic wealth is merely a small part of the wealth of the earth and humanity, most of which has little value to that economy.

    Toffler describes an acceleration effect that makes each unit of time saved more valuable than the last, creating a positive feedback loop — inadvertently identifying the archetypal problem of modern economics — runaway positive feedback loops leading to catastrophe. The fast economy he describes seems to depend on fleets of hypersonic jets racing around the world with the elite and their tonnage of possessions. Telecommunications, transportation, and tourism will accelerate, blithely unaware of their impacts on family structures, biogeochemical cycles, including the ozone layer, and wilderness. Have we learned anything?

Power

Toffler sees revolutionary consequences in new management methods, but not the negative effects. Managerial decisions regarding resources are made often on short-term economic grounds and lead to material shortages and human and environmental degradation. Newer methods seem only to offer a higher degree of impersonality.

Toffler claims that the new wealth creation system holds the possibility of a better future for the vast populations of poor, if their leaders anticipate changes. The new system for making wealth consists of an expanding global network of markets, monetary, and production centers in instant communication with increasing flows of data and information (but not necessarily wisdom or understanding). He argues that the availability of this information flow gives more power to consumers, voters, workers, and small businessmen, taking it away from a centralized few. The potential is there, but Toffler does not go far enough to envision alternate economies and communities. The power still operates under the old assumptions in his synthesis.

Toffler is right to recognize the problems of nonindustrial countries, many of whom depend on cheap labor or strategic military location for foreign investment. But, where does that "investment" go? To the poor or to rich politicians? Wealth could be distributed fairly, depending on many factors, such as synergy, generosity, reciprocity, and cooperation, but it is not. The gaps are growing. Toffler acknowledges that they will keep growing. But, we can redistribute without industrializing. We should try to achieve economic justice before accelerating to new glories.

Toffler concludes that a great technological and cultural wall will separate the slow from the fast, making problems for joint ventures. But, what are the products of these ventures? The debris of advertising fads, such as mink toilet seats, or the tools of real needs, such as evaporative water purifiers?

Local Knowledge

Toffler foresees the emergence of an electronic neural system for a global economy, without which any nation will be doomed to backwardness. What kind of backwardness? Lack of fast things? Lack of professional enslavement? Lack of art, play, or culture? Lack of food, tradition, freedom, or happiness? He describes the fast economies that are forming and concludes that slow economies will have to speed up their responses or risk becoming uncoupled from the fast lane. It might be good for countries to be uncoupled. Uncoupling economically might be a sound option for traditional societies unwilling to make the same mistakes as industrial ones. Local communities are based on traditional cultures, which have long-term lasting power. Traditional cultures often have wealth-leveling properties, absolute property ceilings, fixed wants, and production coupled with need—all of which results in a stable economy. Then, efficiency and productivity are less important than use and appropriateness.

Toffler says that the nonindustrial countries are faced with a shortage of economically-relevant knowledge. Are they? What kind? The knowledge of how to find or grow edible and medicinal plants? The knowledge of how to make appropriate houses and cooking utensils? Toffler touts knowledge-based agriculture as a cutting edge of economic advance; how knowledgeable can it be, if it ignores the erosion of soil and beneficial insects? Traditional communities have lost more knowledge than we will have in the near future. What happened to our rich biological knowledge of animals and plants, to our rich mythical knowledge of animals and plants?

The path to economic power is through the application of the human mind, according to Toffler, and he urges that "revolutionary" forms of education are necessary. What is more revolutionary than traditional education? Learning about plants, animals, families, and cultures is more relevant than theoretical knowledge; computers and economics can be learned after adolescence. We have more than enough information and secondary knowledge.

Economic success is secondary, as is money, the accumulation of goods, and prestige. We are accomplished in the secondary meanings of life. The satisfactions from being in a culture in place, from planting trees, growing apples, watching birds, playing with children, and making love are primary. They are not speed-dependent. We lack the wisdom to act as if we believed this. Is fast technology a necessary part of happiness? Those who are uncomfortable with primary meanings tend to become addicted to power, speed, and possession, as a frantic way to avoid awareness, silence, or responsibility, as a replacement for being grounded in nature.

Nature provides the source, of wonder, of the sacred, of otherness, and of the wild. By submitting ourselves to positive accelerating feedback loops in economics, we distance ourselves from such primary meanings. Nature possesses power that is not speed dependent. Human consciousness has already had a "revolution" — from the wild to the tame — and we regret it. Animals used to be directly experienced; now, they are humanized and domesticated. Humanizing the world has made it tedious, uniform, and dull. Economics is dull! Toffler's assumptions are dull. The needs he describes are transitive wants, and their only measurement is quantitative. For fertile nature, we have substituted a sterile model of production and economy. The model is reductive: trees become resources, people become labor. More is more, faster is better. Although speed is our normal response to dullness, the celebration of speed for itself is ultimately unsatisfying.

What is the result of our fascination with speed in everything? Dismissing nature in disgust, we attempt transcendence through speed. We speed away from nature, from our own bodies, and base our civilization on that momentum, praying, requiring, that it never stops. People's souls die, but secure in their power, they manage the things of civilization and inhabit the treeless flatscapes of the malls of commerce, comforted by the banishment of wilderness and the capture of animals in zoos and of free people in reservations, satisfied that their young are mercilessly tied to televisions and computers, acquiring information without touch and speed without grace.

Wisdom

Nations and communities do not all have to follow the same path and the same rules at the same time and at the same rate. Cultural success is not the "survival of the fastest" any more than it is of the biggest or shallowest or newest. Perhaps if we remain unconscious, there will be a power shift to the fastest that will homogenize and level human cultures. But, we can consciously imagine alternatives and work to preserve cultural and natural diversity and the richness of existence.

We have the knowledge to save cultures, to restore places, to participate in the cycles of the earth, but extra speed and power are not required. The pace of nature is generally balanced and well-established; we violate it at our risk. If we adjust to the pace of the growth of trees and to the movements of animals, we would not be risking extinctions and famines, shortages of water and fuel wood, and the death of humaneness.

We do not need to give our power to faster economies. We need to shift power to local communities through self-reliance and participation. A community protects individual freedoms, guards regional culture (values and identity), and holds groups accountable for their use of power. In communities, people can decide to be conservatively sustainable or to grow and gamble on innovation. Communities can have different economic attitudes, paces, and goals. A community that is balanced and flexible, in tune with natural cycles, based on traditional values — in which industrial production is limited to appropriate goods — can absorb the shocks of change far better than a powerful, accelerating, postindustrial, national vehicle.

CHAPTER 31

Single Vision Science Still Sleeps

(as demonstrated by *Trashing the Planet: How Science can Help ...*,
by Dixie Lee Ray and Lou Guzzo, with special guest references
to *Reality Isn't What It Used to Be*, by Walter Anderson, and to
books by Alvin Toffler and Frank Feather)

Why should we even read these books, much less think about them? Because
ecology is seen as a wet blanket on the fires of industry or as a brake on the
wheel of progress. The extreme statements in books like these, because of
their perceived excitement and blind optimism, may be the ones that prevail
in the public "mind." These economic monomaniacs expect that the increase
in knowledge (or information, rather) is sufficient to suspend the laws of
ecology and the limits of the earth. (Even Barry Commoner states that,
because the ecosphere is not a closed thermodynamic system, the limits to
growth is based on a "serious misconception." Unfortunately, Commoner
overlooks the limits of the system, e.g., net ecological productivity.)

What is wrong with these books? First, ecological problems are
minimized as 'side effects' [sic] or wished away. Second, the scale of the
problems is ignored. Third, the authors discredit conservation movements
by representing them as extreme and then attacking them with incomplete
facts and *ad hominem* arguments. The focus of these arguments is so narrow
that the frame becomes fuzzy. And, finally, fourth, their own solutions are
exclusively scientific, technological, and managerial; they expect to substitute
technology and homogeneity for understanding and diversity.

Ecological Problems as Side Effects?

"Certainly we have to acknowledge that technology has sometimes had
unexpected side effects." says Dixie Lee Ray in her book, *Trashing the
Planet: How Science can Help*—an apt name for this book, especially without
the rest of the subtitle—*Us Deal with Acid Rain, Depletion of the Ozone, and
Nuclear Waste (Among Other Things)*. Sadly, she's in good company; even the
Brundtland report addresses the problem of 'side effects.' As Garrett Hardin
has said, however, there are no such things as side effects, just effects; we
call them side effects because we did not intend or anticipate them, and
calling them side effects lets us pretend that they are less important or merely
accidental.

Ray claims that the side effects cause us to worry to the exclusion of
considering the benefits of a new technology. Her example of the internal
combustion engine is ambiguous. In fact, it has had great undesirable effects
on society, regions, and the planetary atmosphere. Pollution is not a side
effect here; it is an equal effect, along with mechanical power. The costs of air
pollution are staggering: $40,000,000,000 in health care and lost productivity
in the U.S.; $4,000,000,000 from ozone damage to wheat, soybean, and peanut
crops; $5,000,000,000 from acid rain damage to agriculture, forests, and
aquatic systems; and destruction of 20 percent of European forests (figures

from the Worldwatch Institute, 1988). Other pollutions are as bad. The oil pollution of the oceans from spills as well as the continuous discharge of poisonous sludge (up to 17,000 gallons per month per supertanker, including the BTX compounds — benzene, toluene, and xylene) and toxin-contaminated water kills thousands of animals and fish every month (including salmon, ducks, and sea birds with concentrations of metals — zinc, chromium, and cadmium).

The planet's ecological problems are dismissed by Frank Feather in his book G-Forces as "mythical" (would inclusion of cosmic strings as forces require that the book be called 'G-Strings?'). The chaos in the "conduct of our husbandry of the planet is simply due to our lack of long-term vision, long-term planning and adequate geo-strategic management processes." Obviously, Feather does not consider how often the management process is, in fact, the problem itself. Our management of wildlife populations for maximum sustained yield, for instance, has destroyed many local populations and species. Nevertheless, Feather is confident that our problems, "including those of the planet," will continue to be solved as knowledge develops.

No, Frank, no. We have the knowledge now, but not the wisdom to apply it; we have the tools, now, but not the incentive or nerve to use them. Instead of solutions, Ray and Feather offer comforting yet fallacious slogans.

Slogan 1: Poison and radiation are okay! Ray claims that biocides have given us food surpluses new and unique in 6,000 years of history. Not so; from ancient Iraq to ancient Peru, food surpluses permitted great city populations (although poor agricultural practices and overpopulation doomed the cultures to early extinctions). Ray addresses the wonder of agricultural surpluses, but she doesn't even consider the costs in erosion, degradation, and pollution. She simply says that any other way of preindustrial agriculture is "irresponsible."

Ray bemoans the "lyrical hysteria" of Rachel Carson, saying the growing chorus of "self-proclaimed environmentalists" unfairly resulted in DDT being removed from the market. Ray inadvertently identifies many of the problems with DDT, but fails to associate them with science or technology. For instance, she says about DDT that it was overused because of its effectiveness — a common "human failing." She also quotes evidence that the peregrine falcon population was declining before DDT and its fate was more closely related to "the availability of prey and nesting sites than to pesticides." She seems unwilling to consider complex situations; the animals face one form of human interference or another it seems.

Ray notes that PCBs were dumped in waterways with 'proper permits' (as if political incompetence justifies technological blindness). She concludes that "*no harm*" ever came from it. Even PCB-related illness in Japan in 1968 was the result of its conversion to polychlorinated dibenzofurans (which is what happens to PCBs in use). Dioxin poisoning is also dismissed: "No human has ever become chronically ill or died from dioxin exposure in the U.S." Although she admits that it is highly toxic to some animal species, Ray claims that dioxins never claimed one human victim. Perhaps. In that sense, no one has died of AIDS or starvation either, just the 'side effects.' She seems

187

unaware of the 10,000,000 environmental refugees around the world who left their homes because of anthropogenic hazards like dioxins and PCBs.

She asks, with Paracelsus: "What is it that is not poison?" And, she admits Paracelsus proposition: "Only the dose determines that a thing is not poison." Having admitted that everything can be poison, she states unequivocally that DDT, PCBs, and CFCs are *not* poisons, in any dosage. She does not seem concerned that nerve gases and many artificial toxins only need parts per billion to be effective.

She also uses this common sense principle to dismiss the hazards of radiation. Yes, we know the earth is radioactive and that life developed with background radiation. But, we also understand the importance of dosage, and it is the increasing dosages that are dangerous. Missing from her discussion on the happiness of radioactivity is any mention of Chernobyl, not the estimated deaths (1500 by 2000) or the estimated costs (over $120 billion by 2000).

Reversing the attack, she 'reveals' that "60 to 100 million people" are dying each year as the direct or indirect result of anti-pesticide campaigns. Can we see a source or even an argument for that claim? She dismisses activists as naive or misinformed or ignorant. Is it her ignorance or naiveté that lets her not mention the effective and tried alternatives to indiscriminate biocide campaigns? Organic gardeners use good alternatives. Traditional cultures use them.

Feather admits that CFCs "will continue destroying ozone for a few decades" but the process will slow down and stop and ozone will be reestablished by the planet's activities. Feather assumes that we can increase our sunscreen to be safe. He assumes that the trees and crops on which we depend will be safe during this time. He also ignores the tremendous costs of ozone damage.

Slogan 2: There are plenty of trees left! Ray claims that tree-growing areas have increased 18 percent from 1955 to 1977, and forests continue to increase in size, even as we supply much of the world's needs. For responsible timber companies (could this be a qualification?), Ray claims that reforestation is the usual, not "an occasional, practice." First of all, any increase is true only of plantations. Forest land has been decreasing steadily since the 1700s (everywhere except in unfavorable farming areas like New England). Thoreau and Marsh warned of overcutting in the 1800s. Theodore Roosevelt warned of overcutting in 1908. The rate of cutting, for a long time, has exceeded all revegetation, according to R. M. Peterson, former Chief of the U. S. Forest Service. We are mining every old-growth stand we can for short-term economic gains. Although Oregon and other Northwest states are becoming tree plantations (third-world colonies?) for eastern Asia, we cannot supply our own demand, much less all of the world's.

Second, the few areas that are replanted are done with uniformly aged, monocropped (single-species) trees in plantations. These weed trees are harvested for their cellulose; they are not forests, but maintained patches of wood, and Ray attributes this greatness to better forest management. She even states that the main *danger* to forests comes from federal lands, where "no management is allowed, because 'nature knows best.'" Wow! How did the planet ever grow trees without us? With us, much of the earth has been deforested, from North Africa and Lebanon to Greece, Nepal, China, and the Americas.

Feather also recognizes that deforestation is a serious threat and that forests have traditionally been "mined," but he is confident that the ecological deficits are temporary and being corrected. He states that the timber industry (why is everything an industry?) is replanting adequately and that the US Forest Service (USFS) is the "best in the world at forest management." What world is that? The same one where old-growth stands are given away complete with new, expensive, erosion-generating roads? Planned forests are not sustainable, as Weyerhaeuser and other 'miners' have found. So, the USFS and private companies are mining at greater speeds than ever and searching Siberia and Tierra del Fuego for new forests.

Elsewhere Ray patronizes Carl Amory for considering that the killing of a forest is more contemptible and criminal than child slavery. Ecological destruction of forests is a crime! It leads to poverty, disease and many human deaths. What does Ray think is the ultimate cause of any kind of slavery, if not the poverty of the environment?

Deep Ecologists as Nazis?
Ray turns her vitriol on vegetarians, dismissing them as always having "a faddish popularity" among "some religious cults." She links Earth First! (which she calls an "ecology terrorist group") with the Animal Liberation Front and blames them both for sabotage of farms and ranches (no doubt waiting for proof that they also caused the Vietnam war, stock market crash, and recession). She identifies Greenpeace as "zealous adherents" engaging in "physical violence" (possibly by painting seals, photographing waste dumps, and coming between people and their destructive acts). She turns detective, saying, "Richard Whitaker of the FBI ... attributes increasing terrorist attacks to the teaming up of militant vegetarians with radical environmentalists." (Bolt your doors!) Obviously, these terrible radicals are trying to get undeserved media attention from the rational scientific destruction of rabbits to test ground-breaking cosmetics.

She admits that not enough is known about environmental problems, but then concludes that "so-called environmentalists" are wrong, and that she and her industry sponsors are right. The proper approach to ignorance is caution, after all, and not the impulsive foolishness that she seems to recommend. Ray condemns the new environmentalism for incorporating a strongly negative element of anti-development, anti-progress, anti-technology, anti-business, anti-establishment, and anti-capitalism. She says that its only positive side "if that is what it can be called" is seeking a society totally devoid of industry and technology. She includes Thomas Lovejoy and Stephen Schneider with David Foreman, Albert Gore, and Paul Watson, mixing scientists, activists, and politicians indiscriminately. Paul Ehrlich and Kenneth Boulding with David Brower, Prince Philip, and Helen Caldicott, thus mixing scientists with doctors and then economists with activists and royalty (perhaps with a trace of envy that she is not in their company).

Ray starts foaming about Ehrlich, mixing her adjectives and nouns without regard for contradiction or graceful style. She mocks Ehrlich for overestimating the hundreds of millions of deaths from starvation, but she ignores the tens of millions of deaths that did occur, other than a accusatory reference to the government in Ethiopia. What kind of indifferent calculus of misery lets Ray callously dismiss human suffering if it does not fit in her argument for blind growth? She claims that Ehrlich wants to reduce population by "force," but the only forces she mentions are liberalized abortions (surely a choice), tax breaks (how are they different from the ones that fund research in her field of physics?), and deindustrialization (which is more of a cultural trend).

She characterizes the movement as adversarial, punitive, and coercive, as well as elitist, pantheistic, sophomoric, and emotional. "It is quick to resort to force, generally through the courts or through legislation" — if such democratic process is force, let's have more of it! Odd that this hyper-Gandhian sense of nonviolence does not appear elsewhere in her thought. Her characterizations are juvenile and irresponsible. The movement is not against progress, development, technology, or capitalism. It does urge responsibility for business and it does criticize institutions for their violence and waste.

Deep ecology, another of her targets, does reject the tradition of reductionistic and materialistic science. Deep ecology, like social ecology, pan ecology, and mainstream conservation, is part of a large tolerant, nonviolent movement to protect and preserve humanity in its home, the earth. This large movement does not abandon science for mysticism, but accepts them both. It has, like any movement, contradictions and inconsistencies. But it does not claim to be perfect or to be the only way.

Ray uses the words, "finite resources," "limits to growth," and "population control," but without understanding; she associates them with distrust and a rejection of science, technology, and industrialization. What is missing from Ray's vanity piece is knowledge of habitat destruction and species extinctions; she does not exhibit a balanced, reasonable, common-sense point of view. Although she identifies with the rest of us, 'who believe in using scientific data to deal with environmental issues,' Ray shows herself to be everything she labels her targets as being: illogical, unscientific, and biased (is it the result of being an overweight, over-consuming, privileged animal in a rich, wasteful society?).

In his discussion of "totalitarian" religions as "agents of a new Dark Age," Alvin Toffler includes the "antidemocratic fringe" of the environmental movement (based on one issue of New Perspectives Quarterly apparently). He suggests that under catastrophic conditions, a wing of the movement may step up from eco-vandalism to "eco-terrorism." His description of the two wings of the movement is painted in apocalyptic black and white. The good guys, who favor technological advance (within constraints), believe in the "power of the human mind," in imagination and intelligence, whereas the fundamentalists, whose views "dovetail with the thinking of religious extremists," wish to plunge society into "pre-technological medievalism and asceticism." Hopefully, those straw dogs are not the only choices.

Toffler glosses over the issues of the ecology debate, noting only the movement's "deep hostility to secular democracy." Like many others (Richard Neuhaus argued in 1971 that ecological activists used nature to legitimize political power — the same way that Hitler did), Toffler himself then invokes Hitler as the archetypal "eco-fascist," the youth movement (Wandervogel) being equated with today's greens for their spirituality, preindustrial values, organic emphasis, physical fitness, and biological analogies. Never mind that the biological analogies are quite different, the worst of social evolution for the Nazis, the best of cooperation for the greens. Toffler paints the "eco-medievalists" all black, accusing them of wanting a dark ages with all the worst of those times: Cruelty, ignorance, mind-control, force. Of course, any historian knows the dark ages were not dark, especially in Africa, China, Pacific, Japan, South America — in fact, that time in Europe produced the characteristics of civilization, from banking to specialization, that Toffler most admires.

In a fashionable, but silly article in *Audubon* magazine, "Fuzzy-Wuzzy Thinking About Animal Rights," Richard Conniff goes right to ad hominem arguments, implying that animal rights activists are: like "Hitler," "lowlife scum," and "moral fascists." Ironically, Hitler justified the extermination of Polish intellectuals by analogy with the same laws of nature (nature as "slaughterhouse") that Conniff himself espouses. Perhaps Conniff is trying to hide his admiration for Hitler. Conniff's obsession with Nazi themes leads him to dismiss animal rights and environmental protection.

In *Reality Isn't What It Used To Be*, Walter T. Anderson announces that he is calling for people to transform their relationship with nature. Then, he simplifies and rejects the way other groups describe reality. He rejects the biocentric value structure of deep ecologists, claiming that it is a logical impossibility since only humans are capable of valuing. This narrow philosophical use of the word value is worthless for anything other than sophistic discourse. Virtually every living being values those elements that allow it to live. Some beetles value dung and many ticks value blood — even if they construct no philosophies to communicate those values.

Anderson notes that deep ecologists call for a reduction in the size of human populations and the priority of the needs of the ecosphere. He states that the deep ecology platform (of eight principles) does not address the problem of how to reduce populations — in fact, principles, by definition, never state how to do something. Principle 1 does state that nonhuman life has inherent value, and principle 4 does state that the flourishing of nonhuman life requires s decrease in human population — such decrease is compatible with flourishing cultures. In fact, his point 16 is almost identical to the principle 4 of deep ecology. He states that we should seek "harmony with nature" and protect carrying capacity by limiting "growth in human population." Furthermore, his point 17 states that the preservation of the ecosphere is "essential," much like deep ecology principle 1. Could Anderson be a closet deep ecologist?

Although *Earth First!* is a good forum for the exchange of ideas, it is not the only or necessarily the foremost journal of deep ecology as Ray and Anderson mention. It is the journal of practical salvation first, then social ecology, deep ecology, conservation, and humane activities. Many other journals, including *The Trumpeter, Pan Ecology, Environmental Ethics, Inquiry*, and *Wild Earth*, address such deep ecological questions.

Technology Broke It, Technology Can Fix It?
Ray says that our technically advanced society is "based on knowledge and facts." Not on emotion or compassion or sympathy. Wisely, and much earlier, Goethe recognized that all facts are theories, composites of perception, emotion, need, and imagination. Ray doesn't realize this, even though her science is shaped by it. Furthermore, not all facts are "verifiable, determinable" or repeatable. There are nonrepeating phenomena on many levels. For all of her self-congratulation, Ray is not up on her facts. She claims that asbestos is essential in automobile break linings. In fact, it has been replaced by mandate in many states. She states: "The Cuyahoga has since been cleaned up, and so has Lake Erie." The lakes have improved, but then the standard has been lowered; and improvement needs to continue.

Feather's solution for planetary wellness is to provide appropriate levels "of diet and nutrition (that is, money and technology)" to both the overwhelmed and the undernourished parts of the system. Truly, this is a recipe for disaster. The response to gout is not more sweet cream pastries with wine; the response to starvation is not more studies of electric woks. It is the richness of toxics that is poisoning the planetary systems; it is accelerated circulation that is destroying ecological systems. Feather's solution would make things worse.

Ray concludes with a short grocery list of suggestions for thinking about science and the environment. They are: ask for evidence (or "insist on facts"), then put pressure (force perhaps?) on the government not to panic during environmental catastrophes; keep a sense of perspective — after all the dinosaurs died out (a good, unconscious metaphor) and life continued; and, convert everything into a garden, since we're "better" than other species. What a list.

She closes with a quote from Faust, who finds contentment in a land reclamation "engineering" project according to Ray. Alas, Ray does not consider the context of the quote or work (also, the values and functions of wetlands were not understood very well in the early 1800s when Goethe was finishing his work). Salvation for Goethe's Faust was in aspiring and struggling, not in conquest and factual indifference. It was in spiritual improvement, not modernization. From Goethe on, the idea of the novel has reflected the human being as creator of the self and not the master of the cosmos that Ray wants to tout.

We do not need to "master" the cosmos. We do not need better technologies or faster economies.

We do need to create local communities through self-reliance and participation. A community protects individual freedoms, guards regional values and identity, and holds groups accountable for their use of technology and power. In communities, people can decide to be conservatively sustainable or to grow and gamble on innovation (and maybe lose). Communities can have different economic attitudes, paces, and goals.

A community that is balanced and flexible, in tune with natural cycles, based on traditional values — in which industrial production is limited to appropriate goods — can absorb the shocks of change far better than a technologically powerful, accelerating, industrial, national vehicle that is worshipped so unquestionably in these books.

The crisis that results from exponential growth cannot be solved by accelerating growth. The crisis that results from ignorance does not require more information. Paradoxically, the best thing to do may be to stop — stop growing, stop speeding, stop developing new technologies. We have not saved the starving or diseased with our current system. We have not maintained ecosystems or biogeochemical cycles with our current technology. We need to rethink our direction, then restore equity and balance.

Author's Note: Dixie Lee Ray said it was "stupid to argue against her" and then left to pilot a tanker through Puget Sound to show how safe it was.

CHAPTER 32

Cassandra Mocked

Having read Julian Simon's review, "Should we heed the prophets of doom?" (as well as the books he reviewed), I want to state that I prefer possibly "phony bad news" to the kind of cruel phony good news pushed by Mr. Simon in his book, *The Ultimate Resource*. The blurb on the back of his book asks if his book is quackery, stupidity or diabolical propaganda, to which an informed reader must answer: *YES!* All that and less.

His ideas may be accepted by academic economists, but not by rational scientists. Economists have become the priests of the new hunger — the industrial people's hunger for money and convenience, gadgets and fame, that isolates mere physical hunger for food and security behind the glass wall of television. From his comfortable chair *in vitro* (behind glass), Mr. Simon confuses the ubiquitous prophets of doom with the Cassandras of Caution, which in fact is a truer description of the alarmist ecologists (with whom I am identified in fact).

Only by ignoring the overwhelming facts from various sources, from the United Nations to Bell Telephone, can Mr. Simon pretend that Garrett Hardin, Paul Ehrlich, the members of the Club of Rome, and many others appear to suffer from an unfounded hysteria. His own opinions are founded only on a solid rock of ignorance, ignorance of economics, ecology, and of history.

In economics, for instance, necessity was never the mother of invention, as Mr. Simon contents; curiosity shares that honor with leisure time, and invention flows from both. As long as there were several sources of energy at the same time, humans could afford to use some sources to extinction, while developing technology to use other sources. Population growth was probably not the major stimulus for opening new resources; demand was and is more urgent. For example, power demands in the U.S. rose several hundred percent between 1960 and 1980, while population increased only a small fraction of that. The change in quality of life has been negligible, despite the increases. Konrad Lorenz suggests that all our dangers come from overpopulation; he recommends education as the fix. There do not seem to be any natural limits to human growth, so the limits must imposed by an educated populace.

It is true that the Global 2000 report may have had a limited perspective on history, but Mr. Simon's is quite distorted. His "worrisome scarcities" have wiped out hundreds of whole human societies over several thousand years. Those peoples, including Mayan tribes, Saharan tribes, Mesopotamian groups, and Southeast Asian peoples, did not solve their problems; and they perished. Others, such as the Europeans and Mongols, emigrated to avoid the consequences of their failures. And many others, such as the Romans or Easter islanders, were invaded before perishing. Perhaps, with adequate warnings from the Cassandras of their day, some civilizations did change and continue.

Furthermore, Mr. Simon seems to have no understanding or knowledge of ecology. Airborne pollutants have increased world wide, according to a 1982 UN report by Dr. M. Tolba, Executive Director of the UN Environmental Program. Although some forms of pollution may have lessened in some industrial countries, the U.S. and Britain, for example, due to lawful control measures, other forms of pollution, such as acid rain, have increased dramatically everywhere, threatening fish, trees, crops, and buildings. That same UNEP report also states that "the overall state of the environment has not improved." Simon accuses the environmental movement of being "fueled by false information and special interest values," which are "leading the world" to disaster. Not surprisingly, most of the false information and special interest values seem to be his.

UN data on food production does not support Mr. Simon's easy pollyannaisms, either. Across the planet, over 450 million people are chronically hungry. Over 15 million die of starvation every year, and another 15 million die of nutrition-related diseases (many in the U.S., surprisingly). Although these numbers add up to less than the 50 million starvation deaths predicted by Paul Ehrlich in 1977, allowing 30 million people to die is a very strange calculus of misery. The high-yield cereals of the much-ballyhooed green revolution seemed promising for the first year, until many Asian and African farmers realized that they would have to spend thousands of times their annual incomes on complex machinery, fertilizer, and pesticides, which would then have the effects of ruining the soil in several years. The use of industrial agriculture in nonindustrial countries has already resulted in great ruination before the green revolution—Iran, for instance, has lost over 50 percent of its food capacity since 1970 (we wonder how they can angrily call us "devils" when all we tried to do was give them a nonworkable agriculture). Only industrial agricultural countries like the U.S. and Canada can create food surpluses, and only by using great quantities of gas and oil; for many crops, such as corn and dry beans, the energy (in kilocalories) invested exceeds the food energy harvested.

Some countries, such as Pakistan and others on the Indian subcontinent, strip the hills for firewood because of the high price of oil. This practice creates erosion, which results directly in floods that kill thousands of people (and millions of animals and plants) and ruins crops downstream. Creeping deserts, from the complex conditions of overgrazing, deforestation, and climate change, threaten the livelihood of over 700 million people.

Simon states that the sea is not dead, and that fish catches have increased. Observations by Heyerdahl, Cousteau, and many others show that some seas are dying and others are being systematically degraded. Reference to this year's annual almanac shows that the fish take has increased in recent years because "trash" fish are being taken for fertilizer and pet foods. Quality fish populations, including flounder or salmon, are declining or reaching extinction.

Finally, Mr. Simon does not even try to refute the Club of Rome studies. But, neither does he take anything away from them. The main lesson from the *Limits to Growth* was that we are doomed only if no steps are taken to fend off disaster. And, the doom was not the doom of Armageddon, but rather the doom of human hopes linked to unchecked growth; in short, the insulated dreams of economists like Mr. Simon. Assessments of the planetary situation by groups like the Club or Rome and the New Ecologists are substantially correct. The rate of population growth is alarming, but more alarming is the acceleration of the per capita use of energy and materials. These two factors are causing large-scale disruption of ecosystems and the exhaustion of natural resources. In the U.S., we have avoided the consequences of these factors through advantageous trade with poorer countries and the through "drawdown" from the future. That tends to blind many of us to the realities of starvation and suffering elsewhere.

Simon, in "Now (I think) I understand the ecologists better," in *The Futurist* 9/10/87) admits that he shares many goals and desires with ecologists. As a background, he mentions that he chose not to study medicine because of his personal preference not to take medical drugs, except when the need to do so was overwhelming. He was afraid that his preferences would make him "out of sync" with his profession. Yet, that is why doctors prescribe medicine — overwhelming need! Instead, Simon chose to be out of sync with science and now he is out of sync with economics. Recently, Mr. Simon admitted that he had only taken several courses in economics — he is actually a marketeer. As a marketeer (kids, can you say marketeer, m-a-r, k-e-t, m-o-u-s-e — once more, with feeling!) Simon instead recommends the drugs of technology and comfort regardless of need. Simon uses side effects in his arguments — obviously, he does not understand that, for ecologists, there are no such thing as side effects; even the brilliant, eccentric technophile Buckminster Fuller recognized that what humanity "rated as side effects are nature's main effects."

Simon says that the 'vision' of the 'wisdom of the body' is not appropriate for understanding the modern environmental and economic situations. But, he does not even understand the use of his own analogy. Many large, human-made alterations are unnatural. "But more people also bring about greater understanding of the system and increased capacity to bend it to our will," he writes. Sure. With all that understanding, we should realize that trying to bend a self-organizing system to our narrow will may destroy it — and us. Simon is a blind optimist, but he has reason to be; he has been seduced by technological wonder and comforted and confirmed as an over-indulged, over-consuming member of a wasteful society.

Meanwhile, the alarmist ecologists are like Cassandras. The function of a Cassandra, however, is to be wrong. If we heed their warnings, they will always be wrong (and suffer either way). One of the goals of Mankind at the Turning Point, another Club of Rome study, was to force us to act to create the future as a "self-fulfilling prophecy." We can do that only by being aware of the ecology and economics of the earth, not by lying to ourselves and trusting blindly in technology, as Mr. Simon has done and recommends that we do. Clearly the situation calls for, in Mary Midgley's words, "Heaven's cherubim horsed upon the sightless couriers of air, to blow the horrid deed in every eye, that tears may drown the wind."

CHAPTER 33

The Marsh Institute for Research in Ecology

The Marsh Institute is named for George Perkins Marsh, whom Lewis Mumford identified as the "fountainhead of the conservation movement." The aim of Marsh, in his 1864 book Man and Nature, was to show how humans have changed the earth and to suggest the means of conservation suitable to make nature a fit home for humanity. The Marsh Institute continues to apply these ideals through its public research and educational programs, sponsored by a loose association of ecologists who foster the independent practice of science.

Members of the Institute are responsible for their own support, as ecologists, consultants, librarians, or editors. They hold in common that science is a holistic instrument for living appropriately, admitting different truths, considering common sense as well as poetic perception, accepting the limited validity of results and the ambiguity and incompleteness of human inquiry, and being inseparable from ethical and aesthetic concerns. The Institute offers alternate approaches for biological economies, such as farming and logging, while rejecting dangerous options, such as pesticides and nuclear fission, as unacceptable. The Institute is socially sensitive to the needs of wildlife and people. It cooperates with many political and environmental groups.

Members are responsible for designing and funding their own projects. Research includes: habitat study and regeneration, private forest land management (dwarf mistletoe control, selective cutting), natural pest management and nonlethal predator control, animal ecologies (coyote, black bear, mountain bluebird), saving common places (Sonoran desert, Palouse grassland intervention and habitat restoration), the relationship of human populations to ecosystem productivities, frameworks for cultural preservation, the nonviolent defense of habitats and cultures, biogeographical models of wilderness (with attention to quantitative determinations of minimum wilderness areas), and the ecological responsibilities of corporations.

Originally called The Living Earth Concern, The Marsh Institute was formed 21 March 1970, on the original Earth Day, at the University of Delaware for the purpose of promoting the ecological foundations of human communities. Since 1976, it has been a nonprofit Delaware corporation, advised by a board of directors. The research center is situated on twenty acres of coniferous forest between Potlatch and Moscow, Idaho; other research and restoration areas are being proposed. The operating budget is derived entirely from gifts and interest on investments. The Institute does not accept moneys that would limit or prejudice its findings. Please write for further information (ecologists@gpmi.us).

CHAPTER 34

Ecoforestry as Relativistic Science & Crisis Science

Ecoforestry has been defined as selection forestry or restoration forestry. It has been defined as a context-based, community forestry that is based on traditional wisdom combined with scientific knowledge. It could be defined as the pragmatic study and use of genealogical actors in ecological roles (in the evolutionary play in the ecological theater) — the activities of forestry certainly make good theater, with protesters trying to save redwoods, the deafening crashing of trees and roaring saws, and the ant-lines of full logging trucks.

The dominant form of forestry, industrial forestry, is based on the agricultural model of simplify, harvest, and replant. This, and the economic problem of short-term debt loads for forest corporations, causes numerous problems with forests: Wild forests are clearcut; diversity is diminished; erosion increases; and the aesthetic ruin can be seen from airplanes and highways.

Forestry came to be treated as a special form of agriculture. Indeed, forestry has several parallels with agriculture. Like agriculture, forestry uses soil to produce a crop for the purpose of increasing wealth (or perhaps just revenue). Like agriculture, forestry is renewable (unlike mines or oil extraction). Like agriculture, forestry is based on knowledge of many fields, including botany, soil, and meteorology. Like agriculture, forestry deals with vast areas. Unlike agriculture, however, forestry deals with wild plants on wild soils. Furthermore, trees are very long-lived, unlike crops of annuals, and are related in complex cycles, much more so than annuals. Trees are directly responsible for soil fertility and tilling. Despite similarities, forestry cannot be considered a form of agriculture.

As a science, forestry is based on the modern paradigm of science, the study of discrete units, like atoms, in isolation from their context-since the context would increase the complexity of the situation beyond the capacity of the mathematics to describe it. Western science is based on an Aristotelian logic, a Cartesian dualism, and atomism. It is basically concerned with regular, reversible events. This kind of science has never been good with unique, irreversible, long-term, complex, or catastrophic events, the kind that occur in forests, although some new approaches, such as chaos theory, are promising.

Industrial forestry assumes also the current economic system, which is based on winners and losers and on inequality. Economics has not been unsuccessful with its models, for instance of buying behavior, but it has become a highly abstract academic discipline. All its abstractions are applied to the real world without acknowledgment of the high degree of abstraction involved. The philosopher A. N. Whitehead warned that the economic method would triumph if the abstractions were judicious, but even judicious abstractions had limits, and the neglect of those limits lead to disastrous oversights. Considering a fictitious human nature under imaginary circumstances and thinking it is real is the fallacy of "misplaced concreteness" according to Whitehead. Daly and Cobb suggest that the classic instance of the fallacy in economics is "money fetishism," where the characteristics of an abstract symbol, such as limitless growth, are applied to real commodities and values. Misplaced concreteness also occurs in forestry itself. Genetic reductionism is a good example of the fallacy. Genes only function within the organism, which is the creative being; genes are not creative.

European forestry in the mid 19th century was concerned with the entire forest and was relatively holistic. American forestry was going in a different direction. One need only compare the emphasis of two different books on silviculture by two Professors at Yale. James Toomey spent a good part of his 1928 book on forestry describing the reciprocal influence of trees on their environment; he added perhaps only a paragraph on end cuts, such as clearcuts. In his book, written 60 years later, David Smith concentrated on end cuts, with perhaps a paragraph on influence and ecology. Toomey was concerned with the definition of Silviculture — the development and care of forests — where Smith addressed how to cut, spray, and clean them. Toomey allows one page for a discussion of clearcutting — Smith over 40 pages. Toomey spent over 300 pages discussing the relationships of trees, animals, plants, fungi, and humans to the forest; Smith devoted a few pages to climate and fungi.

What happened? Actually, Smith's book was based on Ralph Hawley, whose book editions ran from 1921 to 1962. Hawley addressed foresters who had their "feet on the ground." (Who knows where their heads were?) Toomey's work is based in the German tradition and is written for foresters, teachers, researchers, and ecologists. Toomey's book was still used in the forties and fifties. Toomey is out of print now. The Hawley/Smith approach has been dominant since the 1950s. Smith's book is still used at most schools of forestry today.

Industrial forestry ignores the history of forests as well as the history of use. It assumes the cosmology of the industrial age. Many forests, however, are occupied by people with different (and often better adapted) cosmologies. The Yaruro, for instance, believe that all beings are equal, not just men, as the industrial cosmology assumes; a culture that sees humanity and nature as existing in a mutual harmony will have a different kind of forestry than the modern version. In believing that imperfection cannot be avoided, the Dahomey, for instance, are less likely to try to create a perfect set of circumstances, unlike the industrial cosmology, which holds that 'men are perfectible.' Neither the Dahomey or Yaruro have a formal kind of forestry, but neither destroy their forests.

Western logic is Aristotelian; it is deductive and polar; it is either/or rather than both/and. Ecoforestry considers the morphogenetic logic of the Mandenka or Navajo. Of course, having a different logic means that ecoforestry does not make lists like the one above.

Unlike forestry, ecoforestry includes humans and ultrahuman beings as an integral part of the system. Ecoforestry addresses both large and short-term scales (in time, size, and design). Ecoforestry addresses forests and human societies. Ecoforestry includes humans and ultrahuman beings as an integral part of the whole system. Ecoforestry addresses both large and short-term scales (in time, size, and design). Many forests, however, are occupied by people with different (and often better adapted) cosmologies and economies. Industrial forestry partially uses physics and ecology. It ignores philosophy, ethics, and equity. These are basic tools for ecoforestry.

Principles of Ecoforestry
Forestry is defined by Webster's New Collegiate Dictionary as 1. forest land, 2a. the science of developing, caring for, or cultivating forests, and 2b. the management of growing timber. These definitions are relatively new, although humanity has been using forests for tens of thousands of years. Ecoforestry is the management of the human use of forests for necessary goods at an appropriate scale for the forest. It puts the forest first and considers what to leave. It is also context-based community (ecosystem) forestry that is based on traditional wisdom combined with scientific knowledge.

Contrast the dictionary definition with Buddha's definition of a forest: "a peculiar organism of unlimited kindness and benevolence that makes no demands for its sustenance and extends generously the products of its life activity; it affords protection to all beings, offering shade even to the axeman who destroys it."

Within general topics, such as history or ecology, a number of characteristics, principles, and standards will be presented that should allow you to address every situation (to some extent). Characteristics are qualities that distinguish unique individuals, systems, or patterns; Gregory Bateson calls them differences that make a difference. Principles are fundamental rules or laws that we can use to create images or models. Standards are models or examples of quality or value established by authority or consent. Sometimes these statements will be made explicit in the summary.

For example, one characteristic of a mature forest is its wildness. The corresponding principle is that forest is self-making and self-ordering without human control and management. Our objective for this forest is to allow the foresting process to continue, whether we take resources from the forest or not (forests can be influenced or interfered with by acid rain, pollution, and other industrial effects). We can set standards that are likely to keep mature forests wild: Limit biomass removal to 2 percent of the total forest; use appropriate techniques, e.g., single tree selection, horse skidding; retain mature structure, e.g., 19 snags per hectare, 23 nurse logs per hectare (in mature Ponderosa pine for instance); preserve surrounding landscape patterns.

The principles of ecoforestry are based on a number of fundamental philosophical, historical, scientific, and cosmological principles that were first presented in other contexts by thinkers such as Whitehead and Einstein. The sample characteristics, principles and standards below can be expanded to form the basis of ecological forestry.

Characteristics of Forests: Forestry deals with primarily wild ecosystems. Archaic peoples lived within the limits of forests, whereas as agriculture and modern forestry exceed the limits of forests and land. Changes in scale put pressures on human and natural systems. We ignore very long-term trends in nature and human history that have very dramatic influences on our activities and management styles. Our history of use of forests has been to exploit them to collapse and then move on to unexploited forests. We justify our temporarily successful behavior with myths that allow us to continue the behavior without being responsible.

Principles of Forests: One principle is change. Everything changes (according to Heraklitus). Individuals changes; patterns change. Neither plantations nor old growth forests can remain unchanging. Another principle is uniqueness. History creates unique patterns in ecosystems, especially in forests. Each forest is unique in its parts and structure, in its matter, energy, forms, information, and in its dynamics and history. This principle is related to another, the principle of irreversibility. Forests pass through stages that are never repeated, despite superficial similarities; that is, tree-planting cannot reverse clearcutting (although another old-growth forest may develop in time). The history of land limits or determines its future. Continuity is a principle. Forests proceed through distinct continuous steps in relation to past environments and disturbances; that is, a plantation cannot become old growth without developing through intermediate stages of community. The principle of morphogenesis means that our species is shaped by forests, as well as by other species and ecosystems. Changing wild forests to plantations, fields or deserts will profoundly change our psychology and behavior.

Standards of Forestry: We should consider the consequences of all our actions in their contexts. We should allow wild ecosystems to continue to be self-creating and self-managing. We should modify our myths to provide appropriate images of place. We should apply known ecological ideas to a moderate exploitation of forests.

Many of these statements may be inadequate to address the challenges and problems of industrial forestry, in which case more information is necessary.

Summary
With principles and standards that acknowledge the characteristics of forests, many specific changes would occur in the human use of other species in forests. These would allow us to: Recognize individuality in trees and other beings; Recognize the feelings and emotions of animals and the sensitivity of plants; Promote a noncommodity, drymoperipheral approach to forestry; Minimize the devastation of wild forests and wild ecosystems; and Emphasize habitat loss as a human ethical issue.

Furthermore, in the near and far future, through its basis in ecological philosophy, ecoforestry could: Suggests ways that biosphere cultures can be converted to ecosystem cultures (after Ray Dasmann), characterized by wonder and wildness; Promote ecosystem lifestyles, characterized by frugality and joy; Relate the success of microorganisms to the ultimate success of living in self-sustaining forests; Develop a basis for management techniques for finite ecosystems, especially wild forests; Link ecological sustainability with the richness and diversity of ecosystems, as well as with human pleasure; Work for immediate solutions to inequity and destruction under worsening and thankless conditions; and, Educate with confidence and energy for ecological enlightenment over the long-term, despite short-term wobble.

Ecoforestry is basically concerned with using a new metaphor for forests and forestry. Although ecoforestry sounds like a qualification of the modern, industrial forestry, it is in fact entirely different — it is based on a different metaphysics, a broader ecology, a more comprehensive economics, and it is sensitive to limits, ethics, aesthetics, and spiritual values. Its larger perspective incorporates industrial forestry as a special case, much like the theory of relativity incorporated Newtonian dynamics as a special case.

To some extent, ecoforestry is a crisis science, much like conservation biology, from which it has taken many ideas and principles. Ecoforestry addresses the immediate dangers and catastrophes that are resulting from our destruction of forests for a temporary economic gain for some. The insights from another physical science, quantum mechanics, indicate that participation in the system cannot be avoided — we cannot unparticipate by choice. The larger perspective of ecoforestry requires physical participation and emotional commitment to creating human communities in a context of healthy ecosystems.

Chapter 35

Gigatrends in Human Ecology & Forestry

Some trends are readily evident in the human present; other processes or cycles take many human generations to complete. Many trends in ecology and forestry can be fit into the megatrends identified by John Naisbitt over a decade ago. Naisbitt identified ten larger patterns in society, including the move from industrial society to information society, the economic interdependence of the human world, and the restructuring of society from short-term considerations to long-term time frames, that is, he says, from two-year horizons to a "very long-term" time frame of "six to ten years" — long by economic standards, but not long enough for trees and ecosystems. Some other counter-megatrends (we should resist calling them negamegatrends), such as the denigration of reason and science in the popular press and media, or the incredible explosion in information — without a corresponding increase or spread of wisdom — are ignored by Naisbitt. Negative trends in general are left out of the picture. The things that are most popular to society, such as money, real estate, insurance, and politics, are the things that are treated as the most important. Italian Foreign Minister Gianni De Michelis identifies the most important megatrend of the century as the availability of free time; he claims the US economy will remain the most important economy in the world because its GNP is increasingly geared to entertainment, communications, education, and health care, all of which are about individuals 'feeling well.' The things that will ultimately be most important, such as directing the course of civilization, limiting human activities, or preserving wild ecosystems, are relegated to a sideshow.

There are megatrends in ecology and forestry. Forest companies, for instance, exemplify economic interdependence as they move to countries with weaker regulations to mine their forests. Forest companies have learned to restock some sites, but not to plant well or to nurture the trees. Furthermore, rather than moving from industry to information, forestry is moving backwards, to providing raw resources. And rather than moving towards multiple options, forestry is backsliding into one option: logs.

The real long trends in ecology and forestry — and in ecosystems and forests — occupy the entire human calendar. They are sometimes invisible in the present. We might call them gigatrends, since they are larger and more involving than megatrends. Gigatrends ("giga" from the Greek for very large or giant) are long or very-long-term trends, usually ignored by science and economics, such as atmospheric temperature increases or global deforestation. Some gigatrends may only cover a span of fifteen years, but some over them have continued for thousands of years. In general, the larger the unit of study, the longer the trend; for example, changes in the physiology of a tree occur in days and months, while changes in the forest ecosystem can take decades or centuries. Some gigatrends are beneficial, although most of them seem to be detrimental to human well-being as well as to the health and stability of forests.

Because of their length and scale, most of these trends can be represented graphically with simple lines, which show gradual or rapid (exponential) changes. Several trends seem contradictory or inconclusive in the short-term, but are evident with long-term study. For instance, atmospheric carbon dioxide increases and decreases with seasonal change in vegetation, but has been rising slowly and steadily for at least 35 years, according to Keeling and Whorg. Ecosystem succession is also misleading in the early stages, as pioneer species take advantage of a disturbance—here the short-term abundance of intolerant nitrogen-fixers prepares the site for a mature forest. Other trends are actually complex. For example, wood use as fuel decreased from the 1800s until 1970, when it started to increase again, due not only to the fashion for stoves in industrial countries but to the rising prices of petroleum and coal.

Many of the negative trends have been noticeable for thousands of years, but nothing has been done to halt them. Environmental factors have shaped the course of human history to a greater extent than has been realized. The decline of Rome demonstrates that ignorance of forest ecology can have important consequences. There have been environmental catastrophes in the Tigris and Euphrates valley, Greece, Khmer, Maya, Cahokia, and others. These civilizations were very successful before they failed. Failure from success is tragic. For the Greeks, the operation of tragedy resulted from success taken to great lengths, that is, where successful behavior in one context is applied to all contexts, with the result that the opposite action occurs from the one desired. For example, humans in moderate numbers were able to take what they needed, such as wood, from natural ecosystems without interfering with the processes. Our dominance, once so successful because of our big brains and tool-using hands, has now become self-destructive. When human cultures adapted to ecosystems over long periods of time, the ecosystems also adapted to human cultures; when the human impact has been rapid and intense, as it has been in North America recently, the ecosystems collapsed or stabilized at a simpler state.

It has been argued that humanity is not adapted to live everywhere. Since the human species emerged in a subtropical climate, where it acquired certain biological characteristics, it may lack a degree of fitness to survive in the tropics, the arctic regions, or even temperate forests. The species may remain genetically best adapted to a certain type of subtropical savanna. Rene Dubos presents this development as explaining some of our present behavior patterns: subconscious fear of forested wilderness (where good vision was little use for avoiding danger); the commonness of design features in landscape architecture; the preference for a narrow range of temperature; and the biochemical similarity of nutritional requirements. Perhaps this may explain why people modify their surroundings the way that they do, as well as why forests are less valued than lawns and gardens.

Although many gigatrends are interrelated, they can be discussed in categories, such as general human populations or ecosystems and forest ecosystems. Positive trends, often smaller and more recent, are discussed after negative ones. This list is not meant to be complete or detailed.

General Human Populations and Needs

With the success of the human species has come human domination of ecosystems. Human beings have modified animal and plant associations, simplifying patterns of energy and chemical exchange, and solidifying ourselves at the end of many food chains as a dominant species. Our domination is related to our large biomass, our large annual increase (over 2 percent annually), our high energy use, and our high structural organization (information and matter). These very large-scale effects relate to basic gigatrends having to do with population size and dominance:

• Human populations increase exponentially — at 2 percent per year the doubling time of the entire population is 35 years. The growth of human populations for 500,000 years was minuscule; the agricultural revolution (10,000 years ago), which increased food supplies, and the industrial revolution (in the 1800s), which decreased the death rate, led to dramatic increases in population numbers and the rates of growth.

• Humanity takes over the habitats of other animals and plants, regardless of loss of function of ecosystems, limits to carrying capacity, or deficits. Kenneth Boulding suggests that eventually humans may perform the functions of other species.

• The material goods of human societies have been increasing. Few of us in Canada or the US can carry everything we own on our backs, a bicycle, or a horse. The impacts of a small percentage of people (the wealthy) increase exponentially as the result of heroic consumption.

• With the social avoidance of hard decisions, no one wants to take responsibility for meeting ecological or social limits; no one can say no to their constituents, representatives, or business partners.

• Societies seem to be working toward a minimum for human existence: a box for everyone, with sufficient heat, light, air, and plumbing. Mere existence has become an acceptable option. We have gone from the green forest to the gray box. There is no place to hide, no place on earth that the air cannot stink and the rain not burn, that ugliness cannot reach and misery not touch.

• Human interactions become more violent as a result of competition, inequity, and limits in the distribution of resources.

• There is an increase in the scope of ethics, from family, tribe, nation, humanity, to include reverence for all living beings, identified by Albert Schweitzer. An increase in the scope of ethics to include land and forests, identified by Aldo Leopold. An increase in the scope of law to include legal rights for forests, identified by Christopher Stone.

• Human cultures mutually adapt with ecosystems over time in Asia, Europe, and parts of Africa, resulting in domesticated landscapes.

• Larger areas, such as preservation of ecosystem processes, reservation of archaic cultures, and conservation lands, are set aside from industrial interference.

Human population pressure pushes a lot of trends. The pressure from exponential population growth means that remaining forests will be depleted to meet basic human needs. The existing financial and political resources may not be enough to stop it.

Eric Eckholm concludes that the United Nations must identify, analyze and marshal world resources against negative trends. A scientific method would take a long time, however, and poor countries cannot wait. They must attempt a rural regeneration of some kind, to stop urban drift and ecosystem destruction. Eckholm interprets negative trends as indicating the sinking of marginal peoples on marginal lands into a quiet helpless poverty, later leading to urban deterioration — which may perhaps be less quiet.

Forest Ecosystems

Ecosystems build up information. There are at least three different channels of information in an ecosystem: the genetic (in replicable individuals); an ecological based on interaction between cohabiting species (expressed in changes in their numbers); and the cultural, transmitted through individual learning based on experience. Feedback within the interaction of species is expensive memory with little storage capacity. Whenever succession starts again, after a volcanic eruption for instance, old information in the form of interactions has not been saved. Genetic memory has much larger capacity and is long-term. In higher vertebrates, such as wolves or humans, cultural memory is enlarged. The unconsidered use of information results in still more long-term trends:

Forest destruction was first associated with hunting, when fire was used as a technique to drive animals into the open. Grasslands maintained by burning allowed safer hunting. The technique is still used today by a few small groups of people. Centuries of burning in Africa reduced forest cover to less than 40 percent of its original cover by 1948, according to Eckholm.

Agriculture and forestry have been related in many civilizations. The expansion of agriculture is directly related to the shrinkage of forests; other trends in agriculture, such as opening southern lands, have resulted in forests reclaiming some of their northern territory. Nevertheless, the overall trends have not been affected:

• Forest cover has been reduced since 1100 B.C., constantly reduced since the 1500s, and drastically reduced since the 1950s. No one knows the exact rates of reduction. Overall the world forest cover has been reduced over 35 percent. The actual amount also is very uncertain, since inventories are rare or crude — half the land reported as forest land in many countries is also labeled "unstocked," according to Eckholm and the World Forest Inventory, with the result that grasslands, scrublands, and wastelands are labeled as potential regeneration areas.

• The long-term health of forest ecosystems declines as people fight over access to specific resources for short-term economic gain. Forest health is rarely measured or addressed.

• Natural regeneration is declining, due to interference in the operation of forest ecosystems and the destruction of some of the necessary structure, for example, clearcutting kills mycorrhizal fungi necessary for nutrient uptake.

• Tree planting has been decreasing. According to Robert Mangold et al., 2,419,691 acres were planted in 1993, a five percent decrease from 1992 and almost a 30 percent decrease since 1988. The decrease is in all geographic areas and by all ownership groups.

• The forests that remain are simpler (tending to monocultures), of lower quality, and much younger. Forest ecosystems are kept at early seral stages to benefit from increased production. There is no guarantee that they will ever become mature.

• Ecological limits are ignored, even where they are known. Forests are cut at rates greater than the net primary productivity.

• Ecosystems are simplified and degraded; deforestation, desertification, and exotic take-overs occur on a large scale.

• Vegetation becomes a social artifact. In Scotland, for instance, forest cover was reduced from 55% of the total area to 5% by primitive stock-keeping and agriculture; the moors decreased by half, but meads increased eight-fold.

• Global biogeochemical cycles are disrupted; for instance, atmospheric carbon dioxide has increased since the 1700s.

• Some forests are restored from abandoned fields, anthropogenic deserts, and ruined ecosystems.

Wild forests exist in some kind of balance. Insects damage trees, by attacking different parts of the tree (often during different stages in their life cycle), but not usually at a level more than 10 percent—far smaller than the waste from logging, which approaches 50 percent. After being used and "improved," forests are not being regenerated. They are not being replanted nearly as fast as they are being cut, either in terms of biomass or quality of wood. These trends contribute to a shifting planetary system that has far less flexibility to respond to environmental changes.

Some trends have stopped because planetary limits have been reached. Like ownership, territorial expansion is at an end. There are no "new worlds" to discover and no virgin forests to be logged.

Some of these negative gigatrends are hard to see, much less stop or reverse, because they are based on misunderstandings, fallacies, myths, and psychological blinders. For example, forest planners often treat exponential growth rates the same as linear rates. Thus, if our forests were to last for 400 years at our current rate of use (before extinction), they would only last for 75 years with a demand growing at 3 percent per year, or 50 years at 6 percent (demand is growing now at over 3 percent per year). Also, Dennis Meadows points out the speed with which surpluses disappear with increasing population and increasing per capita demand.

Many predictions about resources are based on fallacies. There is the fallacy of substitution, that states that a substitute can be found for any resource in short supply. This is not always true, especially when cultural preferences are considered. Furthermore, there is a gross underestimation of the length of time that it takes for a substitute resource to attain traditional markets. For instance, the transition from wood to coal as an energy source took about 50 years, despite the fact that coal technology was established and attractive. Andy Kerr has suggested that steel, plastics, and other materials should replace wood in building and other applications. Even if people thought such a shift was desirable, it would probably take many decades.

Meadows identifies another fallacy: The expectation that people and institutions perceive problems and react to them rationally, e.g., with the threat of wood shortages, prices would rise and consumers would value the resource more. Yet, the price of wood is still nowhere near the real costs of production, and wood is used for cheap, impermanent goods. Meadows suggests that the model of addiction might be more appropriate than adaptation for dealing with consumer demand. Futurists are surely right to say that ecologists are premature doomsayers or Cassandras (for spurning the love of Apollo, Cassandra was given the gift of being always right, but never being listened to). The ecologists are certainly right to say there will be shortages and suffering, even if the dates and extents of the shortages are unknown—the function of a modern Cassandra is, after all, to be wrong. The warnings should change our negative behaviors.

Many of these gigatrends are based on myths, such as the economic myth of forestry, which as it is related by Chris Maser, is based on the rationale of "soil rent" theory, a classic economic theory that assumes, fallaciously, that all ecological variables are constant so that capital investment is easily calculated from the rate of growth of the crop species. Due to the uncertainty of natural processes, we should limit our take to far less than a maximum rate.

These trends are partly the result of our unconsciousness of large-scale, long-term events, partly the result of out cultural amnesia about things that make us unhappy, and partly the result of our cultivated indifference—doubtless due to our remoteness from wild nature, remoteness from the forest as a result of our tools, and the general romantic abstraction of civilization.

Summary

What these trends show together is that humanity is using more forests, and more of each forest, to produce more goods that cost more for more of us. These trends are not fated to continue. Some of them can be slowed, redirected, changed, or reversed. Rather than be converted to plantation forests, remaining old-growth could be preserved or harvested at very low sustainable rates.

None of these negative gigatrends can be modified until our remoteness is re-educated into participation and attentiveness. By making long-term trends visible and immediate, we can understand how they shape our use of forests. Combined with other trends in housing, such as arcologies (Paolo Soleri) and ocean arks and bioshelters (the Todds), and agriculture, such as agroforestry, permaculture (Bill Mollison), and tree crops (Russell Smith), further reforestation of some areas appears likely.

The intent of describing large-scale patterns is to fit human patterns with observed patterns in nature; patterns have a form, sometimes repetition, and sometimes regularity, but each of these is caused by some limiting factor. Fitting the pattern can lead to both continuity and predictability, and both of these things are needed to adapt human activities to natural limits.

Thinking we have conquered nature and are omnipotent, we have quit thinking. Satisfied with our comforts, we do not ask enough of ourselves. We seem to be confused between luxuries and necessities. There may be enough forests for necessities, but not for luxuries. We also act confused by the distinction between temporary good and durable goods (nothing is permanent); temporary good are things like cars, entertainment, guns, and drugs (any kind), while durable goods are things like reforested areas, organic farms, well-designed roads, and healthy buildings.

Garrett Hardin used to say that the essence of ecology was found in the question, "and then what?" meaning that everything you did had a primary effect (there were no side-effects) on the system, every action a reaction, or as it has been rephrased by Barry Commoner and others, "you cannot do just one thing." We have to consider the consequences of our actions as much as we can. Even good actions, taken in isolation, can have tragic consequences. For instance, what if, in setting aside forests in North America and reducing the load on them, we put more pressure on forests in Malaysia, causing them to be cut faster and more disastrously? What is the solution then? Social equity with other cultures or peoples? Voluntary simplicity? Global laws?

With these gigatrends possibly ending in tragedy for humanity, we must ask many questions. What kind of ecosystems and forests do we want? Wild or domestic or both? Small or large? Managed or unmanaged or preserved? How shall we use those forests? For wood products? To protect watersheds and maintain global biogeochemical processes? As a home for other beings? Recreation? As some kind of balance for domestic landscapes? How many ecosystems do we (or other cultures or the earth) need? What kinds, in what forms? How many should be wild? These questions lead to new strategies for living with the forests.

We have spent most of our infancy fighting nature. Up until the seventeenth century Europeans regarded untamed nature as a vast, hostile desert. Wild nature still remains unwelcome in our cities and gardens. We might understand the historical failure of cities and walls to lock out the forest or nature. After the Sumerian king Gilgamesh killed the great spirit of the forest, Humbaba, he became possessed with the fear of death and tried to lock out nature with the great wall of Uruk. It did not work. Like any ecosystem, the forest must be wooed. The forest will haunt us until we give it a new life in the heart of modern culture.

Chapter 36

The Promise of Rio (UNCED Uncut)

Why do the advocates of the forestry industry, pro-growthers, and defenders of the blind profit, think that the UNCED was a green platform, all anti-business, full of "Bullying elitists and spoilsports" and "Antihuman ecologists" threatening every modern product as anti-environmental? The earth summit itself has been referred to as a "green tyrant" (*New Scientist* 13 June 1992) attempting to forge a "green empire" by many commentators, including the paid pit bull of any industry, Dixie Lee Ray. Environmentalists are considered blind to alternatives and antagonistic to any other viewpoint. Why are the industry mouthpieces so shrill? An automatic response? Are environmentalists also too shrill, too antagonistic, and unreasonable? Is shrillness necessary to penetrate complacency or does it have the effect of motivating people against the shrill?

Luke Popovich (*Tree Farmer* July/August 1992) expresses disappointment that no one addressed the stated theme of the meeting — environment and development, envying the Greens for having an "unrivaled [sic] platform to state their case for environmental protection." But dismissing the conference as a "Wayne's World for Green extremism," where a few assertions were repeated over and over.

Apparently, Luke did not go to all the meetings (or as an industry workhorse, his blinders were too tight). Not only was development the focus of Agenda 21, but the policies of free trade, deregulation, and massive debt, the very things that created the insurmountable gap between rich and poor nations and between rich and poor people, were reaffirmed. (Nations in the southern hemisphere lose hundreds of billions annually to the north due to established trade practices.) In fact, the industrialized countries manipulated the conventions for their benefit. Agenda 21 did not mention regulation of biotechnology or dumping of toxic materials. Health care and poverty were ignored, also. Instead, the rights of international corporations to make profits were reaffirmed. Human lives and ecological integrity were distant verbal considerations.

Many countries recognized that UNCED was still a license for unlimited development. For instance, Gro Brundtland announced Norway's plans to resume commercial killing of minke whales, thus underscoring her commitment to nonsustainable development.

Luke answers his interpretation of the green case with one of his own. To clearcutting and loss of biodiversity, he answers, "we replant more than we harvest." Not nationally we don't; even less so internationally. To the charge that tree plantations are sterile monocultures, he answers, "we replant after harvesting." With what? Sterile monocultures as in Oregon and Washington? To the charge of clearcutting, he states, "we've preserved more than half ... remaining old growth." Half of what, where? A 10th of one percent in National Forests? To the charge of destruction of ancient forests, he responds only that "business brings jobs ... and higher standards of living."

Jobs that last only as long as the last old growth? His few assertions were repeated with mind-numbing regularity. Doubtless there is a kernel of truth to each of his answers, but he fails to address the real problems with more than one-sentence misdirections.

Finally, Luke accuses the Green movement of living unexamined lives and blaming "impersonal forces" for our predicament (does he want to be blamed personally?). He concludes that environmental protection "without economic growth is self-defeating," echoing George H. W. Bush, who stated (without vomiting) in his prepared talk, that "growth is the engine of change and the friend of the environment." Obviously, they reverse the case, as well as confuse growth and development. Growth is the unconscious increase in size; development is the gradual unfolding of possibility. Growth contributes to the decline of environments; sustained development requires environmental protection. Using a zero-sum game to decide between Spotted Owls and loggers is based on a severe ecological illiteracy. The Pacto de Accion Ecologica Latino-Americana (PAEL, a coalition of over 100 NGOs) recommended recognition of the "ecological debt" of the destructive patterns of Northern development to balance financial debts of the South.

Many greens also evaluated UNCED as a failure; many delegations, including the US, simply ignored most input from the NGOs. The forest convention, to which Marsh Institute members contributed, was reworded as a set of principles, not "legally binding" to countries. In fact, the principles reaffirm the rights of nations to development, with only "periodic" assessment of the principles, doing nothing to diminish deforestation or to protect the indigenous peoples living in the forests. The format of the Earth Summit itself added to ecological problems, through jet travel and pollution, not to mention segregation of the poor, reliance on limousines and taxis, waste of paper on fax machines, press releases, schedules, and drafts. Perhaps, a teleconference would have been more efficient and humane.

UNCED did force the participants to consider the links between environment and development, as well as between growth and degradation, and between poverty and overconsumption. NGOs have now made pledges and agreements within their own numbers. They can work to persuade their governments to approve the conventions. A binding forest convention can still be made. NGOs can monitor the implementation of Agenda 21. UNCED did address understanding of the whole in the public consciousness, healthy people in healthy ecosystems, and that is important.

CHAPTER 37

Strengthening the United Nations

Our many individual national attempts at social improvements have proceeded without adequate reason, without order, without sufficient insight or broad perspective, without enough confidence, without a comprehensive plan, and without a great dream. Our efforts to provide the infrastructure for our civilization are guided by anonymous builders, mediocre designers, minimalist engineers, and rapacious financiers. Our politics have been corrupted by special interests. Historically, however, the creation of states has not been through reason, China, France, Germany, Italy, and the United States, among many others, were united by force.

The notion of a world government seems to satisfy a basic human craving for unity and order. And, an implicit world system has been evolving through economics and science. A global order is necessary to govern the system, but, at the current stage of international relations, there seems to be no agreeable path toward a benevolent world order. The partial adoption of international institutions is insufficient for a world order, especially if those bodies are only advisory.

Many authors have suggested creating a new world authority and abandoning the United Nations (UN). But, the UN is the only existing body with the machinery for constructing a world order; the beginnings of a comprehensive politics can be found in the special services of the UN: UNESCO, FAO, WHO, and the various technical aid services. As long as ecological and political problems are addressed in a framework of nationalism and military power, however, these organizations are treated as peripheral and relatively impotent.

As it is structured, the UN is not capable of handling the responsibility for world order. For example, the UN's solution to economic problems is "sustainable development" within environmental constraints. The Bruntland Report, which proposed that solution, indicated a five to ten-fold increase in world industrial output within the next one hundred years before population stabilization. While the appeal to growth is inarguable, it is really not likely to be sustainable in any meaning of the word, since this kind of growth does not recognize or respect known ecological limits. Considering that the current level of industrial output has imposed severe threats on human society and environmental health, even a five-fold increase should be able to destroy the cultural and ecological diversity of the planet. Other actions of the UN, such as restricting membership in the security council to "great" powers with nuclear arsenals, or its use of the veto principle, indicate that the organization has been captured by the status quo. Furthermore, even when the UN does make good recommendations, it does not have the power to coerce any nation to follow them.

Rather than replace the UN with a new construct, we must revise it. The UN has been a half-hearted investment, but it has historical appeal and wide support. It is a nascent global order, but it must be reorganized and empowered, as first advocated by the Jackson Report (1969) and the Hammarskjold Report (1975). These recommendations are just a continuation and emphasis

New Powers and Responsibilities
The United Nations, an elected body, could have the regulatory powers necessary to maintain a healthy global environment. It could have advisory and regulatory powers to maintain the independence and integrity of its constituent nations and their peoples and places. It could have the regulatory and punitive powers to rectify resource and human rights infringements. This body could have sole international police powers and large-scale weapons. Various advisory bodies could recommend policies and actions to nations. The UN could have several basic functions, such as:
   • To Ensure a Diverse and Healthy Biosphere. This means creating a global inventory of resources and then distinguishing between wild resources (for ecosystem support services) and common resources, to be used for human needs.
   • To manage common resources. The United Nations would have the power to designate areas for conservation, including the oceans and atmosphere. It would regulate all industrial and residential use of common resources, by using new institutions and groups.
   • To Protect Unique Human Cultures with a framework for recognizing and encouraging diversity.
   • To Coordinate the Representation of Cultures through new programs and a representative body.
   • To Provide Services to Nations and Individuals. Especially for civic actions, but also emergency services, educational, and basic security services.
   These functions would be performed by both old and new UN Structures. It could also modify its existing structure to reflect smaller, more peaceful memberships. It could add new structures to perform new functions.

Immediate Steps
The new UN could be based on the old UN, but with immediate responsibilities and powers, for protection and preservation, as well as some temporary powers, such as taxation. Five immediate steps are necessary.
   1. The transference of powers. The major military powers would grant their powers to the United Nations and relinquish their efforts towards global leadership; they also resign from the security council, cease propaganda activities, renounce foreign policy objectives, call back soldiers from foreign countries, and stop giving away produce, factories, or weapons. They put their technical and educational surpluses at the disposal of the UN. If the USA or Russia is to be a world leader, let her lead in tolerance or in trust. Let her be the first to give allegiance to a world organizing body, the UN, the first to divest themselves of nuclear weapons. If they fear for safety, they need only remember the success of nonviolence in India or of guerrilla actions in Southeast Asia and Central America.
   2. Disarmament. Complete disarmament could be accomplished

within a week. Earl Osborn proposes this concept of sudden disarmament in response to the tedious phase-out envisioned by most plans. An agreement would not involve much negotiation. Taking this first step would add to the prestige of the country bold enough to do it. The UN could post a police force to disable all military ordinance. A thousand planes each carrying one hundred trained inspectors could be distributed at all major centers in the nuclear countries within 24 hours.

3. Formation of Nations. Independent cultural areas within nations would have the status of independent nations within the UN. Any culture would be given legal recognition, protection, and full autonomy over their boundaries by application to the UN, which would determine priority of claims (by archaic peoples, agriculturalists, pastoralists, or industrialists). No action would be taken to disband existing nations. Nations could still remain allied with nations as independent or dependent regions. The nations would determine the use of allocated resources. Local economics and technology would provide for populations. Traditional religions and customs are maintained or permitted to develop.

4. Catastrophic Measures. The United Nations promotes a decade of consideration. Starting with population growth, all growth would be suspended. Earth parks, in the Antarctic, Amazon, Arctic, Oceanic areas would be declared immediately. Ecosystem restoration would be begun; massive planting efforts are undertaken. No further expansions are permitted for development in wetlands or other sensitive areas. Destructive searches for resources are suspended, in favor of substitution and recycling. No new building is encouraged until uninhabited ones are restored or recycled.

5. Paths for individuals. Depending on religious, economic, geographic, or personal preference, individuals can join any culture (within ascribed limits); most would probably remain in their native culture. The designation of cultural nations does not involve a major revolution for most people. Revolution is a false dilemma, it does not reflect possibility of thousands of individual actions on farms, factories, and families, all at local levels. Individual actions add up to fate (in Tolstoy's vision). Individuals can work to regain control of their lives. They can make choices to be self-reliant or to limit their impact on their supporting ecosystem.

Taking these steps would solve many of the problems addressed earlier. The satisfaction of physical and cultural needs, as a result of living in stable and small societies, would contribute to the health of people. Fitting economic costs and needs to the limits of ecosystems and monitoring the economic process would reduce wastes and pressures on natural processes. The coupling of agricultural productivity to a solar budget, and the conscious restoration of degraded systems, would contribute to the health of ecosystems. Sufficient wilderness would allow the self-maintenance of global cycles. With the increase in security, wealth, and self-esteem, human populations could be dependent on ecosystem productivities and still be diverse and unique.

With the removal of war capabilities and the equalization of wealth, the remaining issues are not the kind to incite violent passions.

Disagreements over the best way to raise wheat or to maintain a forest may be more easily resolved than deciding the best nation or truest religion. The death of large-scale dogmatic ideology and national idolatry could also mean the end of organized slaughter. Perfecting the art of resolving conflict through social debate would free unprecedented resources to satisfy social needs. Perhaps a planetary electronic referendum would open communication.

The UN works with traditional cultures and realistic planning. The UN is based on the values and forms of traditional cultures. Rational planning can catch up as it develops. The framework is an open, flexible, and partially-planned global relation, a series of Eutopias, instead of a finished, closed, completely-planned society, as imagined in utopias. The UN accepts the imperfect nature of humans and the changing ambiguity of nature. The UN detoxifies cultural rivalries. Racism, sexism, ageism, and speciesism lose their importance in a cooperative society of advanced communication, automation, equality, humane scale, and meaningful preservation.

In designing the world, everyone can participate. We can reduce the violence to nature and ourselves and transmute it to debate. That which has been hitherto left unsaid—-what we want to become, what we could become—could become explicit. Now is the time to define goals in terms of population, quality of life, and preservation of biomes. Goals are not some final state reached once for all time—they are a horizon. The UN offers continuity towards these goals.

CHAPTER 38

Ekogedankenexperiment (Ecological Thought Experiments)

Rather than unconscious, accidental experiments on ourselves and our planet, we need to conduct thought experiments first. Albert Einstein and Leopold Infield, in their book *The Evolution of Physics*, suggest that knowledge of the laws of nature can be gained through the contemplation of idealized experiments created by thought, *Gedanke-Experiment*. For example, to address the equality of intertial and gravitational masses, Einstein imagined an elevator at the top of an incredibly high building, and then imagined what research would be done in this local moving environment. Such experiments might seem "fantastic" he said, but they might help us understand what we want to understand.

Although ecology is orders of magnitude more complex than physical systems, perhaps we could imagine and use such experiments to help us understand what is happening with our complex environment in Bulgaria, which is composed of many interlocking ecological systems. By asking all kinds of questions and then imagining the answers, given what we know about our history and ourselves, we could discuss things that are often taken for granted, or not even thought about. For example, the following thought experiments could be worked out individually in later articles.

- We seem to be replacing natural services with industrial ones. How far can that go? How much will it cost, and what effect will it have on everything else? Should we treat every bit of water? Or avoid polluting it in the first place?

- Bulgaria is keen to join NATO. What would happen if Bulgaria spent more effort supporting the United Nations (in fact, what would happen if the United States and other countries also offered more support to the UN)? What would happen if countries gave up their excess military power to the UN (including nuclear weapons and bombers), and kept only their in-country police forces?

- Bulgaria has a greater percentage of wild nature than most other countries, including the United States — in fact, many European countries are trying to restore what Bulgaria still has. How should Bulgaria deal with the nature in its borders? Create a wilderness inventory and set aside more parks and wildlife corridors? Or allow it to be used by people as they wish?

- Domestic animals are truly loved by Bulgarians, but there seem to be large numbers of them roaming the streets. Many cities are trying to solve this situation by killing all the stray dogs in a city. What should be done? What is the role of these animals in our lives? What effects does it have on wildlife or on people living in the cities?

- Many things are out of place in Bulgaria. Trash is a major problem for any country with ambitions to be a tourist destination. Why does it appear in rivers and fields? Could there be the historical or cultural reason for it? How can it be changed? Or should it?

- Bulgaria has a small and declining population — is this a problem or an advantage? What should we try to do: Grow as quickly as possible? Try to balance the population with the capacity of the land? Or just let things go by themselves. What would be the consequences of each decision?

One expectation that modern life has raised in us is that "there is no right answer." The best response to a situation or a problem or a question may be a hypothesis, that is, a thought experiment. Through that, we can create explanations and discover answers in a dialog with others. Thought experiments could let us examine things and change them without really modifying or destroying the unique systems under study. In practice, this means thinking first, and all the possible connections, before actually making changes.

Ecological Thought Experiments 2
How Many People is Best for Bulgaria?

Bulgaria is said to have a "declining" population. Many politicians and economists are calling for rapid growth. Perhaps because, in the past some times, rapid growth has increased prosperity for some people. But, is rapid growth the best strategy for Bulgaria?

Imagine that the population has grown to 16 million people — every place is twice as crowded. Has the number of good jobs increased or are all these new people unemployed? Where do they live? Some things do not increase: amounts of land, air, and water are still the same. More people need more land for buildings, factories, farms, and roads. That means fewer places for wilderness, recreation, autos, or buses. High populations in many countries, with many different living standards, are only maintained through the constant takeover of natural habitats for farm land, or through the drawdown of fossil fuels, and by economically cheating the poor and powerless. Eventually these things get corrected by war, collapse or adjustments (usually all unpleasant)

What would happen if the number of people keeps decreasing? At some point, maybe half a million, it might be hard to keep the cultural values of the people active and vital. It might be hard to maintain many industrial activities. At very low numbers (5,000-50,000), there might be problems with fertility or social cohesion. The population has been low before, during Thracian, Roman, or medieval times, but it was rarely thought to be too low.

Rather than just let the population grow or shrink, we could try to figure out what a healthy population should be, by relating it to the carrying capacity of the land. The carrying capacity is that number of people who can be supported on a long-term basis, using local renewable and nonrenewable resources for all needs, including clothing, shelter, transportation, information generation, and aesthetic satisfaction. This capacity can be increased or decreased by using different kinds of agriculture or different kinds of technology. For instance, technology can give higher yield crops, but also it hurt crops with unforeseen side-effects (poor uses of pesticides, for example). Furthermore, the capacity decreases as the use of energy and resources per person increases.

The ecologist Eugene Odum suggested using land area as a measure of human carrying capacity. The minimum area requirement per person are is 2.02 hectares to provide all needs sustainably (for the state of Georgia in the US). This number includes land for natural areas, as well for producing food and siting communities and road networks.

Bulgaria and Georgia are roughly comparable in productivity. The size of Bulgaria is 11,091,200 hectares, so if we divide by 2.02 hectares, we get a maximum population of 5,490,693, quite less than the current number. An optimum population might be less than 3 million. Any size human population can disrupt natural cycles and environments, however, so numbers are only the beginning.

All living beings adapt to and change each other over time. They change the climate and environment, also. Wild populations usually exist at far less than a maximum number. They are limited by the productivity of ecosystems. We humans need a similar flexibility to adapt our populations to changes in climate and the productivity of the land. That is why a smaller population may be better ecologically.

Instead of treating a declining population as a problem, why not consider it an advantage? With ecological planning, Bulgaria could become the first balanced nation on earth, by linking its population to its carrying capacity and wild environment. Countries such as China, United States and The Netherlands will have to face these limits soon, but Bulgaria could become a good model for them to follow.

Ecological Thought Experiments 3
The United Nations, Peace and Human Diversity

After the war ended in 1918, there was a popular vision of One World, without borders or barriers, created through reason. Our many individual national attempts at social improvements, however, have proceeded without adequate reason, without order, without sufficient perspective, without adequate confidence, without a comprehensive plan, and without a great dream. Historically, the creation of states has not been through reason. The boundaries of almost all countries, from China, the USA and Mexico, to Germany, France, Italy, Greece and Turkey, are fairly arbitrary, being based on the results of military struggles in the past. These and many other countries were united by force. Our efforts to provide the infrastructure for our global civilization have been guided by anonymous builders, mediocre designers, minimalist engineers, rapacious financiers, and corrupt politicians.

Yet, the notion of a world government seems to satisfy a basic craving for unity and order. And, an implicit world system is evolving slowly through economics and science. A global order is necessary to govern the system, but, at the current stage of international relations, there seems to be no agreeable path toward such a world order. The partial adoption of international institutions is insufficient for a world order, especially if those bodies are only advisory.

The United Nations (UN) is the only existing body with the machinery for constructing a world order. However, as long as ecological and political problems are addressed in a framework of nationalism and military power, the UN is treated as peripheral and relatively impotent. Furthermore, as it is structured, the UN is not capable of handling the responsibility for world order. For example, restricting membership in the security council to powers with nuclear arsenals, or using the veto principle, indicate problems. Furthermore, even when the UN does make good recommendations, it does not have the power to coerce any nation to follow them.

Rather than replace the UN, countries must revise it. The UN has been a half-hearted investment, but it has historical appeal and wide support. It is a nascent global order, but it must have new structures and new functions, new powers and new responsibilities. Two are especially important now.

1. Protecting diversity. The UN must have real explicit responsibilities, such as the protection of biological diversity for the planet, as well as of human cultural diversity. At one time, around 1900, there were over 1000 different human cultures and over 3000 different languages (roughly equivalent to the number of natural biogeographical provinces, subprovinces and habitats on earth). Cultural diversity is necessary to protect biological diversity (the ecological wealth of each country). One critical message of ecology is that if we diminish variety in the natural world, we debase its stability and wholeness. If we wish to advance human civilization, we must preserve and promote variety.

After a long trend of consolidation by colonial powers, new countries are declaring their independence. New countries, such as Kiribati, Liechtenstein, Marshall Islands, Monaco, and Vanuatu, have joined the UN recently. Many more countries want independence based on their cultural and linguistic uniqueness. Perhaps many of the borders drawn by violence, in Bulgaria, Albania, Croatia, and Macedonia, among others, could be redrawn peacefully.

2. Protecting security. The UN must also be given real powers for protecting countries and for policing international terrorism. Most military power, except for local police or national guard, could be turned over to the United Nations (nuclear disarmament itself could be accomplished relatively quickly with complete international support). Wars are being fought over resources and territory, as well as for religious and personal reasons, without an international referee with power or respect. Violence will continue (regardless of how well justified — and justification these days has a very dissolute and tangled history), although the cycle of attack, hatred, and revenge can be broken by an international body

composed of representatives of all peoples.

International terrorism, or criminal actions, such as those recently in Iraq, Afghanistan, Macedonia, and the U.S.A., require an international response, with an international police force, an international justice system, and an international form of punishment, so that all the countries of the world not only have a say, but have a stake in the peace process.

Having the UN address international issues, as a confederacy of concerned neighbors, would do much to diffuse the polarity of one country trying to be world leader and peacemaker (especially when the USA, for instance, is viewed as the power behind many thrones that exist only to protect US economic interests).

The UN could provide support—food, health aid, engineering —to a country such as Afghanistan and it would not be resented as a gift of rich people wanting a market or an ally. This would merely be the act of good neighbors.

Ecological Thought Experiments 4
Ecological Thoughts about Roads

Bulgaria has many roads, although politicians, as well as business people and drivers, are asking for more, wider, smoother, and direct roads. Cities want better roads leading to their centers. Forestry managers want more roads into forests. Resort areas want more roads leading to the sea or mountains. Economically roads can stimulate income, at first. But, there are other ways of looking at roads.

We can look at roads with an ecological perspective. As a science, ecology describes the interrelationships of organisms and environments, that is, the experience of living together in the biosphere. But, ecology is also a way of "seeing" that human beings are participants in nature, as part of the food chain, for example. People, like most mammals, use roads (or paths) to get from one place to another, to get supplies; to visit others, or just to look around.

This is a fine use of our technology. It allows us to increase our horizons and better our lives. Better roads make traveling more efficient. There is less waste of oil and gas, and less wear on vehicles and their occupants. But, every technological innovation in vehicles either requires or makes roads. And, roads have effects that go far beyond the movement of people along them.

New roads lead more people to new places, thus changing the characteristics that often make those places attractive (e.g., being off the road). Roads increase the flow of things between points. But, too much flow (of matter, energy or form) can destroy biological relationships and diversity. Roads are a major force in fragmenting the habitats of plants and animals. Many animals cannot live near roads or noise or human activity. Many animals and plants need large areas to roam and roads cut into their areas (although, highway routes and underpasses can be modified; for instance Britain builds underpasses for frogs). Roads directly affect natural and human communities in many ways, causing:

- changes in populations of animals or plants that cannot cross them (isolation)
- the spread of organisms that use roads to colonize new areas with plant or insect or animal pests (that is, things that are out of place)
- problems with erosion
- problems with spreading trash (other things out of place)
- changes in hydrology and wetlands
- changes in social circumstances. For instance, private cars changed public morality in America. Many kinds of crime are increased, for instance, bank robberies, if there are fast roads nearby with easy access.
- changes in economies, as new roads bypass old routes.

Many countries answered the demands of their citizens, business people, and politicians (and ignored the environment) by building bigger, faster roads. Then as people crowded on the roads, they get more crowded and slower, and the demand for bigger roads rose again. Many countries have found that building more and larger roads does not solve the problems of congestion, accidents, and danger. These problems have a lot to do with the kind of transportation on the roads, that is, cars, trucks or buses.

Maybe Bulgaria should have a new autobahn and maybe all the roads should be upgraded. But, it would be better if it were part of a plan for the entire country that considered population movement, the needs of all the people, and the best forms of transportation. Buses and trains are far more efficient than private cars, and many countries are rebuilding their train and bus routes, from Brazil to France and Japan. By concentrating on a good public transportation system, and limiting the influence of private cars, Bulgaria could become a good model for them to follow.

Ecological Thought Experiments 5
Why Bulgaria is Wealthy

Recently I was asked what I thought about poverty in Bulgaria, the lack of cash that makes Bulgaria poor. I did not know what to say. From what I see, Bulgaria is rich. Perhaps we misunderstood each other's ideas of wealth.

Economics has always been concerned with measuring wealth. Wealth once meant tangible things, such as land, ships, houses. Later, it was measured by labor and production. Now, it has come to mean negotiable symbols such as cash and stocks. Lately, information is considered a form of wealth.

The first economies depended on their own food and minerals. The mass economy rose after the industrial revolution, when networks of governments and institutions were created to hunt and acquire vast quantities of material in order to manufacture products that could be sold at profit. This economy is being altered by increasing populations and by increasing difficulties in finding cheap resources. Some of our perceived wealth and assets are disappearing in the process. But, symbols such as cash are considered new forms of wealth. And with computers, information itself is valued as the ultimate resource and source of wealth.

Information is apparently boundless. Yet it can be manipulated. It is information that defines the use of resources by people. For example, hydrogen is worthless unless technologies exist to transmute it to helium and manage the released energy.

Land and resources are considered less important. But information without "form" is nothing. Information lets us use resources and land more efficiently. Land and resources still are part of the basis of wealth (a *material dimension*) — as many native peoples have found out when coal or pharmaceutical plants were discovered on their lands.

The narrow definition of wealth (as just one thing, resources or information) means that it can be increased only by producing a bigger supply of goods or reducing the demand for goods. Wealth is defined as supply divided by demand. If supply is limited then wealth can only be increased two ways: reduce the expectations of individuals (smaller pieces of the pie) or reduce the number of individuals (fewer larger pieces of the pie). Supply may be mostly material things — but not status, for instance — while demand has the more psychological dimension. Wealth has a *psychological dimension*. This dimension is not limited by strict logic. Wealth can therefore be expanded without being limited to supply or demand for materials.

The assessment of personal or cultural wealth, for instance, is mostly psychological; wealth may be measured by how many valuables one has, which may be physical, like feathers or salmon or gold, or by how by much status one has, which may be behavioral, as when enjoying deference or a good reputation.

Rich sensory experiences can be derived from direct contact with nature. But economists rarely mention these values. Light, wind, dirt, plants, birds, all act on us — but not with the meaning of crops or vehicles, which is for their utility — they just are. People do not live without these things. They are valuable to us.

Until now, economics has required growth to increase wealth. Growth has been a substitute for equality; it seems to be necessary to avoid revolt — even after 400 years, growth has not brought wealth or equality to most people. Ecologically, the goal of economics should be mature development, not growth. Development means the introduction of an innovation. Economic development will still require technology and new forms.

A mature economy would be like an animal, or plant, or like a mature ecosystem, like an old beech forest. In its early stages, a beech tree can still be stable. Growth in trees can delay the onset of senility by ridding it of waste products in more diluted form. However, too much growth produces a strain on tissues and early decay. Later stability must result from limits and metabolism. When it reaches a size that fits its genetic and environmental limits, it is mature. It continues to change, but that is development of new relationships and forms.

It seems that the wealth of a country is a function of its physical attributes (resources including information) and its culture (application of information by people). In fact, the attributes are only possibilities until appropriate cultural perceptions and technologies exist. The inclusion of nature as a source requires an ecological dimension to wealth.

Bulgaria has wild lands with all kinds of unique wildlife and plants. Bulgaria has fertile fields that could grow all the food that her people need. These things are an important basis of true wealth. Bulgaria also has an educated and exceptional people, who have the knowledge and information to make sure that some of the natural wealth is transformed to human wealth. By this most advanced definition of wealth, Bulgarians are indeed wealthy.

Ecological Thought Experiments 6
Communism, Capitalism or Community

Bulgaria has accepted much of the help offered by capitalist countries, and seems eager to replace the old communist command model with the old capitalist market-oriented economy. Is this a good idea? Will capitalism automatically improve things? Is there a third alternative, after communism and capitalism, that Bulgaria should use?

The communist model provided many things for many the people(while maintaining a large war and research establishment). The philosopher Karl Marx thought that socialism was inevitable, that public ownership of the means of production would provide equality and social security for all people. But, in practice the distribution of wealth was very inequitable, as the result of historical trends, old economic rules, and cultural confusion. Some people were treated as more equal than others.

As a result of central planning, the patterns of life under the communist model were pressured to be uniform and efficient. Yet, the strengths of this kind of economy — especially the planning of production and the control of resources — were not admitted. Instead, military competition ruined these command economies and socialism is considered a failure..

Capitalist (or market or free enterprise) economies are exploiting the resources, such as the forests of Siberia, of failed economies, while at the same time trying to rehabilitate other resources, such as the forests of East Germany and Poland. This contradictory behavior is due to its own economic myths. Capitalism is based on erroneous assumptions about nature and culture, for instance that "nature is a resource to be exploited" or "anything can be substituted for anything else." Marx thought that free enterprise was the most efficient and dynamic system, but he — correctly — identified its basic flaw as the accumulation of wealth without the capability of using it wisely. Misery has increased as fast as wealth. The capitalist model has proved to be a limited, flawed and self-serving system that has many negative effects, such as unemployment, poverty, homelessness, environmental ruin, and wild inequity. Americans are criticized by the

French, not unreasonably, for having a "frivolous" culture based on "savage" capitalism. Capitalism also increases the pressure for uniformity. The patterns of life have become the products of market forces and stylish transportation operating in a sterile abstract order. It is as sterile and poor as the communist model!

The choice is not necessarily between communism or capitalism, two old flawed systems. There are other models for economies, such as the Scandinavian model, or the Ecological models (described by Herman Daly and John B. Cobb, Jr.), or community anarchism (described by Paul Goodman). These models argue for strong communities first.

We do not need to surrender to fast, giant, national economies. We need to shift power to local communities, through self-reliance and participation. Community economics has many advantages over communism and capitalism. For instance, community economics allows barter and other informal exchanges. It also offers more services far cheaper than a formal modern corporate system. A community protects individual freedoms, guards regional culture (values and identity), and holds groups accountable for their use of power. In communities, people can decide to be conservatively sustainable or to grow and gamble on innovation. Communities can have different economic attitudes, paces, and goals. A community that is balanced and flexible, in tune with natural cycles, based on traditional values — in which industrial production is limited to appropriate goods — can position the community for a long, sustainable future in a healthy environment.

Bulgaria has numerous small businesses, e.g., clothing, food, and hardware stores, automotive repair, and lumber yards and construction companies. But, it also has large business that allow money to flow to foreign countries. Certainly, the country could implement ways to keep money circulating in local communities; sometimes this can be done by simply buying locally, but more ambitious solutions, such as local barter "bucks," are also possible. Bulgaria could still differentiate its unique products, within a regional partnership, such as the European Union. This would benefit Bulgaria, as well as the larger regional community. Bulgaria is an optimum size for a country to start an international experiment on the economy of community.

CHAPTER 39

The Role of Radical Ecology in Making Humane Places

Summary
This paper presents an outline of the tenets of radical ecology. The concern
of radical ecology is to ensure the survival of human communities in place
on earth, which is, in fact, the goal of politics. Because survival is in nature,
politics must rest on the foundation of ecology. As a science, ecology
describes the interrelationships of organisms and environments, that is,
the experience of living together in the biosphere. As a philosophy, radical
ecology investigates the normative aspects of living together, that is, ethics,
and the maintenance of the affairs of a community, which is, on fact the
function of economics and politics. As a noetic discipline, radical ecology
provides information on the state of nature, but it recognizes that human
beings are participants in nature, as part of the food chain, for example,
as well as participants in the societies that are trying to survive. Radical
ecology offers a new perspective of humanity in the total field of nature and
defines balanced relationships with ultrahuman beings and species. Radical
ecology addresses the determination of separate wilderness areas necessary
for a healthy ecosphere, and an optimum human population, based on net
ecosystem productivities and modified by appropriate technologies within
ecological and cultural restraints. It urges local, self-reliant cultures with
adaptive cosmologies and natural values.

Modern science has provided us with the means to manipulate natural
processes for human advantage. But we distance ourselves from processes
that are uncontrolled or unowned. This detachment is the greatest threat
to the welfare of nature and, ultimately, to ourselves. The vivisection of the
world depletes our ability to feel compassion for it. We are "destroying the
voices of existence," in the words of Neil Evernden,[1] without even knowing
most of them.

Analytic science has reached its limits. Data and information
developed by hard studies have undercut the paradigms that guided their
investigation. The compartmentalization of scientific fields has exposed
the complex connections of the subjects. Science does not need to be based
on logical positivism and reductionism, though these have allowed great,
insensitive changes. A. N. Whitehead thought that what had been missing
during the formation of science was a sense of relatedness. Early science
saw the world as mechanism; modern biology is seeing it as resembling an
organism. Organismic trends can be seen in sciences, from relativity and
gestalt psychology to ecology.

Ecology deals with the relationships of organisms to environments. It
is not a reductive discipline, and not readily amenable to quantification. Even
scientific ecology is an integrative discipline that extends beyond the bounds

of science.[2] Ecology is an amphibious discipline, with the authority of science and the force of moral knowledge. Ecology, studied through its components and relations, is a perspective, a way of "seeing," according to Paul Shepard.[3] It is a perspective of the human situation in its interconnection. For Paul Sears, ecology is a "subversive subject."[4] It is normative and sensible, it also offers a "sacramental vision" of nature.[5]

Radical — from the Latin word meaning "rooted" — ecology forms part of a new metaphor that is more appropriate to the unity and interrelatedness of the earth. It is part of a movement of consciousness, concerned with equality, diversity, health, with humane methods, and with a holopoetic cosmology.[6] And it affects them simultaneously.[7]

Radical ecology emphasizes biological equality, like the "deep ecology" of Arne Naess.[8] Charles Elton transformed the Great Chain of Being into a chain of eating. And, the result of Elton's food chain was the realization that the bottom link — plants — is the most important. Humanity is part of the chain. The exploitative competition of humans in ecosystems is an important part of their cycles. It is "interference" competition that is so destructive and should be limited. Humans need to recognize that they automatically participate in everything, and that they cannot unparticipate by choice.

And, like John Rodman's theory of ecological resistance, Radical ecology argues for diversity.[9] It is a world of action directed at preserving and enlarging diversity. If we wish to advance the continuity of nature, we must preserve and promote variety. In nature, variety emerges spontaneously, as the capacities of new species are tested by the environment. Ethical, aesthetic, and utilitarian reasons all support the efforts to conserve the diversity of nature. A critical message of ecology is that if we diminish variety in the natural earth, we debase its wholeness and stability.

Radical ecology incorporates a broader scientific method that might be called patient practice. There are ways of dealing with the earth that are not scientific or technological; they are aesthetic or ethical. These alternatives are not incompatible with a whole science. The methodology of traditional science is limited and wasteful. Radical ecology considers the method of Goethe. His natural philosophy incorporates a world view[10] of organic dialectics; its methods are contemplative nonintervention and the primacy of the qualitative.

1. Contemplative Nonintervention: Goethe rejected analysis; his approach was passive attentiveness. Knowledge comes of itself, in quantum leaps; a gestalt is perceived. Thought experiments and computer simulations can supplement observation.
2. Primacy of the Qualitative: The qualitative cannot be subordinated to the quantitative. Qualities must be evaluated. Intuition works toward deeper sensory participation. Applied systematically, Goethe's method of analogy, supported by empirical evidence and observation, is complementary to the method of difference.

An alternative science considers every-day observations, unique occurrences, short-lived phenomena. Goethe recognized that different people are sensitive to different aspects of a thing. Any investigative effort should incorporate the observations of many others.

To examine nature in general, we must shift to a taoistic approach, asking rather than telling, observing rather than manipulating; receptive and passive, not active and forceful; "nonintruding;"[11] and noncontrolling. It stresses observation rather than manipulation; it is receptive rather than forceful. Classical objectivity may be contrasted with taoist perception. Caring perception provides kinds of knowledge not available to remote researchers; this is especially true in ethological literature: Maslow cites his own work with monkeys; Lorenz, Tinbergen, Schaller, Van Lowick-Goodall, and Fox have found it to be true. This is the way a good therapist, teacher, scientist, parent, or friend functions.

What is necessary is not a primitive animism or a single-vision science, but a scientific animism, to understand our animalistic nature and use it as the foundation for a sound human ecology. A scientific animism would consider the relations of humans to vegetation and the human attitudes toward ecotypes, like open plains or dense forests; it would consider the need for sacred places, and open, quiet or wild landscapes; it would consider territoriality, aggression, and the aesthetic reaction to the wonder and beauty of life.

Radical ecology is, in a sense, a scientific animism, a soul science ('Anima' is from the Latin word for soul). Nature is a feeling system. We need an animism to approach nature.[12] This animism would allow us to behave "as if" nature were intelligent and sensitive, with the proper reverence.

A scientific animism would be concerned with far more than the anatomy and taxonomy of animals. It would be concerned with mutual experience between human and nonhuman animals, with the need for touch and the phylogenetic possibilities of animal empathy — dogs, for instance, exhibit strong physiological changes when they are petted and human blood pressure drops.

Radical ecology considers the vast scope of ecology, including the economic and political behavior of human beings. The global character of its approach permits the creation of an ecological ethic and the valuing of nature, within a formal economics and politics, based on an ecological education.

Economics. Ecology started out as bio-economics, a division of economics, the application of human economics to nature, where cooperation is defined by functions of production and consumption. Ecology and economics are close etymologically. The study of the house precedes the management of the house.

Both ecology and economics attempt to understand and predict the behavior of complex, interconnected systems where individual behavior and flows of energy and material are important. Although economics provided the model for ecology, few ideas on environmental limits and interdependence were taken from ecology by economists. Human economies are based on natural economies. Economists, however, fear that "letting things alone" will lead to stagnation, poverty, and chaos. The technological vision is "life under control." The technological imperative is to strive for better efficiency. But nature is not perfectly efficient. Looseness is necessary for any open system to function—a lesson that economics needs to learn.

The idea that everything should be managed is based on the extreme belief that nature is a resource to be processed. As Aldo Leopold pointed out,[13] the weakness of relying on economic motives is that most members of biological communities, such as wildflowers and songbirds, have no economic value. Yet, all members of a community contribute to the integrity of the whole, which is vital to maintaining what we consider important. Those beings with no economic value are either ignored or labeled as weeds or vermin and destroyed so that crops and animals with short-term advantages for human ends can be substituted. We will never understand nature unless we dissociate the wild from utility.[14] Our indifference toward nature comes from our judgment of its uselessness—that is has no value apart from the human.

Economics has always been concerned with measuring wealth. The basis of wealth has been variously described as labor, resources, production, net plant production, and information. Yet, no single basis is adequate. The redefinition of wealth in an ecological framework would promote human enrichment and natural preservation. For instance, diversity is a form of wealth. Differences do not necessarily cause conflicts because each fills other's needs. Other species fit where humans cannot. In nature as a non-zero-sum game, many groups can gain at the same time, ultrahuman and human. An ecological economics dictates that some resources are just not ownable—air, water, silence.

Economic and ecologic systems interact. Because human and natural systems are interlocked, there must be a common framework for ecology and economics. Economic decisions are based on human reference and not nature. Human reference is not large enough. Economics is the study of budgets, material and energetic. Ecology is the study of natural budgets, material and energetic. Ecological and economic processes and values are often the same. Economics must recognize that ecological health is vital to its own continuation. An ecological economics is survival-oriented, not profit-oriented.

Nature itself, as wilderness, has meta-economic values. Wilderness areas have values, as mirrors of existence, examples of natural, complex processes, expressions of love for nature, and wild places. But the most important value is as a vital organ for the life of the earth; the generator of hydrological, geochemical, and atmospheric cycles. It is where ultrahuman species live for their own purposes.

Radical ecology tries to define the human relationship to the earth or wild beings. As fully conscious, self-limiting beings, rather than pandominant species, lord and master, or good stewards. When we understand our roles and relationships in nature, then we will not be managers or stewards, but participants and sharers in experience.

The descriptive value of radical ecology is reverence for life. Letting live. Humans have no need for the entire earth. Vast parts of it, especially those that also act as aquifers and nutrient cycling, should be set aside with virtually no human interference. Perhaps forty to fifty percent of the land area of the planet should be set aside in wilderness. Foundation areas, that should be left entirely without any human use or visitation; preservation areas, which may be visited for research or inspiration; reservation areas, for nonindustrial native peoples with traditional ways; and conservation areas, for true multiple use, including forestry and grazing — in short, a limited commons.

Radical ecology strives to change the direction of the economy, to stop growth and expansion. It refuses to compromise on survival issues, to use cost-benefit analysis and resource management to accede to a shallow, temporary economic domination. The current human population is maintained by the take-over of habitats, the drawdown of resources, and by cheating the balance of humanity. We must consider not only resources and products but their effect on the environment and on human life.

Politics.
The scope of economics was narrowed to production and the scarcity of means by J. S. Mill, who considered distribution a political process, since it depended on laws and customs that varied widely in different cultures and ages. Political theory is a symbolic idiom. Misguided politics arise from the wrong relationship of worlds and symbols. The desire for order without a scientific search for the real nature of order can have disastrous political consequences. The Aztecs, for instance, used the wrong symbols to interpret their universe; they believed that the sun needed human blood to survive and sacrificed great numbers of lives to ensure the sun's life. Their political policy was based on continual raids for victims, and this policy contributed to their overthrow and decline with the arrival of the Spanish.

Ecological, economic, social, and religious phenomena are part of the broad definition of politics. The basic goal of such a politics is the "survival of the community" as William Ophuls identifies it.[15] Politics is the interactive means of providing the basic food and necessities of a community. As survival is survival in nature, politics rests on an ecological foundation. The organization of a community must be in accord with natural laws. Political participation depends on information, much of which can be provided by ecologists.

There can be no separation of politics and ecology. Every political act has ecological consequences and every ecological decision is a political demand for control over use of the environment. Ecological consciousness must be identified with political consciousness.

The ecological, social, and political problems of today do not have simple disciplinary solutions. The problems are cosmological and must be solved on that level. But a single cosmology cannot solve all problems in all places. Where human understanding is still underdeveloped, humanity cannot afford to suppress the diversity of thought necessary for adaptation to the diversity of environments, or to eliminate ecosystems and the societies adapted to them. Therefore, radical ecology recommends a framework for local cosmologies, a holopoetic cosmology, as a means for the preservation of human diversity. Modern technological cosmology, beyond being another kind of order, more linear and abstract, is wrongly considered the evolutionary successor to traditional cosmologies, and is displacing them rapidly.

The most pressing condition is the effect of human pressure on natural systems. It is important to limit the number of human beings to a maximum, optimum, or minimum number. How many humans are necessary? One million, one billion, one trillion? As far as a minimum for genetic variation, creative mass, or species advantage, probably only a million at the least. An optimum number would be safest. Maximums in nature are dangerous and unstable. Desired substances have an optimum value, not a maximum one. Salt or calcium is necessary, too much is deadly. More lemmings result in mass migrations. More rats in social disorder. An optimum population can be calculated from the net ecosystem productivity of wild vegetation or from the net primary productivity of agro-ecosystems. It is less than a billion.

An optimum population within ecological restraints should consider the quality of life, nonrenewable resources, appropriate technology, genetic variety, cultural framework, wilderness areas, and intangible values, such as adventure or beauty. To expand the methods of science, the understanding of economics and politics, to increase the quality of human life while lowering the population and preserving wild areas, will take education.

Education
It was the belief of Frederich Schiller that human society could be improved by political means. But after studies on the Thirty Years War in Europe, he became skeptical of the ability of politics. After reading a work on art,[16] Schiller considered it historical proof that art can achieve what violence and law cannot — art educates and liberates the individuals of society in a gradual and peaceful process. In spite of the cultural forces dominant at any moment, an individual has the potential to determine a different course. Unlike the classical humanism committed to lessons of the past, the aesthetic humanism of Schiller was open to possibility. Humanism started as a revolution against scholasticism in the 15th century, but it has become every bit as dry and reactionary. In its search for a philosophy of universal values in the classic literature of the ages, humanism ignored the otherness of the cultures of the past.

An ecological education based on Schiller's ideas could have a whole image of the place of humanity within nature, and not a transcendent view. It could confront the past without the baggage of sentiment and the future without the paralysis of dread. The appreciation of the differences of other cultures could allow us to transcend our present identities. Art would broaden the mental worlds of observers and encourage tolerance and wonder. Education in aesthetic humanism embraces three concepts: liberation, play and community.

1. Liberation. Humanity has taken its own opportunities. These opportunities have been codified for centuries as rights. Now, we must allow animals equal opportunities. The interrelatedness of life dictates the interrelatedness of rights.[17] And these rights are necessary to the integrity of the whole planet. Humanity developed in a community of animals and plants. The quality of human life has always depended

on the quality of animal and plant life. The extension of rights to animals and plants does not deny any traditional human rights. Their intrinsic worth is independent of the instrumental values imposed on them by humanity. Symbiosis is a basic attribute of wild communities. Symbiosis means "living together." The word "ethics" is derived from the Sanskrit word meaning "doing together." Life is more than competition; it involves cooperation, play, and competition. Rights are simply rules for living together.

2. Play. Play is the creation of meaning for an animal that can experience it. The play of life includes youth, health, and challenge, as well as sickness, aging, and death. Science and philosophy are forms of play, attempts to solve the basic puzzles of the universe. Intellectual play is thinking just to think, without having change or revolt as a purpose. Play is imaginative experience, natural learning entered into freely — education should be more like play than work. Most human activity is play, in place within a community of nature.

3. Community. Human beings gravitate into groups to live, but communication across the barriers is necessary for a world community. The wholeness of humanity needs to be affirmed, but from a firm cultural base. The complete surrender of cultural identity is as dangerous as too little openness. Every culture needs its own local, sacred center, which cannot be broken if the group is not to perish.

Education could enlarge or alter the perceptions of all human beings on earth with the selection and presentation of relevant information, to form an ecological consciousness. The survival of society now depends on a consciousness of the global system in its complexity and connectedness. The spirit of humanity depends on a consciousness of its proper relation to the wild places of the earth. Radical ecology is a basis for a science and politics adequate to deal with the creation and maintenance of good places on earth.

Chapter 40

Deep Anthropology: Ecology and Human Order

Summary
A form of radical ecology, called "deep ecology," has been criticized for being anti-anthropocentric and utopian. Most of the arguments against deep ecology, however, are based on an uncritical use of the concepts of anthropocentrism and utopianism. Deep ecology places anthropocentrism, as well as anthropomorphism and anthropometrism, into a proper perspective, which permits understanding of the human relationships with other beings in nature, without accepting unhealthy extremes. The principles of deep ecology are concerned with creating good places (eutopias), rather than nonexistent places (utopias).

Recently, deep ecology[1] — or ecosophy, Naess' neologism — has been castigated by Richard Watson[2] for its 'utopianism,' by Ariel Kay Salleh[3] for its ignorance of feminism, and by Henryk Skolimowski[4] for its 'anti-anthropocentrism.' Yet what are the alternatives? Anthropocentric realism? Reductionist femocentrism? Or, evolutionary anthropocentrism? Each of these sesquipedalian structures hides a limited perspective.

The arguments for and against various degrees of anthropocentrism are based on different interpretations of the term. The misunderstandings can be cleared up with an examination of these sibling terms from a larger perspective. The human place in nature can be explored through the concepts of anthropomorphism, anthropocentrism, and anthropometrism in a framework of anthropology.

The Shape of Knowing

Human images of order (worlds) were wedded to shape (anthropomorphism) rather than to position, measure, or language. Archaic peoples[5] saw human forms in every form of nature. The order of nature was a human order. Natural events, like lightening or rain, had needs and reasons. And these events could be controlled by human rituals, which satisfied the needs or influenced the reasons.

Anthropomorphism gave human beings a place in nature. Other beings were seen as relatives. Anthropomorphism is the only way to understand mother and father and kin. Anthropomorphism leads to an understanding of the ultrahuman.[6] We understand other beings by expanding ourselves, not by shrinking them.[7] This application of anthropomorphism is concerned with the projection of experience into other beings, not with making other beings into humans. The social life of humans and other beings is not separate. Symbolic associations and transformations are made between diverse entities.

Because other beings are related, they have proper places in the order of nature, also. Richard Reichel-Dolmatoff claims that most rituals of the tropical American Indian tribes are concerned with ecological balance (though not necessarily self-consciously).[8] The cosmological myths of the Amazonian Tukano, for instance, do not describe their place in nature in terms of mastery of a subordinate environment. Instead the Tukano learn that they are part of a larger system that transcends individuals. Survival and maintenance of the quality of life are possible only if all other lives are allowed to evolve according to their specific needs, which are described in myths and traditions. Likewise, the ritual cycles of the Tsembaga of New Guinea regulate their impact on the supporting environment.[9]

Toynbee concludes that worship of nature unites people because it is not "self-centered." Anthropomorphic thought increases the dimensions of the human intellect. Language itself has an anthropomorphic base. Yet paradoxically anthropomorphism is limited. It is nonabstract. Humans project human need and thought patterns as guiding forces in nature. Nature, however, is not a human creation, as is the image of nature. Anthropomorphism is a personal interpretation of an order that encompasses all human orders. Anthropomorphism is limited by the range of human experience.

The Center of Existence

An increase in knowledge from the neolithic to classical periods brought about a realization of the vastness and strangeness of nature. Concentration on human interests, anthropocentrism, resulted in greater success for the species. Humans became successful competitors. Then they became instruments of change in ecosystems. With Plato and Aristotle, nature became anthropocentric; it turned around a center: humanity. Humans were the most important beings, at least through the Middle Ages. By fifteenth-century European and Arabic standards, the universe was a rational order. The human place was prominent, and human life had meaning and purpose. Then the Copernican revolution transformed the universe from geocentric to sun-centered and then centerless. The biological universe, however, was still great chain of being where humanity was a link between the beasts and the angels. When Darwin linked humanity too closely with the beasts, cosmology became even less meaningful.

But the industrial, scientific revolution restored human importance by showing the power of human reason. The success of this revolution in modifying the environment to fit human needs almost filled the vacuum of meaning. Nature is seen exclusively as anthropocentric, as a human resource (especially by philosophers such as John Dewey). Even the Biosphere Reserve Program is justified according to anthropocentric use.

Much modern thought is concerned with inflating the uniqueness and position of humanity in the universe. Yet much of the justification for the centrality of humanity is based on a naive a posteriori reasoning, where the explanation proceeds directly from the conclusion.

An anthropic principle was introduced by R.H. Dicke,[10] who thought that a causal connection might be founded on Mach's principle. Mach had proposed that the inertial mass of a particle was determined by its gravitational interaction with distant matter. According to this principle, the weakness of gravity is related to the enormous amount of distant matter;

thus the whole determines the parts. Dicke sees that man has appeared at a "privileged moment," that certain conditions are necessary for human existence: The earth that is necessary for life depends on solar evolution, which depends on the state of the galaxy, which could not have condensed without a high recessional velocity from a primordial 'big bang.' Dicke invokes the anthropic principle, that humans are necessary for the prehuman stages to exist, that the size and complexity of the mathematically improbable universe exist to support human life. This anthropic principle implies the existence of a divine plan. This argument entails a misunderstanding of necessity and existence. Although the current conditions are necessary, it is not necessary that they had to occur. That humanity exists on earth shows only that they did occur, not that they had to occur. If humanity did not exist then the universe would probably be much the same with different species.

Ecophilosophy is also enchanted with the importance of humanity. Skolimowski believes that the universe is as it is, in its size and age, "in order to enable life to evolve."[11] Humans are the "crowning glory" of the universe. "Man is of utmost importance," he asserts.

To assume that evolution necessarily progressed to human beings ascribes an anthropocentric purpose to nature. But for environmental effects, dinosaurs, birds or whales could be the dominant species. Humans are not the unique end or goal. In fact, like new dinosaurs, humans are good competitors, suppressing other species and creating their own pseudo-species.

All fields of study are trying to confirm that man makes himself, according to Paul Shepard,[12] no matter how the world is made. Geography endorses economic determinism; history studies the rise of Promethean civilization; the arts separate abstract qualities from content; socio-anthropology encourages the theme that everything is possible; the sciences posture with value-free facts. Ideas are no longer connected. All aspects of life have become interchangeable, including soil, water and land. All concepts of the natural seem to turn on the definition of human. Humans desire to make everything conform to images of themselves; the "stink" of humanity clings to everything.

All of nature is not human nature, however. There are many other sentient species. All animals, 'two-legged and four-legged,' are equals in the view of Black Elk.[13] Science is only beginning to support this idea. Adolf Portmann shows that every form of life appears as a gestalt, developing in a specific place.[14] All living forms create an image of their environment. Genetics provides the proper image choices for some — frogs, for instance, focus most closely on objects that have the same size and trajectory as flies. Others must learn what is valuable.

Jakob von Uexkull suggests that the unfamiliar world of animals can be represented with bubbles to denote the self-world or phenomenal world of an animal.[15] The world — life-image — is what has meaning for an organism. It is a focus. The first principle of the life-image theory is that all animals from the simple to the complex are "fitted to their unique worlds with equal completeness." A simple world corresponds to simple animal; a well-articulated world to a complex animal. Von Uexkull implies that the human world is only one of the many possible. Animals are not suboptimal beings relegated by evolution to second-rate habitats. They are optimally fitted to places that humans are not.

The theory of life-images is a basis for a new, genuinely nonanthropocentric metaphysics.[16] Natural processes take on an expression of significance of their own without reference to humanity. All things have an ultrahuman value of their own. There are other life-images that are measuring parts of habitats. There are other centers. The center is everywhere.[17] These centers are equals. Humanity is not the center of all. It is not the reason for the existence of all. The entire earth cannot be controlled for only human benefit. Ecosophy is not anti-anthropocentric; it provides a proper perspective for anthropocentrism: humanity is one 'center' among many, and that center is placed in context.

The Measure of All

The universe is not anthropomorphic, in the image of man. Nor is it anthropocentric, centered around man. But it is measured and valued by man, as, indeed, it is measured and valued by all beings. Protagoras recognized the relative basis of mathematics and philosophy, when he stated that "Man is the measure of all things." Indeed, the first physical measurements for building were expressed in human body lengths. The body was a standard.

Measurement applies to all spheres of activity. The human mind filters information from the coevolved environment, from which working models are made. All mental models of the environment or universe (cosmologies) are human models, from that of the Kalahari Bushmen, to the medieval German hierarchy of heaven, to the Newtonian dynamic view.

Measuring social interactions leads to moral codes. The word 'morality'[18] comes from the Latin word for will of the people; the singular meant the 'will' of a person. It was probably derived from the verb 'to measure,' as to measure one's way, to go one's way. Morals means the 'way of going together.' And, in an anthropometric universe, it is entirely appropriate. Measure implies limit. An ecological ethics can address the limited relationships of all beings without becoming entangled in the fuddle of reciprocity or sentience.

Maurice Merleau-Ponty grounded philosophy in the human bodily dimension.[19] The body is the basis for metaphysics, an access to being as well as an expression of it. Philosophy (or ecosophy) is tied to a comprehensive anthropology — everything perceived and expressed is anthropocentric and anthropomorphic. All philosophy becomes human first. Ecosophy can expand the narrow anthropocentric evaluation and see things from the perspective of the whole. Arne Naess offers a biospherical egalitarianism, where all beings have an equal right to life and fulfillment, but without denying necessary human exploitation. Total egalitarianism is impractical, even for Jainists. But ecological egalitarianism considers the beings of different species in context. Complex beings may be more valuable on a one-to-one basis, but have less value on a regional basis.[20] For example, the Snail Darter was certainly more valuable than the excess power generated by the

Tellico dam, a temporary luxury (dams only last 10-200 years). Microbes are more important to ecosystems than tertiary predators.[21]

Certainly, nature is too subtle and intricate to be reduced to variables in the human calculus of development. The answer is caution, preservation of other beings in place, preservation of natural areas. Naess understands that human reference may be the rod that measures, but what is measured may be greater than humanity and its survival.[22]

Putting It All Together

Anthropomorphism is a necessary human way of knowing; all knowing is based on it. But the knowledge is not limited to just human experience. Anthropocentrism is the natural centering of human experience. But humans are not the only centers of experience. Anthropometric behavior is the statement that humans are the measure of all things. But humans are not the value of all things. Not everything can be measured. But everything can be put together in a metaphorical language. Metaphor permits language to carry beyond direct reference, to extend meaning. The word 'logic' means putting together.[23] Anthropology is the putting together of diverse human worlds embedded in unique places.

All three concepts dealing with shape, center, and measure are needed for human knowledge. By rejecting anthropomorphism, the experience of others is restricted, and the scope of self-knowledge is reduced. Narrowness of experience is a source of human insecurity. By rejecting humanity as its own center, the experience of selfness is suppressed. The self is the basis for exploration and success. By rejecting measure, perspective is lost. And if humans claim all value for themselves, the term is meaningless. Deep anthropology includes the full spectrum of human knowledge and interaction with all. The proper study of humanity is all beings.

Humanity is embedded in an ecological world. Ecosophy attempts to preserve the balance of humanity with other diverse species. Balance is an ecological value, as is flexibility and richness.[24] If we diminish the richness of the earth and let its balance readjust as a desert planet, we will wipe out ninety-nine percent of the human population as well.

Richard Watson emphasizes that humans have the right to thrive, as do all animals. But he does not distinguish between thriving and the interference pandomination of all ecosystems for human interests. Being powerful may permit electric lights over miles of deserted malls, but being smart does not mean converting every possibility for power into a temporary technological marvel. Possibly Watson is correct in saying that few ecologists have a good comprehension of economics, but virtually no economists have any comprehension of the ecologies upon which economics are based, and industrial civilization is only just learning the costs of such ignorance.

Both Watson and Salleh complain that ecosophy maintains a "hands-off" position that sets humanity aside from nature to allow nature to go its "separate" way. They misunderstand. Humanity does not belong in all ecosystems. Many plants and animals cannot cohabit with humanity. This does not mean that humanity is apart from nature, just restricted to some ecosystems. The limitation of human population and impact on wild ecosystems is not contradictory to life-affirming values, as Salleh argues. It is necessary for the development of ultrahuman species. Species on which humanity depends. The adventure of life has always been the creation of new forms, new experiences. It is foolish to say that humanity cannot be self-limiting, as Watson does, because it never has limited itself before, just as it was once foolish to say that humanity could never dominate every ecosystem, reach other planets in the solar system, or invent weapons that can destroy the biosphere, just because it had never been done before.

Skolimowski has presented the return to hunting and gathering societies as a 'gospel' of ecosophy.[25] This is an exaggeration. Hunting and gathering should be an option for some cultures and some people. But it should not be the only one. Human populations are plastic and could probably be decreased without fascism, by economic, religious, or cultural means.

Watson judges the ideals of ecosophy to be utopian. He objects to 'factual incorrectness' in the presentation of the ideals of ecosophy and finds practically no evidence for the view that the human species has ever refrained from disrupting the environment or killing for benefit. But he ignores most archaic cultures. Some cultures, including the Aranda of Australia and the Mbuti Pygmies of the Ituri Forest in Africa, have lived in ecological balance for tens of thousands of years. Indeed, many are fighting industrial cultures for the privilege of continuing to live in balance. His realistic view is severely limited to modern, twentieth century.

Surprisingly, Watson has not seen that his list of modern 'ologies' confirm that most humans live in and prefer small, low-technology communities (there are just too many small communities). Large cities have passed the level of complexity where they can provide a good environment or exist in harmony with the surrounding land. Not to realize this is to project a dream world populated by machines and moorlocks, which is worse than "Peter Pan" and the choir of angels that Watson attributes to the ecosophical vision. It is true that not everyone in an overpopulated world can live in small, self-sufficient communities. But many do, on every continent. There is no commandment in ecosophy that every human group should live in small communities. Skolimowski, Cobb, and others have praised the arcologies of Paolo Soleri as an appropriate solution for large groups.

Placing humanity at the center of the universe makes a cosmology unsustainable. The modern industrial cosmology that dominates most human cultures is creating flatscapes, 'no places.' And that is the true meaning of utopia, no place. With changes in consciousness, with understanding of the ecological relationships of human cultures to wild habitats, and the moral relationships of all beings, real eutopias, good places, can be created.[26] It is the role of ecosophy and other synthetic disciplines to encourage these changes.

Chapter 41

Metaphysical Implications from Physics and Ecology

Summary
Metaphysical implications from physics and ecology are contrasted and
compared through two concepts, the field, which was developed primarily in
physics and borrowed by ecology, and wholeness, which was first postulated
in ecology and borrowed by physics. Several implications from physics are
found unacceptably reductive or erroneous. An old and a new ecology are
identified. Metaphysical implications from the old ecology are quite different
from the new ecology, as well as from Quantum or Newtonian physics.

Recently, in an article in *Environmental Ethics*, J. Baird Callicott[1] addressed
the principal metaphysical implications of physics and ecology in his
presentation of a historical outline of ideas. Some of his conclusions,
however, are erroneous. This paper examines several areas of convergence
between physics and ecology, and then discusses how the metaphysical
implications of ecology are different from those of physics.

Callicott states that ecology has emerged only recently as an
independent, quantitative science, and that it is less foundational than
physics or cosmology. In fact, ecology is one of the oldest disciplines, not the
youngest. Early in their development, human beings realized the value of
recognizing edible plants and animals and their interrelationships, and they
built up traditions of knowledge. This practical ecology has been obvious for
a long time, but because it is subtle and complex, it is not easy to quantify,
and its development as a science is recent. This practical ecology enabled
some cultures to achieve long-term stability in a natural environment; it
also embodied teleological and holistic concepts expressed in qualitative
terms rather than in mathematical forms. As a young science, these practical
concepts were formulated as various principles: wholeness, the relationship
of complexity and stability, succession and climax states, and the balance
of nature. As ecology became more quantitative, that is, mathematical
and reductionistic, its methods and topics more reflected the old physics.
The new ecology, according to C.H. Waddington, places emphasis on the
discreteness of individual genes, the randomness and nonrelational nature
of the process of mutation, and the unimportance of the experience and
reaction of an organism to its environment. Old ecological principles were
rejected. D. Simberloff, for instance, even argues that the ecosystem model of
A.G. Tansley is only another way of formulating the `balance of nature' and
must be rejected. Similarly, holism is unacceptable to the new ecology and
is replaced with an individualistic view. Thus, the new ecology rejects the
ecosystem model that Callicott places at the center of his argument.[2] Many
of Callicott's arguments are taken from the old ecology and not the new,
quantitative ecology. Henceforth, it is basically the old ecology that will be
treated. Leopold, Naess, and others belong to the old tradition and not to the
modern science.

Physics and Ecology

Callicott observes that ecology and the new physics converge on some similar metaphysical notions, such as field and wholeness, the former from physics and the latter from ecology. These notions are examined in more detail.

Field

The field concept was introduced by Michael Faraday into studies of electricity and magnetism. Albert Einstein states that "fields are physical conditions of space."[3] He identifies the underlying unity of nature as space. The field concept is central to the unification of theories of light, electricity, and magnetism. Einstein's general theory of relativity replaced the distinct categories of space, time, energy, and matter with multiple components of a unitary field characterized by Riemannian geometry. This four-dimensional model accounted for the phenomena of mechanics and electrodynamics. The field is still the fundamental primitive concept in the quantum electrodynamics of Paul Dirac. This Space/Time/Energy/Mass (STEM) field has general characteristics of discretion, participation, connection, consistency, limitation, wholeness, self-making, self-ordering, individuating, and developing.[4] No one component is ontologically subordinate to another. Clearly, Callicott is mistaken to present matter as subordinate to energy.[5]

One of the most important properties of the field in physics is the participation of the observer. John Wheeler suggests that the universe is brought into being by participation — the sum of an infinite number of elementary acts of observation (although he unnecessarily reverses the causation). Quantum mechanics eliminates the notion of the neutral observer behind glass. The observer participates in the act of measurement and the universe changes by the act of measurement. A similar principle of participation can be postulated for ecology. Not only do organisms participate in the field of nature by virtue of their existence, but the experiments of ecologists alter the system being studied, often degrading it for a period of time.[6]

Biological field theories dealt with problems, such as regulation and reproduction, thought to be insoluble in mechanistic terms. In the 1920s, Aron Gurwitsch and Paul Weiss independently advanced the idea of developmental fields to account for the properties of wholeness and directedness. In Weiss's theory the field became a system of organizing factors that proceeded from organized parts to developing regions, resulting in typical patterns. Growth and pattern became emergent field effects. For Weiss, field was a symbolic term for the dynamics underlying the ordered behavior of a "collective;" it denoted the properties lost in the process of analysis.[7] Waddington extended the field to an epigenetic landscape, but regarded his own use of field as a descriptive "convenience." Living fields have form and impose restrictions on the probabilities of nonliving fields.

Paul Shepard describes living natural "objects" in terms of events which constitute a "field pattern." Callicott concludes that relations are then prior to things and that the characteristics of species result from adaptation to a niche in the environment.[8] But, relations are not prior to objects; they

239

arise together. The wasp and the yucca coevolve; they are not co-linked by prior relations. Furthermore, a specimen is more than a sum of its species' relationships to an environment; it is an intentional being that, with other members of the species, can create niches, as well as adapt to them. Because the STEM field produces life, the qualities of life cannot be separated from its physical qualities. While it is true that living subjects are at a different level of description than events in field patterns, they should not be treated as ontologically subordinate. All of the aspects of the field have equal status. The ecosystem model, as a reaction to "superorganismic" metaphors of early ecologists, attempted to be a field theory, but has been limited by its parentage, thermodynamics, and rejected by new practitioners.

Some exponents of the new ecology, misreading the new physics, consider energy to be a more fundamental reality than material objects. These ecologists apply classical thermodynamics to living systems. Unfortunately, entropy is an incomplete explanation of living systems.[9] A living system is a natural phenomenon and all natural phenomena are constrained by the laws of thermodynamics. As W. Gibbs stated, the generality of entropy extends well beyond the border of thermodynamics. Entropy, as R. Clausius defined it, is the constant transformation of motion into heat. Classical entropy is a thermodynamic variable of a system, commonly a measure of the unavailable energy in a closed system. The classical concept is concerned only with macroscopic states of matter, the qualitative change of free energy into bound.

Ludwig Boltzmann applied statistical mechanics toward interpreting the properties of the microscopic systems that make up the macroscopic ones. Statistical entropy is defined as a measure of the number of ways in which the elementary particles of a system may be arranged and determined by counting the number of microstates. Statistical entropy became a measure of disorder, described mathematically in six-dimensional phase space. This application not only reduces the level of explanation, but changes the coordinates from heat and temperature to space, time, and force; it explains thermal equilibrium as the result of an undefined universal shuffling process. A process opposite and complementary to the shuffling of entropy – a sorting process – has been proposed independently a number of times, from Georg Hirth to Schrodinger, Szent Gyorgi, Woltereck, and Whyte, but not developed adequately.[10] With the connection of entropy and information by Claude Shannon, the concept of entropy has been extended to applications in engineering, art theory, Gestalt psychology, and cosmology. The assumptions of statistical thermo-dynamics – chance motion, the independence of particles, time reversibility, ahistoricity, continuity, and equilibrium – are problematic for physics, much less for ecology and human civilization. Time reversibility, for instance, has never been observed in macroscopic systems.

This use of entropy seems to be little improvement from the mechanics of the old physics (and presumably the new ecology) for understanding organisms, which are reduced to "energy moments" rather than "atoms." In wanting to combine quantum theory with ecology, Callicott describes organisms as being configured by energy through time. But, organisms are material patterns in space as well. The focus on either frame permits subtle differences and limitations in interpretation. Even if energy is considered primary metaphysically, organisms are still composed of the atoms and molecules that energy forms under certain conditions of temperature and pressure, and they act differently than just `energy vortices' or `patterns of energy.'

Reliance on physical explanation impoverishes the complexity of ecological reality. Morowitz's[11] portrayal of each living thing as a dissipative structure is reductive and distorting. Ilya Prigogine defined a dissipative structure as one of two[12] types of organization, whose order is governed by amplified fluctuations; his examples include walls and slime molds. However, Prigogine misuses[13] the concepts of order and complexity by making them dependent on random events; furthermore, his concept of stability assumes a reversibility of biological time (the result of its basis in quantum mechanics). Dissipative structures are more applicable to pans of boiling water than to black bears. The reduction of ecological patterns to dissipative structures ignores observed behaviors like communication and intention.

In describing a general concept of nature, Callicott follows Prigogine in emphasizing process over structure. Each assumes that energy flow is more primary than matter, that energy is a more fundamental reality than discrete entities. Although an organism may be characterized as a `configuration of energy,'[14] that is an artifact of the quantum perspective (and, perhaps, of the desire for an absolute reality). There are philosophers and ecologists, such as Ramon Margalef,[15] who consider `information' to be more basic than energy or matter and more in-line with patterning. Thus, organisms are reduced to information in a cybernetic perspective. These perspectives are useful to an understanding of complex behavior, but sometimes ecologists and philosophers simply take over the vocabulary of a paradigmic trend, new physics or information theory, and apply it uncritically to the epistemology of the older paradigm. Thus, 'efficient cause' can disguise itself as a `genetic program.'

As the new physics has transformed the mechanical picture by placing atoms in a field that accounts for the qualitative emergence of properties from simple quantities, so the new ecology has placed living `objects' in a field. This field determines the limits of any ecological field of activity, and no field of ecological activity can be described without taking the physical field into account. But physics and ecology cannot be equated, as Callicott, Morowitz, and others try to do. Ecology has principles that cannot apply to particles. The field is living and intentional, as well as physical, and an ecosystem model based exclusively on physics cannot account for intention or other emergent properties.

Wholeness

Gilbert White characterized nature as `one organic whole' and influenced Darwin and subsequent generations of biological scientists. E. A. Birge's early work on the heat budgets of lakes, for instance, was holistic.[16] J. C. Smuts, in trying to synthesize the evolutionary theory of Darwin and relativistic physics of Einstein, presented the whole as a powerful organizing principle inherent in nature. L. von Bertalanffy and E. Laszlo extended holism with general systems theory. David Bohm suggests that the universe as a holomovement carries an implicate order that is undefinable and immeasurable.

The concept of nature that emerges from ecology and the new physics is more holistic, as Callicott recognizes, but he does not sufficiently explicate the levels of organization of wholes.[17] Each level is real as a whole; it is a whole, or a `holon' in Arthur Koestler's term.[18] A number of ecologists use the concept of holon to describe the organizational levels of hierarchical systems.[19] Given that nature is a structured and differentiated whole, the character of the organism or particle is determined as a subwhole. According to Koestler, all complex structures and processes of a relatively stable character display hierarchical organization. Levels of a hierarchy tend to be contained in subassemblies. Each subwhole behaves as a whole to its components, as a self-contained whole, and as a dependent part in context. Wholes and parts do not exist absolutely. There are intermediary structures on a series of levels in an ascending order of complexity; each subwhole faces in opposite directions. These Janus-faced subassemblies are holons; a holon is any stable subwhole in a hierarchy that displays rule-governed behavior and structural Gestalt constancy, to paraphrase Koestler. The rules lend order and stability, as well as flexibility.

Wholes are mutually defining, but also self-defining or self-making. Nature is a self-making system; species and organisms are self-making. The ontology of any living system is the history of the maintenance of its identity through continuous self-making, or autopoiesis.[20] The evolutionary stability of the subassemblies — organs, organisms, species — is reflected by the degree of autonomy (self-government) each has, according to F. Varela. The system develops through a continuous dance of autonomy and control; autonomy represents generation, internal definition, internal regulation, and self-assertion, whereas control represents consumption, instruction, assertion of other identity, and external definition. Furthermore, the holistic nature of the STEM field eliminates the unsatisfactory notion of the priority of relationships to beings or of wholes to components — a notion Callicott is inclined to use.[21]

Discussion

The new ecology is a central dimension in biology and overlaps many specialties. It deals with different levels of a hierarchy, focusing on organisms, populations, communities, and ecosystems, but with attention to genetics as well as to geological and evolutionary events. Although ecology is a scientific newcomer, it is certainly a foundational science (especially in view of Callicott's own statement that the essence of a being is determined by its relationships). Any science that studies those basic relationships is foundational.

If the old ecology were truly linked to the new physics, at least in the terms of field and wholeness, it would have a much different flavor than that presented by Callicott. For example, the field has a historical character; a whole is self-making. The implications of these attributes are neglected. Ecological principles are not the same as physical laws. Physical operations necessarily apply to biological systems, and ecological theory must be consistent with physical laws. But biological systems exhibit unique regularities that are not reducible to lower levels of activity or understanding. Although the ideas of ecology to some extent parallel the ideas of physics, ecology lacks the laws and constants of physics.[22] Physics lacks the high-level predictability of ecology. The ideas of ecology and physics can benefit from cross-fertilization, but ecological ideas are not reducible to physical ones.

A metaphysics inspired by ecology would be larger and more comprehensive that a metaphysical consensus from the Newtonian paradigm or from quantum physics. Such a metaphysics would provide a place for rarity and diversity, as well as perhaps a reason to value it. It would include ultrahuman and human interests. Because of his indiscriminate use of old and new physics and ecology, Callicott's implications for moral psychology are anthropocentric and teleological. The use of Shepard's arguments for the preservation of species — as educational devices — is anthropocentric and incomplete. Similarly, Callicott's statement that the environment becomes "fully actual" with humanity implies that human consciousness is the goal of evolution.[23] The environment may be extended with consciousness (human and other), but not completed. Callicott has made a good, tentative start. But, a more rigorous list of the qualities of ecology is needed, before anyone can begin an exploration of their implications for human behavior.

Chapter 42

Metaphysical Principles from Ecological Foundations

Summary
The processes and levels of ecology are discussed in terms of individuation, interaction, and evolution. Ecological principles are distilled and linked together in the concept of biological maturity. Hypotheses and norms inspired by these ecological principles are presented and described within the framework of ecological philosophy. These hypotheses and norms for the basis of an ecocentric metaphysics, epitomized by self-realization.

The Levels and Processes of Ecology
Ernst Haeckel's 1869 definition of ecology as "the total relations of the animal to both its organic and inorganic environment" was notably broad. Ecology deals with the highest levels of biological integration, from organisms to the ecosphere. Autecology, for example, is the study of individual organisms in an environment. A group of individual organisms of the same species in a particular place is studied as a population. The assemblage of populations of different species in a habitat is studied as a community. The community in its biotic environment is studied as an ecosystem, and ecosystems comprise the ecosphere. There are emergent properties at each level of organization, such as the diversity of species or the structure of the food web. Furthermore, each level of integration has distinct attributes. A population has a proper "density," the number of individuals per unit area, which is not applicable to individuals; a community has "species diversity," which is meaningless at the population level; processes such as homeostasis or homeorhesis,[1] which involve a relationship with the environment, occur on an ecosystem level. In sum, each level acquires additional characteristics. Three processes — individuation, interaction, and evolution — are addressed in more detail.

Individuation. Even if individuals can be described in terms of vortices, as the poet Pound did before the physicist Prigogine, they do exist materially, and they participate in the field in which they exist. The organism must adapt to the environment, which implies having a memory and being capable of learning; the organisms must also reproduce, that is, duplicate its pattern in a separate being. Organisms are goal-seeking, and often stability is sought above change or complexity. The individual is a subject centered in a milieu. Because of this implied point of reference, Rodman[2] concludes that ecology is teleological. Often, organisms strive for well-being beyond just survival. Their goal is to come into the fullness of being; A. N. Whitehead[3] considered that all organisms have three urges: To live, to live well, and to live better. Living better is being more attuned, stimulated, receptive, flexible, spontaneous, and integrated in a milieu.

Each participant in a field creates an image of nature — or world — from what is meaningful to it. J. Von Uexkull[4] suggests representing these unfamiliar worlds with a bubble model. The life-image, or umwelt, of an animal is what has perceptual and operational meaning for the animal. All animals are fitted to their unique worlds with equal completeness — simple animals to simple worlds and complex ones to well-articulated worlds. Each is optimally fitted to its habitat.

Interaction. The individuals in a community engage in interactive behaviors, which can be considered positive (+), negative (-), or neutral (0) in effect for each individual or species in relation to another. These interactions are described as:

neutralism (0,0)	charity (0,+)
competition (-,-)	
amensalism (-.0)	altruism (-,+)
parasitism (+,-)[5]	
predation (+,-)	interference (+,-)
commensalism (+,0)	
protocooperation (+,+)	
mutualism (+,+)	

Most interactions are not simple, but are complex and paradoxical because of the integration of levels. For example, parasitism that is detrimental on an individual level may have benefits on the species level, by influencing reproduction rates or resistance to diseases. Some of these interactions are poorly defined in the literature. Predation, for example, is regarded as the adverse effect of one population on another, while being dependent on it. Yet, predation can benefit both species; it may be more mutualistic than parasitic in character. The predator/prey are not excluding opposites, but generate a whole unity on the community level, where there is stabilization and survival for both species, according to M. W. Fox.[6]

Predation increases the survivability of two species. In Caswell's open nonequilibrium model,[7] the incorporation of predation results in an indefinite coexistence of species — extending the extinction of either indefinitely. Predation opens up cells for colonization by inferior competitors (gaps caused by physical disturbance can also have the same effect). Predation increases the diversity of species in a community. Steven Stanley[8] argues that predation, or cropping, may have controlled the evolution of metazoans. In communities of primary producers, a few species can monopolize a place. A predator dominates its favorite species, usually the most populous, thus limiting it so that other species can develop to claim a niche. Stanley explains the Cambrian explosion of life through the evolution of cropping herbivores, which "opened space" for a greater diversity of producers, which in turn permitted more specialized croppers, for which specialized predators evolved. Each new level of the trophic pyramid broadened the one below; the pyramid became wider and higher.

Competition was once considered the basic interaction between individuals and between species, from Darwin to Birch, Rodman, Lehman, and others. But, cooperation is seen now to be as effective a strategy as competition and as necessary. Survival of the fitter is correct only to a certain point, then it becomes survival of the more cooperative. Both old studies (Reinheimer and Kropotkin) and new research (Lorenz, Fox, and Schaller) stress the primary use of cooperation both within and between species instead of unrelieved struggle. As Naess says, "live and let live" is a more powerful principles than "either/or." Cooperation creates communities of many species in which competition is necessary but limited. Neil Evernden[9] notes that organisms often go to extreme lengths to avoid direct competition; species form a spectrum of attempts to share the life base without risking their health. This diversity enhances the potential for survival.

Many species in communities live closely together; this constant, intimate relationship between dissimilar species was identified as "symbiosis" by Heinrich De Bary (1879). Biological symbiosis results in a greater store of genetic information — in a new species. Lichen, for example, is composed of fungus and algae; each is part of the milieu of the other in a necessary and beneficial relationship. In this paper, symbiosis is also used in a philosophical, etymological sense, to denote living together, which involves all kinds of interactions. from competition and conflict to cooperation and mutualism. No one interaction can dominate others without unfortunate consequences. If all plants were mutualistic and none were competitive, it is unlikely that trees and flowers could have evolved.

Evolution. The process by which species reorganize their structures to adapt to their environment is called evolution, an integrated, partly open process that selects whole individuals in whole environments. Evolution can be said to flow upward and outward, as well as inward and downward, from the simple to the complex, but also back again. For moths that mimic bark patterns, there are others in the same area that are conspicuous; for herbivores with complicated stomachs, such as deer, there are those with simple stomachs, such as elephants or horses. It cannot be shown that a particular evolution took place by necessity, only that an adaptation had value and the species survived. The consideration of observations can lead to the conclusion that evolution is converging to a single end. Darwin himself did not want anyone to consider evolution a purposive movement towards a goal. Rather he regarded evolution as a bush, growing where it can. Evolution can be considered as a building up of complexity, or an unfolding of patterns, in Merleau-Ponty's term,[10] a "pattern mixed-upness" of styles of living , as beings radiate through time and space. Evolution does not seem to be a hierarchical ladder or an up escalator, but a history of forms adapting to changing environments.

The adaptation of beings to a changing milieu cannot be perfect. Over-specialization reduces flexibility and the ability to change, but underspecialization reduces efficiency. Beings that survive tend to have a satisfactory level of specialization. Beings that are not optimally (or satisfactorily) adapted are eliminated through competition and stress. Evolution can increase the levels of complexity through the operation of natural events.

Species are defined by their position in the environment and thus are in internal relations within the environment,[11] but it is also true that they define the environment through positive and negative feedback. While some species adapt to a niche, others create new niches. J. Baird Callicott presents species as too passive (in his article in *Environmental Ethics*),[12] regarding a specimen as "a summation of its species' historical, adaptive relationship to the environment." The specimen is much more than this — it is intentional and flexible, sometimes stress-seeking and maladaptive. Species in their milieu are in dynamic relationships.[13] While relationships are as real as the organisms,[14] the relationships are not necessarily or logically prior. The whole part, or holon, creates the whole as the whole creates the part. The organism creates the ecosystem as the ecosystem creates the organism. The multiplicity of beings and relationships create and are created by the field.

Ecological Principles
Ecological principles are not the same as physical laws. Physical operations necessarily apply to biological systems, and ecological theory must be consistent with physical laws, but biological systems exhibit unique regularities that are not reducible to lower levels of activity or of understanding. Although the ideas of ecology to some extent parallel the ideas of physics, ecology lacks the laws and constants of physics.[15] Ecologists, such as Odum, Hardin, and Margalef present sets of basic principles and concepts that can be said to represent the science of ecology. These principles are incorporated, expanded, and synthesized here on four levels: The individual, population, community, and ecosystem.

I Individual
  • The individual is the unit of experience and reproduction
  • The size of an organism is related to its metabolism
  • The organism is inseparably related to its habitat, the place where it lives
  • The niche of an organism depends on what it does in its place
  • Many organisms identify strongly with place; the bond can mean life or death
II Population
  • The population is the unit that evolves in nature (according to Krebs)
  • A species population has unique properties, such as density, mortality, natality, potential, dispersion, age distribution, growth form, and structure (isolation , territoriality)
  • Populations interact in neutral, positive, or negative ways
  • Competition limits the number of species in a niche (the competitive exclusion principle)[16]
  • Cooperation can increase the number of species in a niche (within limits)
III Community
  • The community is the level of survival
  • Diverse organisms live together in an order
  • Communities are named by structural features such as dominant species
  • Communities are stratified
  • Communities have a diversity of species
  • Communities are characterized by rhythmic changes in the activities

of organisms which produce regularly recurring changes in the community (periodicity may be daily, lunar, seasonal, genetic, or climactic)
- Communities replace one another in an area in sequence by an orderly process of change called succession[17]
- The final community in a successional series is self-perpetuating and in equilibrium with the physical habitat (that is, the energy/material budget is balanced in the mature community)

IV Ecosystem
- The ecosystem is the level of integration and the unit of organization undergoing a directional development (after E. Odum)
- In an ecosystem, energy is bound into organic matter, measurable as productivity
- That quantity of matter and energy no longer of use to the ecosystem is wasted
- Chemical elements, especially those used by living organisms, circulate in the ecosphere in characteristic paths known as biogeochemical cycles
- Living organisms are limited by elements and physical factors, e.g., light, water, gas, salt; too little of an element limits an organism (Liebig's law; and too much of an element also limits an organism, Shelford's law of tolerance)
- The transfer of energy and materials through organisms is referred to as a food chain
- The interaction of individuals in a food chain results in the trophic structure of communities (as ecological pyramids)
- The energy required to maintain an ecosystem is inversely related to the complexity of the system; succession decreases the flow of energy per unit biomass (Margalef's concept of maturity)

Discussion

The concept of maturity incorporates many of these principles and is important to the understanding of complexity and diversity. Ramon Margalef proposes maturity as a quantitative measure of the pattern in which the components of an ecosystem are arranged. The life-form communities and physical elements are related in a definite pattern, which is a real but untouchable property (structure). In general, this structure becomes more complex as time passes (or rather the entire STEM changes), as long as the environment is stable or predictable. The structure acquires a historical character. Maturation, as a function of historical processes, increases the levels of complexity of an ecosystem.

The structure is based on material and energetic exchanges. The matter present is biomass (B); the material output is primary productivity (P). Their relation (P/B) is the flow of energy per unit biomass. More mature systems have a richer structure and a lower productivity per unit biomass. There are more steps in the trophic pyramid.[18] There is higher efficiency in every relation. The loss of energy is less, so less energy is needed to maintain the system.

Any ecosystem not subjected to outside disturbance changes in an orderly and directional way: The complexity of structure increases and the energy flow per unit biomass decreases. The physical environment limits the type of change. Homeostatic (or homeorhetic) mechanisms protect the system from many disruptions. Thus, maturity is self-preserving.

This concept of maturity, as an attribute of a community, is related to structural complexity and organization. Maturity increases in time in an undisturbed community. The species diversity,[19] that is, the information content, of a community also increases with maturity, leading to a more complex spatial structure. The energy in a mature system goes more to the maintenance of order and less to the production of new materials. In general, diversity is higher, and life cycles are more complex; symbiosis between species increases, and nutrients are conserved. Complexity and diversity offer advantages for living forms. Complexity allows increases in size, which allows the colonization of harsher environments. Diversity allows more effective behavior through specialization; for example, a specialized organelle may digest less common molecules.

But, Odum points out, as some communities age, Wisconsin forests for example, there is a decrease in diversity (in the understory anyway). Also diversity can decline with productivity, as in the eutrophication of lakes, for instance. While it is meaningful to speak of an optimum diversity, as the result of limits and the interaction of many factors, a maximum diversity may never be reached in an ecosystem.

Conventional wisdom[20] holds that increased complexity in a community leads to increased stability. But, in the 1970s, work with mathematical models tended to support the reverse, that complexity leads to instability. Robert May constructed simple mathematical models[21] concerned with local stability, in which an increase in complexity lead to a decrease in stability. His connection, however, may have been a mathematical artifact, since his food webs were randomly assembled and sometimes unreasonable. May admits that his arguments are only true for mathematical models and that "things may be different in the real world." In the real world, ecosystems are the result of historical processes that are mathematically atypical. Furthermore, real communities are not randomly structured. A system drives to a nonequilibrium state as a mature ecosystem. The adaptively reorganized system is not necessarily more stable, but it is optimally resistant to the outside conditions that elicited the self-organization, a natural normalization process. The ecosystem learns the changes, periods and seasons of the environment.

The structure of food webs may enhance stability. May suggests that communities are more stable if they are compartmentalized (as holons perhaps), that is, if there are subunits where interactions within a unit are stronger than interactions between units. Ecologists have observed that complex communities have existed for thousands of years in stable environments, although many of them are now vulnerable to human interference. The idea that diversity promotes stability has been attacked and defended elsewhere in detail.[22]

Ecology and Philosophy

Arne Naess suggests some of the metaphysical implications of ecology when he called the metaphysical dimension of ecology "ecosophy"[23] — a philosophy of ecological harmony not to be equated with the Deep Ecology Movement, despite the desire of many commentators to do so. Ecosophy is a "total view" inspired by ecological principles and ideas. This total view is a larger philosophical view than that implied by the new physics or the new ecology (movements). Certainly it is larger and more comprehensive than the metaphysical consensus from the older Newtonian paradigm or its equivalent in natural history. Like any new view, it includes the old paradigm as a "special case," leaving it useful in special circumstances. The acceptance of an ecological view has few effects on a physical level of explanation, for which the mechanical model is still adequate.

The ecological principles listed here can serve as a source of inspiration for the philosophical hypotheses and norms of an ecosophy. A complete formulation of an ecosophy is impossible, perhaps meaningless, due to the complexity of living organisms and to the uncertainty of chaotic events. A general model may be made, however, in the form of a truncated pyramid, whose top of hypotheses and norms is supported by a broad base of individual actions and decisions. From a logical point of view, decisions are derived from norms and hypotheses. A small number of abstract formulations result in many concrete, practical actions in unique situations. Norms are derived historically from motivations and impulses.

The pyramid is a stable form. The rejection or modification of a lower level norm or hypothesis does not destabilize the pyramid but results in modifications or adoption of a different specific interpretation. Individual actions can be inducted to norms from which other actions can be deducted. Following Naess's format, hypotheses are indicated with a period, and norms are indicated with an exclamation mark to designate an imperative mood as a special case.[24] The exposition of this particular form of ecosophy, which we shall call Ecosophy A, using this model, results in the following hypotheses.

Hypotheses
- Individuals are related to places by basic physical factors
- Places provide limits
- Places and organisms shape each other through feedback
- Organisms, as well as higher levels of organization, are self-making
- Life-images off meaning and value to individuals
- Organisms strive to live well
- Species interact and are interdependent
- Communities develop historically
- The diversity of life increases the potential for self-realization
- The complexity of life increases the potential for self-realization
- Symbioses increase the potential for self-realization
- Higher self-realization results in deeper identification with others
- Higher self-realization depends on the self-realization of others

The following set of norms is derived from the preceding list of hypotheses.

Norms[25]
- Nature!
- Levels (hierarchical)!
- Individual organisms (whole beings)!
- Organisms have intrinsic value!
- Organisms participate in populations, communities, and ecosystems!
- Limitation (local/global discontinuity)!
- Mutual historical adaptation (evolution)!
- Interdependence (interactions)!
- Pattern unfolding (complexity)!
- Diversity of species (of experience)!
- The diversity of organisms has intrinsic value!
- Self-realization for all organisms!
- Self-realization!

The precise formulations (the first eight hypotheses) refer directly to ecological principles. The hypotheses of self-realization potential are biologically colored but have a metaphysical dimension because of the inclusion of value and self-realization. The last two hypotheses are primarily metaphysical in character. The norms are assertions based on the hypotheses. They also proceed from ecological theory to metaphysical principles. Decisions from hypotheses and norms let us act accordingly to preserve ecosystems and species, encourage ecological diversity, set aside habitats, plant trees to restore habitats, and help others.

Organisms and places shape each other. The life-images of organisms limits the goals of organisms. Organisms interact and develop in time. As populations change, complexity increases. Networks of activities increase interactions; relationships become more complex. Species adapt; sometimes their structures become physically more complex. But, complexity is limited by effectiveness. Often, in the case of gripping limbs, for example, simplicity allows more flexibility and generality.

The concept of complexity is often used to support the judgment of higher and lower animals. Higher functions in "higher" animals result from complex differentiation of tissues. Increases in physical complexity often confer advantages to species in terms of added functions or adaptive values. But, there is an optimum of complexity at each level of being that has nothing to do with importance or being "higher." As categories, higher and lower are anachronistic with their connotations of superior and inferior.[26] Darwin acknowledged,[27] "Never use the words higher and lower." The Aristotelian hierarchy of high and low on the scale is gradually being altered. The classification hierarchy itself (modified by Linnaeus) is also being modified.[28]

The middle hypotheses introduce the refinement of self-realization potential. Self-realization is a generalization of the psychological and sociological potentials of individuals, groups, or institutions, but restricted by norms. It has a metaphysical character. The variety of organisms with

251

different capacities add qualities to the whole. Each individual organism contains an indefinite number of potentials that can be released. From the diversity of life comes a diversity of potentials.

Self-realization is a norm formulation in a metaphysical sense. The conceptual bridge from Self-realization to a positive evaluation of diversity, complexity, and symbiosis, as well as other ecological principles is furnished by self-realization potentials; the realization of potentials increases the Self-realization of the Earth — life-images, not just man-images. Each species has an equal right to live and develop, free from interference.[29] Diversity, complexity, and symbiosis are all necessary norms for the accomplishment of self-realization. Furthermore, the realizations should be qualitatively different. The numerical abundance of one life-form, such as rats or humans, is not equivalent to diversity. A single being, even a human being, cannot realize the goal in itself. The plurality of potentials is crucial and introduces plurality into unity.

Each being also mirrors the whole of life, much as microcosm mirrored macrocosm in the Renaissance and much as a hologram can be produced from any part, although at a reduced level of resolution. The part is not separate from the whole; it is essential for the existence of the whole. That more potential can always be realized implies a continued evolution at all levels of complexity from protozoan to human.

The hypotheses and norms dealing with self-realization are ultimates, since they are not logically derivable from others in this exposition. The Self-realization norms are logically derived from the self-realization hypotheses, however. Self-realization is a logical ultimate in this exposition of an ecosophy. The term, capitalized, includes personal and community Self-realization, but is generally conceived to mean an unfolding of reality as a totality. Thus, it is a process as well as a goal of perfection; self-realization may be expanded to ego-realization, self-realization, and Self-realization.

The prevalent usage in utilitarian thinking equates self-realization with self-interest and self-expression (stressing the incompatibility of individuals in a sort of competitive exclusion), here labeled ego-realization. By comparison, other trends, such a Spinoza's idea of self-preservation,[30] are based on a hypothesis of increased compatibility of individuals as a result of their maturity. Maturity allows the development of the narrow ego of a child into the comprehensive structure of an adult human being. The capitalized concept refers to the development of a deep identification with all life forms; this concept is known in the history of philosophy under various names: the universal self, the Atman, or the absolute.

Maturity is linked to the increase of identification with, and care for, others. Albert Schweitzer noticed the expanding circle of care from family to humanity to animals, although different cultures have different emphases. Realizing higher levels of potential for the self favors Self-realization in others. As a corollary, increased Self-realization is dependent on, and internally related to, the Self-realization of others — giving "high self-realization results in deeper identification with others," which is important for the conceptual development of ecosophy — its assertion reflects an attitude opposed to an unconditional Verherrlichung of life and nature in general.

The development of life since the Cambrian era displays a diversity of forms in an expansion of life into places that can only be described as self-realization, since it is far more active than the passive adaptation of self-preservation. Self-realization as a result of maturation and the natural inclination to engage in beautiful action, as opposed to the moral actions distinguished by Kant, condenses certain social, psychological, and ontological hypotheses. The hypotheses, "the diversity of life increases self-realization potential" refers to all living beings — all beings that are in principle capable of self-realization. And, the norm "pattern unfolding!" is derivable from the norm "diversity of species!" and the hypothesis "complexity of life increases self-realization potential" — it is instrumental only in relation to the norm "Self-realization for all organisms!" so it is not a purely instrumental norm.

An unconditional "yes" in response to the norm "Self-realization for all organisms!" implies that Self-realization is something of an intrinsic value — it could never be a purely instrumental norm. Agreement with the norm "The diversity of organisms has intrinsic value!" implies the intrinsic values of all beings. The platform of the deep ecology movement, including an epistemology and ethics, can be derived from these metaphysical statements, which are based on ecological principles.

Conclusion

Ecology has prompted questions of philosophy. Does knowledge of ecology inspire us to create a metaphysics large enough to encompass ultrahuman beings? Using nonscientific hypotheses inspired by the science of ecology it is possible to escape the presuppositions, probably from the use of physics as a model, that have contributed to the predicament of classical philosophy?

Science accepted the banishment of metaphysics from its boundaries, but kept the tradition of substance as an explanation of phenomena. Substance metaphysics is still a western tradition. But, the ecological view suggests that events and patterns are as primary as substantial objects. Basing a metaphysics on ecology provides a place in philosophy for fluctuation, irregularity, uncertainty, rarity, and diversity. In the ecological view, humanity is an integral part of nature. The proper relationship is symbiosis, which includes competition and exploitation as well as mutualism, but not interference. In exploiting nature, humans are interfering with other species rather than just competing with them. Humans, like other organisms, are limited by environmental constraints; they have life-images and goals; they depend on other species for their existence; and they are capable of self-realization.

Author's Note: This paper and the previous one on metaphysical implications were written in Spring 1987 in response to a challenge by Arne Naess, who asked what a metaphysics based on ecology, rather than on physics, would look like. The author thanks Professor Naess for his discussion and encouragement.

Chapter 43

Ecophilia: Animal Welfare, Wilderness Preservation, and the Metaphor of Home

Summary
Humanity is exploiting nature recklessly, without attention to the minimal health of ecosystems. Yet various societies are working to preserve animals, species, and habitats. Their efforts are described, according to three levels: individuals, species, and habitats. An ecological philosophy is outlined as a basis for the united effort of these societies. Ecology supports the uniqueness of individuals in their life-worlds and the interrelatedness of species in communities. Psychological and geographical studies support the importance of healthy places for human beings. The concept of earth as home is proposed as a metaphor for the development of appropriate attitudes and participation in appropriate ways of living.

Death and Metaphor
The list of animal deaths in the United States (Hoyt, 1984) in 1984 reads like a doomsday book of atrocities: 22,078 North Pacific fur seals clubbed to death; 17 million mammals trapped for fur (303 million throughout the world); 12 million unwanted pets put to death; 70 million laboratory animals used in experiments; 3.5 billion chickens killed for food; 700,000 cattle dead from transport-related injuries; 598,757 animals shot for sport on wildlife refuges. Although species are still being identified at the rate of 8,500 new insects species and 100 new fish species per year (Cousteau et al., 1984), probably 400 species are driven to a premature extinction and 1,000,000 species are threatened every year (Myers et al., 1984). Statistics for habitats are almost incomprehensible. 3 billion cubic meters of wood are consumed annually. 12 million hectares of forest are cleared annually and 10 million hectares are degraded. Marsh lands are filled in; coral reefs are mined; and grasslands are paved over. The human impact is not negligible. 450 million humans are chronically malnourished and 40 million die annually from hunger-related disease (Myers et al., 1984).

Yet people are ignorant of these facts and detached from the consequences of their personal actions. In a national study of American attitudes towards animals, Stephen Kellert found that the prevalent attitudes were humanistic (anthropocentric) and negative (either the affection for or avoidance of individual animals, usually pets), and moralistic and utilitarian (concern for the treatment or use of animals). The natural and ecological views were less prevalent.

Results regarding recognition and knowledge of animals indicated that Americans were most knowledgeable about animals known to inflict injury and about domestic animals. Endangered species were least known. On wildlife issues, 32 percent of the respondents had never heard of the baby seal controversy. More that 70 percent of the people questioned about the Tellico Dam/Snail Darter issue were unaware of it.

We are distanced from what we cannot see — whales, otters, microbes — and from what we cannot understand — the Amazonian forest, the arctic plain. So much is outside our experience. We are incapable of responsibility. Helpless, afraid, we become more detached. We cannot express what is wrong.

Language has always had difficulty describing actions and things in the world. If each unique thing or action were named, speaking would become a burdensome impossibility. Paradoxically, in being spoken, language avoids being an inert catalog; it progresses outward from the body of the speaker toward the world, metaphorically. The concept of metaphor has been defined and used for over twenty centuries. Metaphorical attitudes toward nature were changing. Where Wordsworth saw consolation, joy and wisdom in nature, Tennyson saw nature "red in tooth and claw." If, for Shakespeare, our bodies were gardens tended by will, many saw their bodies as machines.

The advent of the machine made processes of order more amenable to description. Although only a closed system, the machine was a fruitful metaphor for living systems. The theory of the living organism as a mechanical contrivance explained biological phenomena from the physiology of an organism to the processes of cells. From the mechanical machine to the cybernetic machine, the metaphor was successful at explaining detailed processes without answering fundamental questions of meaning.

According to Thomas Kuhn, there is no methodological evolution of science; rather, normal science progresses by a succession of paradigms, which he described as noncompetitive and open-ended. He states that paradigms are the traditions described by historians under rubrics such as: Ptolemaic or Copernican astronomy; Aristotlean or Newtonian dynamics. These examples include law, theory, application, and instrumentation together, and provide models from which the traditions of scientific research spring. In his view, science proceeds by working out problems uncovered by each current paradigm. Should problems occur that can not be ignored, suppressed or resolved, then a revolution occurs to replace the paradigm. The new paradigm has to include all the old data as well as the new problematical data. Metaphoric systems are the core of structural coherence. For Kuhn, a metaphor is the vital spirit of a paradigm.

Science makes use of the metaphorical process to construct its models. Bacon referred to true metaphor as "the footsteps of nature." "Man is an animal" (Pribram). "Man is a system" (Laszlo). "Man is a computer" (Arbib). Kenneth Boulding offered the perfect machine metaphor for the operation of the earth: as a spaceship. As a metaphor, it suggested the limits of earth and the value of a limited life-support system. Machine metaphors for the body, animals, the earth, and the solar system, were illuminating for a while. But they have been extended too far. The body is not a machine; animals are not devoid of consciousness.

The technological paradigm has reached its limits. Data and information developed by hard studies have undercut the paradigms that guided their investigation. When a paradigm shifts, perceptions change. There was a paradigm change in metaphor from machine to organic system that undermined atomism and animism alike in developmental biology. The notion of organicism can be traced from taoism through Leibnitz, Goethe, Whitehead, to Naess and the deep ecology movement.

Deep ecology forms part of a new metaphor that is more appropriate to the unity and interrelatedness of the earth. Lack of a proper metaphor can lead to illness (Shepard, 1982). Deep ecology can provide a healthy metaphor for living, the metaphor of the earth as home. Humane and Conservation societies have already pointed the way.

Societies and Care

Humane Society and the Care of Animals

Throughout the evolution of the species, humans have killed animals for food and clothing. Comparatively recently, animals and then plants were domesticated; animals were followed, herded, corralled, tamed, and finally bred. As human technologies developed, relationships with animals changed. Hunting, grazing and agriculture produced large ecological disturbances. Early domestic animals were revered, but nondomestic animals became competitors or at least nuisances. Lately, animals have been treated as commodities processed in factories. Human overpopulation and emphasis on economic efficiency have resulted in utility hens in batteries and milk production from overcalved cows whose calves are converted to veal. Wildlife is considered useless, and hence valueless. Hunting activities persist, but mainly as recreation. Technology developed a human world apart from the natural world, which became only the source of raw materials.

As animals have been exploited, the humane movement has sought their protection from cruelty and 'needless suffering.' All animals – domestic, captive, and wild – have been defended by the Humane Society. It defends the 'welfare of pets, livestock, laboratory animals, and wildlife.' Its members speak for creatures that otherwise would be victims of abuse 'in the arenas of sport,' research, and farming. The traditional humane movement is based on the assumption that when an animal comes under human domination, it is entitled to fair treatment. Because of the 'pandomination' of humanity, humane ideas have been extended to wild animals. The society extends its umbrella to animal hunting, experimentation, and neglect, as well as the human poor and war victims.

Argument for Animals

Paul Shepard presents an argument that caring for pets progresses to care for all domestic animals and then to all animals. He considers the pet-attitude, stewardship, to be an adjustment to modern life, a grasping for ecological connections in a humanized world. Furthermore, he judges that the humane movement "is marred by its faulty ecology and its unwillingness to accept death as the way of ecosystem life." Shepard presents four arguments for preserving animals:

- economic – animals are useful for their products (milk, fur, oil, hormones).
- ecological – they have functional roles to play and the 'well-being' of all depends on their existence.

- ethical—they are part of the human ethical system.
- educational—humans need animals to develop properly.

He concludes that the first three arguments are not sufficient to save animals. Animals are inefficient; substitutes for their products can be created chemically. "Ecologically, all the creatures in ecosystems are not equally necessary to it ... although it cannot be proved that their presence does not add a little to the efficiency or the stability of the whole." Shepard says the ethical argument is not new; its application is ambiguous because 'unlimited rights' will conflict with human interest.

Shephard's defense of nonhuman life is "minding animals". It is not dependent on changing technology or idealistic ethics. Animals present us with related otherness. The human mind needs animals in order to develop and work. Animals are code images for ideas; they shape cognition, self-identity, self-consciousness. This is a good argument, but incomplete. Plants, rock, and water shape us; and wind and fire. As for economic value, it is a function of our state of knowledge; penicillium was just mold before Flemming amplified antibiotics from it; wheat was a natural before hybridized with a weed, goat grass. Animals and plants are the source of economics. The ethical argument will be expanded shortly.

Many humane magazines present animals abstracted from any natural setting. They concentrate on animals that are cute or are symbolic of human virtues. Animals perceived as ugly, microfauna and microflora, responsible for organic recycling, are ignored. Animals need to be considered in communities.

The Defenders and Care for Species

Through speciation, orders of animals and plants probe the environment. A species is thought of as a morphological extension of its niche, but the niche extender enriches nature. The species that enlarges its niche also enlarges the ecology as a whole, it expands the environment for itself and others. An expanding whole is created by diversification and enrichment of the parts.

Paul Colinvaux states that species result from the process of avoiding struggles for existence. Yet they are still related. Alan Watts developed a metaphor of species on earth as heads of a hydra; each has some autonomy and finite life, but is part of a longer-lived whole (referred to in Klopfer). The notion of human separateness may be an illusion of imperfect senses or an underdeveloped brain. Speciation is a great invention of nature. Although individual species, and certainly individual animals, may not be necessary to its functioning, the richness of nature depends on complexity.

The Defenders of Wildlife organization sponsors an endangered species campaign. In a recent issue of their regular magazine, the Defenders presented the plight of the red-cockaded woodpecker, a native of the pine forests of the Southern United States. This bird is a highly territorial, highly social bird, with a vary narrow range of habitat. Colonies of the bird are so territorial that relocation after nest abandonment occurs in less than one percent of observed cases. Unfortunately, the habitat is coveted by developers and military bases alike. The species cannot be saved unless the habitat is saved.

Argument for Species

Shepard argues that we cannot show that large mammals and endangered species are really indispensable to their ecosystems. "Elephants could be removed from the productive lands of Africa and bred in zoos." He continues by pointing out that the habitat does not die; the communities are intact. "Ecology cannot prove that the whole requires any one, or ten, species of large animals for its continuation." He claims that species rework their reciprocities others vanish.

That argument is fallacious. For example, elephants are grazers by choice, but their browsing activity, which breaks or tears up bushes and thickets, actually opens up the bushland in a way that is beneficial to grass. Without elephant activity, a diverse mosaic would become thick scrub (Eltringham, 1979). Elephants can also destroy trees, making them more susceptible to fires, and prevent regeneration of woodland. Elephant relationships with plants are highly cyclic, also. Without elephants, the whole range of species would shift to those preferring scrublands.

Perhaps it would be better to say that all species are necessary but not sufficient to a system. That is, the system can survive as a system without large species, but it is reduced accordingly. Hippopotami and crocodiles are necessary to theirs also. Hippopotami support fish populations in lakes and rivers with the minute animals in their excrement. As hippos are eliminated, the fish population dwindles, and native human populations have less protein. Similarly, alligators play an important role in the equilibrium of the Florida Everglades. The alligator creates pools by digging in damp soil; these pools become lairs of fish that eat mosquito larva. The pool also serves as a refuge to more species, including birds, in times of drought. Every species is "useful" in nature: as expressions of variety, niche makers, and feeling beings. According to Eugene Odum, any heterotroph that consumes autotrophs and excretes matter (with inorganic ions) contributes to the circulation of nutrients and minerals in proportion to its respiration. It contributes to the energy flow of the system. It is an interlink in the chain or net or web. By killing off select species, humans are changing the character of ecosystems, possibly reducing stability and diversity. Fox (1980b) states that "No species is more — or less — important than another, even if the ecosystem were to remain relatively unchanged in the absence of one species ..."

Shepard considers a human association devoted to each species of wild animal, so that every creature on earth would have a human constituency. If land and water use were required to face review by survival committees for each organism, no other environmental safety system would be necessary. Leagues dedicated to single species would accomplish little, however. Animals live in communities and are parts of food chains. Emphasis on endangered species does not address habitat; often, it does not consider other species or individuals.

International Union and Care for Habitats

Earth is a mosaic of cells of communities; the cells have boundaries like rivers or climates that occasionally break down and allow invasion and transformation. It is possible to define close but not exact subsystems, that is, ecosystems. The vast number of interrelationships between systems keeps them open. For example, grassland is affected by climates, soil conditions, fires, surrounding communities, and human agents. Each locality supports a segment of the total species population in a unique context, with a particular set of predators, competition, food, physical habitat.

The goal of the World Conservation Strategy of the International Union for the Conservation of Nature and Natural Resources (IUCN) is to identify important biological areas that do not have adequate protection. Priority habitats include tundra, desert, Antarctic ice cap, tropical deciduous forest, and islands. Large areas of representative ecosystems should be preserved to insure natural processes and diversity. Recent *IUCN Bulletin* Supplements have addressed the conservation strategies of Zambia, Nepal, and the Philippines. Yet national conservation has a nagging reputation as an elitist plot by industrial powers.

Argument for Habitats
Unlike a cat or dog, or panda or coyote, a desert or rain forest does not evoke sympathy. Worse, many habitats are perceived as useless or dangerous, composed of sand or wet leaves, harboring leaches and snakes. This is a public image difficulty, not a cosmological constant. The problem is more typical of industrial peoples, who draw raw materials from a variety of systems (Dasmann makes a distinction between ecosystem people versus world people, which is typifies the two approaches to resources and place). The majority of people are still directly dependent on an ecosystem, and their view of it is different. The Mbuti pygmies view their Congo forest as generous and friendly. The Pueblo Indians see their American desert as a providential home, because of attachment and knowledge. Saving habitats means saving large areas of land. It means placing wilderness, which is support for cultivated and industrial areas, as well as natural communities, off limits to development and perhaps any use.

Figure 6. The Palouse Grasslands (where the author proposed a Palouse Wilderness area in 1984).

Shepard concludes that the consequences of technological civilization are either exile or sanctuary; and sanctuary is the only solution available to the humanitarian ideology. The idea of sanctuary recognizes the multiplicity of factors necessary for viable populations. But, although there are sanctuaries for frogs and ferrets, it would be impossible to establish one for every creature. Shepard discounts the humanitarian objective as considering only "worthy' species. If all species were considered, the whole planet would end up as a sanctuary. He considers a 19th century political solution, when space was unlimited. He describes sanctuary as an unfeasible arrangement, in evolutionary terms as allopatric (life not occurring together). Shepard states that exile (extirpation or extinction) and sanctuary are allopatric choices — Allopatry from the base word for fatherland. Throughout, the etymologies of words trace traditional thought — is consistent with the tradition of personal property, domination of nature, and model of the nation state.

What Shepard has overlooked is that humanity was never sympatric. It never lived together (the meaning of sympatry) with major carnivores. So allopatry is consistent with the nonexploitation and nondomination of nature as well. Sanctuary as personal property or as nonhuman property is a moot point if habitats are saved from destruction. Domestication and enslavement are sympatric forms. A conservation program, despite Shepard's argument, cannot be based solely on sympatry, although at the planetary level, all beings are sympatric. Not all species occur together in the same place. Sympatry describes only those that do. At the habitat level, allopatry is "intelligent" use of available resources by animal communities. Large herbivores, such as elephants and rhinoceros, may choose poorer quality food and avoid competition with smaller animals and exploit an untapped food source. Many interactions between different species contribute to the mutual benefit of the members of the community, as well as the community itself. Humanity must be allopatric with most wild species and allow them to develop independently in their own places.

Ecosophy and Ethics
Interaction of Beings
A community is a collection of individuals engaging in interactive behavior. Many interactions are positive, negative, or neutral in effect. These include neutralism, competition, amensalism, parasitism, commensalism, protocooperation, and mutualism. But not all are properly defined. For example, predation is regarded as being where one population adversely affects another, but is dependent on other. This is not so; it benefits both. Predation could be considered a mutualism and not a parasitism; nor is it an entirely separate category. Predation can increase the diversity of species, at the cost of perfect fitness. Darwin noticed the predation principle. S.M. Stanley has argued that the cropping (predation) principle may provide biological control of the evolution of metazoans. Intuitively one would expect the introduction of a cropper to reduce the number of species in given area; but the opposite occurs. In communities of primary producers (photosynthesizers), a few species will be superior and monopolize space.

A cropper dominates its favorite prey species, usually the most populous, limiting it so others can develop. A new level on the pyramid broadens the one below it. Stanley explains the Cambrian explosion through evolution of cropping herbivores, which opened space for a greater diversity of producers, which permitted more specialized croppers. The ecological pyramid became wider and higher.

Competition was once considered the basic interaction between individuals and species. The better adapted organisms survived and reproduced; others died, after violent struggles. Darwin also perceived nature as violent. "Thus, from the war of nature, from famine and death... the production of the highest animals, directly follows." Nature is seen as a pyramid of death, a slaughter of all by all, even by contemporary philosophers and biologists (Rodman, Lehmann, Birch). The processes of nature considered inhumane. They see nature clearly, in Tennyson's words, 'red in tooth and claw'. But it is not so.

Cooperation is as an effective strategy as competition, and as necessary. Survival of the fitter is correct only up to a point; beyond that it is survival of the more cooperative. Neil Evernden notes that organisms, unlike the standard Darwinian description of them, go to absurd lengths to avoid direct competition. What are recognized as species are a rainbow of attempts to avoid competition, to share the life-base. Diversity enhances the potential for survival. Arne Naess states that "live and let live" is a more powerful principle than either/or.

A number of old and recent studies in ethology have stressed the use of cooperation instead of struggle. Herman Reinheimer characterized organisms as bioeconomic traders, that put cooperation before competition. P.A. Kropotkin, in his work, concluded that the element of cooperation in animal life, even between different species, was more impressive than instances of competition. Many more recent studies stress the primary use of cooperation as opposed to struggle: K. Lorenz, M.W. Fox, and G. Schaller. L.L. Whyte considered even genetic inheritance to occur by rules of cooperation.

The mycologist Heinrich deBary coined the word "symbiosis" in 1876 (or 1879) to describe "living together." (from the Greek word *sumbioein*, meaning 'to live together.' Etymologies of words are used only to trace traditional thought.) It defined a constant and intimate relationship between two dissimilar species. Such economy of effort is very common in nature. Living together is convenient for food and protection. Symbiosis has considerable ecological importance. Herbivores, such as cattle or sheep, depend on intestinal microflora for digesting grass — for survival. Two species that live together closely enough are biologically symbiotic. Symbiosis acts as a higher level store of genetic information; a lichen is fungus plus algae, where each is part of environment for the other. It may benefit the individual (Lincicome, 1969). There are elements of symbiosis in wolf-deer predation. In one sense, all interactions are positive and symbiotic on the planetary level. Symbiosis is a necessary and beneficial relationship.

Cooperation is a natural rule. It creates communities, living entities in ecological balance. There is a wisdom of life, a "biosophy" in Stephan

Lackner's words. In spite of the emphasis on competition and survival from predation, Lackner notes that there are peaceful habitats; the predatorless islands that Darwin knew, African lakes filled with flamingos. Lackner notes that the largest mammals — elephants, hippopotami — are vegetarian; the smallest insects — gnats, mayflies — are too small to be preyed upon. George Schaller estimates that less than ten percent of the grazing animals on the Serengeti die of violence. Karl Popper conjectures that besides the theory of the hostile environment (passive Darwinism), there is a complementary theory of the friendly environment. Many organisms are active explorers, searching for new, friendly environments. Birch and Cobb consider that the richness of the world and the freshness of response are matters of novelty.

No interaction can dominate without sad consequences. If all plants were symbiotic and none were competitive, there would be no trees or flowers. Most of these interactions are not simple, but are complex and paradoxical. For example, parasitism on an individual level may have benefits on a species level. The predator/prey pair are not excluding opposites, but generate a whole unity, an autonomous domain where there is complementarity, stabilization, and survival values for both. There is no real opposition in natural systems.

Life-images

C.H. Waddington said that the "general anagenesis of evolution is towards what may be crudely called richness of experience." The goal of all creatures is to come into the fullness of being (spontaneous flowering). For Whitehead, all living things have three urges: 'to live, to live well, and to live better' (in *The Function of Reason*, p. 8). Living better is being more attuned, stimulated, integrated, receptive, and spontaneous.

Adolf Portmann shows that every form of life appears as a gestalt with a specific development in space-time. All living forms develop an image of their environments. Genetics provides the proper image choices for some — frogs, for instance. Others must learn what is valuable using their senses. Animals have their own universes that are strange and fascinating. Reality is immeasurably greater than the human idea of it. Jakob von Uexkull suggests representing the unfamiliar world of animals with bubbles to denote the self-world or phenomenal world of an animal. According to von Uexkull, perceptual and effector worlds form a closed unit, the umwelt. "Figuratively speaking each animal grasps its object with two arms of a forceps: receptor and effector. With the first it invests the object with perceptual meaning, with the second operational meaning."

The world — life-image — is what has meaning for an organism. It is a focus. The first principle of the life-image theory is that all animals from the simplest to complex are "fitted to their unique worlds with equal completeness." A simple world corresponds to simple animal; a well-articulated world to a complex animal. Von Uexkull implies that the human world is only one of the many possible. Animals are not suboptimal beings relegated by evolution to second-rate habitats. They are optimally fitted.

Equality of Experience

The human mind usually gives direction to the process of evolution. Evolution is considered a building up of complexity, when it should be regarded as an unfolding of patterns. Evolution is not a hierarchical ladder or escalator going up and up, but a series of adapted forms; the goal of each is fulfillment. Darwin wrote a note for himself (in Singer, 1981), "Never use the words higher and lower." He did not want anyone to believe that evolution was a purposive movement toward a goal. He considered it the result of natural forces. Recently, Hobart Smith has considered the categories of higher and lower, with their connotations of superior and inferior, as anachronistic. Although there may be chronological differences in the development of species, each surviving species is well-adapted to the environment.

Animals and plants radiate through the environments. Evolution does not have just one direction. It can be regarded as varieties of styles of life, as the "pattern mixed-upness" of Merleau-Ponty (1968, perhaps based on the "mixed-up-ness" of physicist Willard Gibbs). Rather than relying on evolution for metaphysics, Merleau-Ponty called for a kind of "phenomenal topology" of things as they loom upward bodily around us. All forms fit themselves into a changing environment. In a strange and complex earth, other species fit in places that humans cannot. Naess offers a biospherical egalitarianism, where all beings have an equal right to live and blossom. Rejecting human superiority entails an egalitarian doctrine of species impartiality.

Ecology: Knowledge of the Home

The study of life in place is ecology (from the Greek word *oikos*, a dwelling place or house. 'Economy' derives from *oikonomia*, household management.) Ecology is knowledge of the house, as economy is its management. Although economics provided the model for ecology, few ideas on environmental limits and interdependence were taken from ecology. Ecology is not a division of economics; if anything, economics is a division of ecology. Ecology deals with the relationships of organisms to environments. It is not a reductive discipline, and not amenable to easy quantification. The joy of ecology is variety. In nature variety emerges spontaneously, as the capacities of new species are tested by the environment. Its basic premise is interrelatedness. The interpenetration of boundaries makes humans less discrete, less alone.

Human Symbiosis with Earth

Our bodies contain the ashes of stars; human cell structure is shared with trees; human brain patterns are shared with reptiles, birds, and mammals. We share our bodies and land with hundreds of species of bacteria, fungus, insects, that are beneficial. As Lewis Thomas shows, our human bodies are living communities, hosting amoeba in the blood, mitochondria in the cells, bacteria in the intestines. We are connected to the largest and smallest beings. We are part of a food chain (according to Soleri, 1983). Naess rejects the image of man-in-environment for the relational, total-field image. He characterizes organisms as knots in the biospherical net, a field with intrinsic relations. The relationship with other beings is part of the basic constitution of a being. Fitness has more to do with cooperation in complex relationships than with the ability to kill or suppress.

Under a broad definition, human activities such as planting trees, growing corn, raising camels, qualify as symbiosis. Without suitable places for livestock, or with cruelty to animals, the symbiotic element disappears, and the human relationship is parasitic. With care, domestic animals benefit from association. The same can be true of pets, bats, owls, coyotes, rats, and cats. Humanity is part of the system. The earth is a loosely formed spherical organism, with all its working parts in symbiosis. Leopold described conservation as an attempt to harmonize civilization with the land towards a "universal symbiosis" (Leopold, 1949). Rene Dubos sees "humankind and Earth as constituting a diversity of systems of symbiosis that constantly undergo adaptive changes and thus contribute to a continuous evolutionary process ..."

Humanity is embedded in the earth (states Merleau-Ponty, 1968). From the oldest language we know, the Indo-european tongue, we took the word for earth and turned it into humus and human (earth-born, *dhghem*=earth—>*humanus* in Latin—>human in English) Yet the word for man was shaped into man-image, world (Indo-European *wiros*=man—>*weorold* in O. English—>world) One word progresses from earth to human, the other from human to earth. We cannot be any closer to the earth and its processes, since the parts are combined in us. We are indissolubly one with nature. The German poet Novalis (in Bly, 1980) equated man and metaphor, as the blueprint of the world. "Man=Metaphor," stated Novalis. "We are looking for the blueprint of the world — that is what we are ourselves."

We mistakenly conclude that our skin is the boundary to ourselves. But intuition also senses the interdependence of nature. We extend the boundaries of personality to other things and people. The human skin is like a pond surface, according to Paul Shepard. The skin's interpenetration ennobles and extends the self — the beauty and complexity of nature are continuous with ourselves. We know subjectively that we are not separate from the earth, that wolves are capable of love and tenderness, that trees are beautiful. Edward Wilson argues that the essence of humanity is inextricable tied to life on the planet. Biophilia is the natural affinity for life, and central to the evolution of the mind.

Our species has been shaped by the earth. The desire to save forests, wetlands — all natural ecosystems is a expression of deep human values or perhaps a more basic survival instinct. Experience of wildness lets us capture some of our own wildness and authenticity. Our emotional response to the unfathomability of the ocean or luminosity of the desert is an expression of aspects of our fundamental being that are still in resonance with these forces. We know these things. John Fowles judged experience "quintessentially wild," irrational, uncontrollable, and incalculable. It corresponded with wild nature.

Perception of the body as landscape and of natural terrain as a body is as fundamental to psychology as it is to mythology. We depend completely on the natural environment, physically and psychologically. D.O. Hebb has conducted experiments that show the effects of a limited environment. Cut off from external stimuli, the mind becomes strange. The external world is needed to keep us alive and sane. This world is composed of remote occurrences, on polar icecaps and distant stars, as well as immediate personal events. The person is inextricably woven with the world. This bond of betweenness constitutes the foundation of an understanding of other species. What is impaired in the absence of a rich ecology is the individual's knowledge of himself, not only as a person, but as a member of a species.

Mental health can be related to the quality of the landscape, as Dubos, Passmore, Shepard, and others have done. The ecological health of civilization depends on an environment that is healthy, that is, has sufficient wilderness to renew air, water, genetic resources. For Skolimowski, Ecophilosophy entails health-consciousness. Humans are complex fields of force that are maintained through effort. To be healthy is to be on good terms with the earth. In the taoist view, also, sickness is a symptom of disharmony with the universe. Taking care of health means taking responsibility for the focus of the universe that is the self.

Participation in Existence
John Rodman offers a participatory image of humanity, as an integral part of food chain and part of an organic cycle of birth and death. Humans need to recognize that they automatically participate in everything, and that they cannot unparticipate by choice. We know that from the quantum level, through the ecological and cultural. Human nature does not find meaning in an absurd world, but discovers its structure through interaction with the ultrahuman order. Human identity exists partly in relation to nature; the destruction of one involves the other. An act of 'ecological resistance' is an affirmation of the integrity of the naturally diverse self and world. The meaning of such an act is not exhausted by success or failure in linear sequence of events. One is aligned with the ultimate order of things by ritual action, by the affirmation of value.

Ecocentric Value
Humans use an incomplete source of value for nonhuman beings; the source is human need. Human need shapes facts. As Goethe recognized, all fact is theory, a blend of perception, imagination, and needs. Abraham Maslow established a hierarchy of human needs, beginning with food and continuing through social acceptance to self-actualization. Vital human needs are based on the health of the earth. Human needs could be extended to include a foundation of wilderness. Nature, which is self-supporting and self-managing, is a life-support system. Human systems depend on natural ones, for recycling of wastes, water, and air.

Skolimowski identifies a new moral order to address values. One should behave to enhance life, as a condition of evolution, to enhance the ecosystem, and to enhance capacities of the highest form: consciousness, creativity, and compassion. Arne Naess presents a deep movement alternative to the shallow fight against resource depletion and pollution. Deep ecology is inspired and fortified by ecological knowledge derived from experience, not logic. Its tenets are normative. Its value system is only partly based on scientific research. The movement is ecophilosophical; Naess offers the term ecosophy, which means 'wisdom of the house.' Ecosophy is a philosophy of ecological harmony. It contains both value judgments and hypotheses.

Ecology can expand the narrow human-centered evaluation and see things from viewpoint of nature. Lovejoy's Great Chain of Being traces the deductive order in classical nature from Greeks to German idealism. But Lamarck inverted the chain in his theory of transformism; mind is immanent and can determine transformations. Although the hypothesis of inherited characteristics was rejected by Darwin, who shared that hypothesis but denied mind as an explanatory principle, both Lamarck and Darwin inverted the value of life. By inverting the great chain of being, Lamarck escaped the directive that the perfect must precede the imperfect. The result of Elton's food chain was the realization that the bottom link — plants — is the most important.

Francis of Assisi was the exception to the general attitude of Christianity: compassion to man only. He tried to unite the compassion of Christianity and the animistic sense of union with the natural world. Natural processes take on an expression of significance of their own without reference to man. All things have an inhuman value of their own. St. Francis tried to depose man from his monarchy and set up a democracy of all God's creatures. In parallel, the taoists saw that we were indistinguishable from other creatures; if we seemed distinguishable, it was through our feelings of self-importance. Lao Tse turned pyramid of human values upside down. As there are more commoners than aristocrats, there is a net gain to the success of the community, according to Holmes Welch.

Humans have stripped the world of qualities and significance and claimed them for themselves. By valuing humans alone, we make value subjective and end up without value. Ecological philosophy reclaims value by placing it at the center of life. Whitehead has stated that existence is the upholding of value intensity; for itself and shared with the universe, from which it cannot be separate. Everything that exists has two sides: its individual self and its signification for the universe. Each aspect is a factor in the other. Whitehead finds: "Remembering the poetic rendering of our concrete experience, we see at once that the element of value, of being valuable, of having value, of being an end in itself, of being something which is for its own sake, must not be omitted in any account of an event... " Human values come from knowing what is valuable in nature. Values usually encode information having survival or prestige importance. Perhaps the most valuable thing is living time. Then experience of life — aesthetics, from the Greek meaning perception — is also valuable. This may be why humans value walking in the woods or observing the production of art. Natural processes are their own purpose and constitute their own value. A growing tree is; it does not have to demonstrate or prove.

Perhaps we should not argue that things have value in the human system. Let us just respect the nonhuman system. Bees have bee value; wolves have wolf value. Wolves are not efficient at binding nitrogen; neither are humans. Lichen are poor predators, but they break apart rock better than bighorns. From a functional point of view, all beings are equal. In ecocentric perspective, all beings have intrinsic value and are equally important.

Every being has an intrinsic value, before any utilitarian value to humanity. Associations of plants and animals are just as unique as their components. The value of wild nature is its independence and wildness. If we can admit the independence of nature, that things continue in their own complex way, we may feel more respect. We can contemplate with admiration, sense as well as manipulate. The emergence of new moral attitudes depends on a more realistic philosophy of nature.

Reverence for Life

Albert Schweitzer thought that ethical thought had been developing since prehuman history and that it culminated in the principle of reverence for life. Schweitzer challenges ethics: "Let it dare, then, to accept the thought that self-devotion must stretch out not simply to mankind but to all creation, and especially to all life in the world within reach of humanity. Let it rise to the conception that the relation of man to man is only an expression of the relation in which he stands to all being and to the world in general." Our attitudes are grounded in a belief system that constitutes a particular world view. The system constitutes a coherent whole. With Schweitzer, the system began to shift toward a biocentric outlook. The concept of reverence (ehrfurcht, meaning honor-fear in German) offered some respectability for nature through a proper attitude. Schweitzer proposed an ethics derived from Christian ethics (but really larger) that affirmed the world. But reverence for life sometimes conflicts with the Christian paradigm, which is just a particular manifestation. And it can lead to an instrumental ethic.

But Schweitzer's reverence entailed a constant effort to make excruciating decisions. His attitude was a noblesse oblige toward lesser species, based on a Christian idealism and on a mistaken image of nature as brutish and dumb. During human evolution, the circle of responsibilities widened, from the family, to tribe, nation, humanity, and now toward all life. Although Schweitzer noted that the circle of knowledge was widening, also, he felt the streams were divergent, that ethics had nothing to gain from understanding nature, and furthermore, that there was no hope of finding meaning in natural phenomena. His ethic was not based on ecological knowledge. His reverence for life principle acquires a new aspect when it is restored to ontologically and ecologically firm ground.

Ethics and Knowledge

The effectiveness of ethics, however, depends on a scientific knowledge of the way the world works. The Aztecs based an ethic on imperfect knowledge of the sun, and suffered disastrously. Julian Huxley stated that knowledge of ecology is necessary for responsibility for the future of the earth. The correction for ignorance of ecology is knowledge of the home system. The correction for nonacceptance of death is a generalized concept of reverence for life. Death is a part of life, but not a driving force. The reverence for life has to be based on an ecological ethic, that understands the necessity of predation as well as altruism. Killing as well as saving. Reverence for life must include awareness of natural laws. Wolves need deer and mice to survive, as much as they need wolves to be healthy. Human intervention into natural communities must be responsible, not sentimental. The concept of reverence allows the center to be everywhere.

Knowledge by itself, however, merely permits a more efficient utility. That utility can be seen in Singer, the animal rights philosopher, as well as in Hoyt, the Director of the Humane Society, who referred to animals as "one of our nations' most precious natural resources..." The problem with utilitarian ethics is that it permits the use and exploitation of any natural object, including human beings. Based on a limited science, the ethic failed to see those beings and communities for which no use was known.

Knowledge cannot be the sole basis of decision making. It is always incomplete and therefore cannot describe all aspects of the earth that bear on human life or environmental quality. Knowledge must be humane. Abraham Maslow saw the organism as having biological wisdom; it can be trusted as autonomous, self-governing and self-choosing. To examine organisms, and nature in general, we must shift to a taoistic approach, asking rather than telling, observing rather than manipulating; receptive and passive, not active and forceful; "nonintruding," and noncontrolling. It stresses noninterfering observation rather than controlling manipulation; it is receptive rather than forceful. This is part of the paradox of duality; it is detached yet concerned; free yet committed; and independent yet responsible.

Classical objectivity may be contrasted with taoist, which is another path to objectivity with greater perception. Loving perception provides kinds of knowledge not available to nonlovers; this is especially true in ethological literature. Maslow cites his own work with monkeys. Lorenz, Tinbergen, Schaller, Van Lowick-Goodall, and Fox have found it to be true. This is the way a good psychotherapist, teacher, scientist, parent, or friend functions.

Ecological Ethics

The word ethics is derived from the Greek word meaning 'custom' (Greek *ethos*, Sanskrit *svadha*, Latin *mores*, plural of *mos*, from *meare*), which itself came from the Sanskrit word for one's 'own doing.' Since it was used in the plural, it meant 'doing together;' it also meant 'abode.' The word 'morality' comes from the Latin word for will of the people; the singular meant the 'will' of a person. It was probably derived from the verb 'to measure,' as to measure one's way, to go one's way. Morals means the way of going together. Ethics means doing together, which of course one does living together.

Ethics are assembled inductively, from experience in living places. Because of the uncertainty of human actions, ethics has to encompass the far past and distant future. No one knew that when DDT killed mosquitoes, it would concentrate to kill birds. Values are time dependent, and ecological time can be very long indeed. The futures we invent are viable only if compatible with constraints imposed by evolutionary past. An ecological ethic recognizes all human endeavors as part of nature. We have a moral obligation to leave the world habitable for future generations of humans. An ethics that requires a long-range responsibility also requires a new humility.

Our technological power exceeds our ability to foresee its consequences.

Aldo Leopold has proposed a conservation ethic, dealing with human relationships to land, plants and animals. The land ethic Leopold had in mind was a sense of ecological community between man and other species. When we see land as community to which we belong, we will use it with love and respect. Such an ethic would change the human role from master of earth to plain member of it. Leopold describes the extension of ethics as "actually a process in ecological evolution. Its sequences may be described in ecological as well as in philosophical terms. An ethic, ecologically, is a limitation on freedom of action in the struggle for existence. An ethic, philosophically, is a differentiation of social from anti-social conduct. These are two different definitions of one thing. The thing has its origin in the tendency of interdependent individuals or groups to evolve modes of cooperation."

An evolutionary ethic suggests that humans avoid tampering with complex evolved systems, not because they are good, but because they are the basis of life at this stage of development. Ecological ethics is situational. Because ecology is the study of a changing system. The morality of the act is determined by the current state of the system. Adaptive modes should conform to ecological patterns. An ecological ethics is based on attributes of ecosystems and human compliance with ecological laws.

The aim of an ethic must be harmonious to the total ideas of the world's population of living beings. Fox proposes a biospiritual ethic as a unifying set of principles, ethics and values that will bring about a nonconflicting state of one earth, one mind. The ethic is based on the biological fact that all humans and living beings are kin and that life is spiritual — love is stronger than violence. It arises from seeing humanity in an ecological perspective.

Reverence for Being

The world would not be a better place without sharks, silverfish, cockroaches, rats, hyenas, or whales. Their existence has value; they have functions. Humanity has upset the balance of nature in favor of itself. Civilization will have to correct future activities to reduce the margin between bare survival and social development. The recovery of implicit natural values can be expressed as a reverence for natural systems. Ervin Laszlo proposes a social ethic for the age of humanity. Laszlo calls for reverence for the level-structure of the microhierarchy, including all systems on all levels, from atoms to an emerging planetary culture and ecology. "We can express the recovery of our implicit natural values in requesting a reverence for natural systems." This reverence expresses the insight that humanity is in nature, a part of an embracing network of dynamic, self-regulating and self-creating processes. Marveling at humanity is marveling at nature, the matrix (the word shares the same root as 'mother') from which we arose.

Reverence must include all natural and artificial beings and things and fields, from atoms to weeds, computers to galaxies. Atoms, molecules, and organic cycles are parts of humanity. We must revere all the arrangements of earth stuff. The greatest human dignity follows from respectfulness of everything as meaningful as ourselves. Such a reverence would treat all substances of the earth as precious, to be used carefully, if at all — and certainly not for the flood of mass-produced consumer items. It would include all human artifacts, manufactures, and societies. It would promote a society where individuals would live in close contact with natural support systems. It would provide each with the aesthetic necessities of life, to develop all capacities. Reverence can only be felt at the alienness of nature, not its comfortable conquest.

A new metaphysics, that is genuinely nonanthropocentric, would be the best foundation for ecological concern. St. Francis calls for a radical rejection of anthropocentrism. This implies a rejection of technological domination and fosters an attitude of letting beings be. Rather than ask the purpose of existence, we should accept that it exists, in place, in an ecological whole. Every being has a right to be and express itself and to seek fulfillment.

All organisms, including human ones, are continuous with nature. In ancient philosophies, being was the ground of ethics. Modern philosophy divorced the two. But a revised image of nature reunites them. Ethics is ultimately grounded in the order of things, the ordo creationis. Being is universal; everything has value in itself by virtue of its existence. Therefore every human action that affects the ultrahuman world has significance. There is nothing that does not have value. For Plato, knowledge and value were combined. Therefore, every human action has ethical consequences.

The basis of all value is being (the verb form). It is reality undistorted by human needs. Of course, humanity can still enhance nature with its presence, in a nonexploitable manner. The reverence for beings as they are is the law of noninterference. In nature, the law of noninterference means 'letting be' (Heidegger), 'letting alone' (Wilson), and 'not killing for pleasure' (Fox). Noninterference is not indifference, which is diffuse. It is caring. Noninterference will not lead to chaos, poverty, and stagnation. The technocratic vision strives for "life under control," but the earth is self-managing, productive, efficient, and orderly.

Ecophilia and Home
In Place
The human ordering of the world makes places from wilderness. A place changes qualitatively; it becomes structured. Natural complexity decreases as the human increases, although the two are not mutually exclusive. Fitness is achieved after slow progressive reciprocal adaptations; it requires a stability of relationships between societies and the place. Human places are complex integrations of nature and culture that develop in particular locations. The place precedes knowledge of it. The knowledge of place is one of the first links in chain of knowledge. This knowing is essential to our existence. Being human is having and knowing a place. Only learning flowing from hospitable presence can promote life and enhance human existence.

Paul Shepard suggests that for each individual the organization of thinking and meaning is intimately related to specific places. Experience focuses on a place, which acts as the background for specific events. The features of world are experienced meaningfully. The place is a matrix for ordering experience. The specificity of place is important. The earth extrudes itself into particular plants and animals; flexes mountains; and sweats weather. Places animate (from the Latin *anima*, meaning 'inspire') people. The inspiration of the sentiment of dependence is called impregnation. Animals and humans are imprinted early in life to particular places (philopatry). Each difference in the landscape has meaning, as when the aborigines of Northwest Australia perceive physical differences and even a symbolic landscape. In fact they structure space according to myth, where Europeans use buildings and roads. Every place has a unique identity, a persistent sameness as a result of combinations of factors. Topophilia (Tuan's word, 1974), love of place, is the recognition that all human beings have affective ties with the material environment.

The attachment to a place is rootedness. Von Uexkull describes the importance of rootedness in his concept of life-world. Simone Weil regarded rootedness as "perhaps the most important and least recognized need of the human soul." Rootedness arises from participation in a place. It is the need for order, liberty, security, status, and responsibility. A deep relationship with a place is necessary. Without it, existence loses much of its significance. Caring for a place involves concern and responsibility. This attitude is similar to one described by Martin Heidegger as caring (Sorge). Care is the recognition that a human being is a participant in the world. It is tolerance for the essence of a place; absorption in a place; concern, the willingness to not change or exploit.

One ecological benefit of rootedness is that people will take care of a place if they realize they are going to be there for a thousand years. Having a place means that the inhabitant has stock in it and participates in its unfolding, through planting and caring. Detailed understanding of plants in a locale allow gathering of food and medicine. People cultivating a sense of place are people in place. Their work can be appropriate; appropriate growing, logging, mining, or building. People in place acquire a sense of community, nonhuman and human; a shared set of values and concerns; health and spiritual benefit.

At Home

Adolf Portmann observed that insects and animals displayed a powerful attachment to places; that it was best understood as home. What does it mean to be at-home? The fundamental ambiguity of existence is that humans have different capacities for feelings and awareness. Some feel strongly about a place or home; others never do. Several metaphors have been used to describe the human place on earth. The earth is a storehouse, property, a spaceship. But the earth is not a spaceship or storehouse; it is home. Victor Ferkiss (in Fox, 1980b) proclaimed that: "The world and humanity are one entity, one system in equilibrium. Earth is humanity's only home; humanity is one people in relationship to the earth."

There are a wide variety of meanings of home. It is a place of family residence, the family social unit, habitat, and place of origin. The word 'home' comes from the Middle English (word *hom* and Old English *ham*,

Old Norwegian *heimr*, Greek *kome*, and Sanskrit *kayati*) meaning village or home. The Old Norwegian word for home meant village or world. The word can be traced through the Greek to the Sanskrit, which meant 'he is lying down.' Its spectrum of reference is enormous. The word is used to describe house, village, city, bioregion, cultural world, and the earth. Its content is also ambiguous. Home can hold a single person, family, relatives, pets, domestic food animals, neighbors, and others.

Home living is simultaneously on different levels; the importance may shift from city to nation, or nation to state, or house to bioregion, or state to habitat. Each level is a metaphor for the next. There are parallels between nature and a house, as the basis for home. Solar space is like the landscaping; wilderness is the foundation; conservation areas form the shell and provide services; and each bioregion is a unique room. The analogy cannot be carried too far. But it shows that a house is not, as Le Corbusier said, "a machine to live in." It is a matrix for home. Home is not just a house, either; it is a complex of significant events centered in place. It is the foundation of our individual identity on one level, and our role in the community, on another. What makes home different from house? Participation in making, commitment. People invest parts of themselves in a place, to make a home.

The concept of home has a mixed reputation. Paul Shepard finds humanitarians obsessed with the 'homelessness' of stray pets and wild animals. He points out how the fixation on shelter is taken over by the advertising of wood industries, who describe their meager reseeding efforts as 'creating a home for wildlife.' He sees protective organizations swaying to the tunes of propagandist lullabies. Perhaps he is partially right. But that is a misunderstanding of home. a home is not the house, not an undifferentiated place. Animals accustomed to rich woodlands are not at home in a replanted clearcut. The use of the word is a cheap advertising device; yet, it shows the importance of the meaning. A home is living, a house is not.

A home is a part of the environment claimed by feeling. Emotion creates an 'in-place'. A place must be found and made. Humans, like plants and animals, identify greatly with local environments. Maybe this is a function of the limbic system of the brain, a function we share with territorial mammals. Human emotion creates an 'in-place.' So far, no psychologists have studied what happens when a person sees her/his place, their very context, destroyed. These catastrophes may be the basis for diseases, depression, or cancers.

The word nostalgia was coined by Johannes Hofer a Swiss medical student, in 1678 to describe an illness characterized by insomnia, palpitations, stupor, fever, and persistent thought of home. The disease could result in death. For the Northern Aranda in Australia, as well as for émigré Russians, it is not possible to stay away from home indefinitely and still live. Nostalgia can be a fatal disease. Thus far, the sense of place cannot be gleaned from an analysis of the nervous system. Yet a place shapes the nervous system, somehow.

In English the term for dwelling is to stay. This is the symbolic opposite of moving or changing. It means to withstand time. Dwelling resists and persists. Permanence is important element in idea of home. The Royal

Commission on Local Government in England and Wales found that people's attachment to 'home area' increased over time (in Kaplan, 1983). Gaston Bachelard has written much about the significance of home: "For our home is our corner of the world. As has often been said, it is our first universe". The home was a springboard to understanding the universe.

The Poetic Species

Humanity is the poetic species, according to Richard Rorty: "the one which can change itself by changing its behavior — and especially its linguistic behavior, the words it uses." Sewell claimed that biology needed poetry rather than mathematics to think with. Ecology requires a combination of aesthetic perception and disciplined thinking, as does poetry. Elton remarked in his work on animal ecology that there is more ecology in Old Testament or Shakespeare than in zoological texts.

Ecology contains a secret: attention to detail. A metaphorical ecology has more than a scientific or political meaning. The whole can be seen by the part, because it is implicit in every part — this is why hologram can be reconstructed from a small piece. Blake wrote of seeing the world in a grain of sand; that is the secret expanded. Rilke sees that if we leave our locked-in interior and use imagination to see a tree, we grant the whole world its being (in Robert Bly, 1981). Intimacy does not necessarily require details.

Poetry is communicative of the quality of things. Like philosophy and science, it discriminates the unsuspected in the commonplace. Poetry has an ancient, ontologically mythic function. Not different from science, but more diffuse; not better than science, but more comprehensive. It pulls in; showing is not simply mirroring. The poet accepts ontological parity; aspects of the world are not negated or reduced by one another. The poet is willing to grant consciousness to trees and hills or other living creatures. A poetic language could go a way to including a view of the infinite interpenetration of all existence in a sublime ecology.

A conquered world is no good for humanity; conquest is boring — there would be nothing to live with. The connection to otherness has been severed. In the living universe there is no boredom, because everything is alive and active; danger is inherent in every movement. Contact between things is wary and keen; wariness is a kind of reverence. Nothing can be taken for granted. Life itself consists in a live relatedness between man and animals, flowers, rocks, and stars.

Poetry is a tool for comprehending partially what cannot be totally known: feelings, aesthetic experiences, moral practices, and spiritual awareness. People need to be made aware of the power of self-determination, and the possibility. The poet celebrates the unity of existence by sharing it with everyone. The poet reconnects the real world and the thought-world (or perhaps heaven and earth) in a poem. The poet creates new forms and myths to make the earth sacred again. The richness hidden in thoughts, places and beings is revealed. We need to feel the immensity of nature. People need to feel things, before they can act on them. Poetry can help them feel themselves as part of the web of life, or on an oasis in space. Poetic ecology is home-making on earth. As Bachelard stated, the poet's occupation makes the universe into a habitable cosmic house. Only when we are comfortable can we know the house and manage our part. A creative language can express a new vision of humanity and nature, based on ignorance and feeling, as well as knowledge and thought—a poetry of the earth.

Conclusion

Animals, plants, and habitats are being destroyed because of short-sighted, short-term economic interests. Animals do not need to be saved from natural death, a great regulator of life, but from suffering, experimentation, and premature extinction. A reasonable place needs to be created for animals in human society or left for them outside society. The safety of the environment is too important to be left to scientists, even ecologists. The crude history of science shows that scientists fall as willingly as others under the influence of money and power. Education will provide more knowledge, eventually. But participation in a humane society is the most effective way of convincing people of the importance of conserving their larger home. The humane movement could educate people so that they understand their long-term interests. The benefits of natural ecosystems are not trivial or dispensable, just little known.

Humans arise from the matrix of the earth. They make places that are home. Poetry is making. Yet this is done together with other species, in a broad symbiosis. This coevolved activity is an ethos, an abode together. Living together creates an ecos, homes. The survival of society now depends on an expanded ecological consciousness, an awareness of the global system in its complexity and connectedness. The spirit of humanity depends on an ecological consciousness that would place humanity in a proper relation to the wild places of the earth, taking what it needs, but letting the rest be.

Theodore Roszak has identified the needs of the person and those of the planet as one. We need a wild universe to live fully. We need wild animals and plants to be fully conscious. When we understand our roles in nature, then we will not be stewards or managers, but participants and sharers of experience. If we are not sure how, then we will have to act as if we were wise, as was recommended by Hans Vaihinger. Let the first step be an acknowledgement of fellowship with animals. There is a prayer in the Liturgy of Saint Basil that reads (in Niven, 1967): "O God, enlarge within us the sense of fellowship with all living things, our brothers the animals to whom thou hast given the earth as their home in common with us. We remember with shame that in the past we have exercised the high dominion of man with ruthless cruelty, so that the voice of the earth, which should have gone up to Thee in song, has been a groan of travail."

Chapter 44

Nature as Self

Many ideas of nature are not objective or scientific. They underlie thought. Their insights are embedded in language. For example, the root word of both ecology and economy is based on the Greek word for house; ecology is its study, and economics is its management. The word house is used as a metaphor for nature. Similarly, other metaphors are used for nature, so that nature is seen as mother, father, sister, brother, and self. These words are applied in different ways to describe nature. Understanding attitudes towards nature, and the bases for them, could lead to a more benign ethics, to healthy human beings in healthy environments. Such an ethics could be based on love and respect, that is after all the proper attitude towards family, relatives and the self.

Nature as Home
The biologist Adolf Portmann observed that insects and animals display a powerful attachment to specific places, an attachment best understood in human terms as home. Human beings feel strongly about places. Several metaphors have been used to describe the human place on earth. The earth is a storehouse, property, a spaceship. But the earth is not a spaceship or property or storehouse; it is home. Victor Ferkiss proclaimed that: "The world and humanity are one entity, one system in equilibrium. Earth is humanity's only home; humanity is one people in relationship to the earth."

There is a wide variety of meanings of home. It is a place of family residence, the family social unit, habitat, and place of origin. The word 'home' comes from the Middle English word (*horn* and Old English *ham*, Old Norwegian *heimr*, Greek *kome*, and Sanskrit *kayati*) meaning village or home. Its spectrum of reference is enormous. The word is used to describe house, village, city, bioregion, cultural world, and the earth. Its content is also ambiguous. Home can hold a single person, family, relatives, pets, domestic food animals, neighbors, and others.

Living at home occurs simultaneously on different levels; the importance may shift from city to nation, or from nation to state, or from house to bioregion, or from state to habitat. Each level serves as a metaphor for the next. There are parallels between nature and a house, as the basis for home. Solar space is like the landscaping; wilderness is the foundation; conservation areas form the shell and provide services; and each bioregion is a unique room. The analogy cannot be carried too far, but it shows that a house is not, as Le Corbusier said, "a machine to live in." It is a matrix for home. Home is not just a house, either; it is a complex of significant events centered in place. It is the foundation of our individual identity on one level, and our role in the community, on another. What makes home different from house? Participation in the making of it, commitment to it. People invest parts of themselves in a place in making a home. A home is a part of the environment claimed by feeling. Emotion creates an 'in-place'. All beings find and make a home. Humans, like plants and animals, identify greatly with local environments. (Maybe this is a function of the limbic system of the brain, a function we share with territorial mammals.) Being away from home results in nostalgia, a 'disease' identified by Johannes Hofer ,a Swiss medical student, in 1678 to describe an illness characterized by insomnia, palpitations, stupor, fever, and persistent thought of home. The disease could result in death. While nostalgia is not considered in organic etiology by current medical science, the symptoms are still manifested psychologically and physically. The connotations of place have been stripped from the meaning of nostalgia, the use of the word has been trivialized, and the symptoms have been reassigned, but the disease still erupts, unnamed, and its effects are everywhere. For many, it is not possible to stay away from home indefinitely and still live. Thus far, the sense of place cannot be gleaned from an analysis of the nervous system. Yet a place shapes the nervous system, somehow. The relationship of human beings to nature is deeper than just home.

Nature as Mother and Father
Gary Snyder discerns an undercurrent in civilization since the late Paleolithic. He considers Buddhist Tantrism to be its finest and most modem statement: "that Mankind's mother is Nature and Nature should be tenderly respected; that man's life and destiny is growth and enlightenment in self-disciplined freedom; that the divine has been made flesh and that flesh is divine; that we not only should but do love one another ... these values seem almost biologically essential to the survival of humanity."

Homer sang "of Gaia, universal mother,/firmly founded, the oldest of divinities." Goethe found that deep knowledge could only be sought in "the realm of the mothers." The idea of nature as mother forms the basis of a modem, scientific hypothesis, to explain how the planet exerts a living control of the atmospheric and hydrologic processes to maintain minimum conditions for life over long periods of time. James Lovelock believes that there is a collective global mind (however unconscious) immanent in the cybernetic structure of the global system. He calls it Gaia, after the Greek earth goddess, as suggested by William Golding. When ecology strives to think of the planet dynamically and holistically, it returns to personification, "Mother Nature". Lovelock's Gaia is a metaphor designating a field of atmospheric study; technical analysis follows from the metaphor.

The hypothesis notes that: the average surface temperature of the earth has moderated, despite a gradual rise in solar energy; the concentration of the atmosphere is improbable, compared to the composition of Venus and Mars — it should be mostly carbon dioxide; each atmospheric gas is in the optimal proportion for a life-supporting function; the salinity of the ocean is far lower than it should be from runoff from land--the present percentage of salt in water could have been achieved after a mere eighty million years. Lovelock concludes that the chemical and climactic properties of the earth have always been optimal for life. Since this could not happened by chance, the explanation is Gaia, who he defines as: "a complex entity involving the Earth's biosphere, atmosphere, oceans, and soil; the totality constituting a cybernetic system that seeks an. optimum physical and chemical environment for life on this planet."

The earth's biomasses, air, oceans, and lands form part of a giant system which is a single organism. Life exists as a consequence of the right material conditions. Life defines the material conditions needed for survival and then tries to maintain them. The earth's biosphere controls the temperature of the surface and the composition of the atmosphere, from major constituents to trace elements. The system has maintained control over instabilities for millions of years through a variety of responses. (Just as the human body maintains homeostasis.)

At a scientific level of thought, the Gaia hypothesis extends the fundamental ecological doctrine that all things in nature are densely, subtly, and systematically interrelated until they include humanity, ethically and mentally, as well as physically. The entire earth is envisioned as a unified entity, actively shaping the material conditions of the planet for the purpose of maximizing the survival and variety of living beings. Nature is the fundamental matrix (from the Latin *mater*, meaning mother, maw, or void) for human development. Nature is the source of life, on which humanity depends. The matrix is historical. It has duration and it extends from the past into a future of following lives.

The metaphor of "Mother Nature" enables bonds of kinship and responsibility in human communities directed to the earth. However, the feeling of subordination to an indifferent mother could have a negative effect; Paul Shepard suggests that resentment, violence, and guilt might result. Effective metaphors require a level of maturity in individuals as well as in cultures. The metaphors are not literal. Nature may be indifferent, but permissible. The metaphor of mother does imply that nature is peaceable and nurturing. Often, cooperation contributes as much as competition in shaping species.

But, life is violent as well as peaceable, and this is reflected in philosophies and myths, and violence is often associated with the image of 'father'. Heraklitus regarded conflict as the father of all things. Many archaic cultures, such as the Tukano Indians of the Northwest Amazon, addressed the sun as Father, the creator of earth and all life. Many peoples refer to their home territories as fatherlands. On local levels, animals and plants have 'fatherlands;' they live together and are sympatric (from the Latin word for father, used to mean occurring together). Not all species occur together in the same place. Sympatry describes only those that do. At the habitat level, allopatry (occurring apart) is "intelligent" use of available resources by animal communities. Large herbivores, such as elephants and rhinoceros, may choose poorer quality food and avoid competition with smaller animals and, thus, exploit an untapped food source. Many interactions between different species contribute to the mutual benefit of the members of community, as well as to the community itself. Humanity must be allopatric with most wild species and allow them to develop independently in their own places.

Nature as Sister or Brother

Saint Francis of Assisi, in "The Canticle of Brother Sun," addressed the Sun, Air, Fire, Wind, and Water as his brothers, then the moon and stars as his sisters; he praised the earth as his mother. American Indians, such as Black Elk and Seattle, also referred to animals as brothers and sisters. All animals, "Two-legged and four-legged," were equals. The phrase, "all our relatives," was used in prayers and rituals referring to plants and animals as well as to human kin. Science is only beginning to support this kind of closeness. Adolf Portmann shows that every form of life appears as a gestalt, developing in a specific place. All living forms create an image of their environment. Genetics provides the proper image for some. Frogs, for instance, focus most closely on objects that have the same size and trajectory as flies. Others, such as coyotes, must learn what is valuable, what is edible and preferable.

Animals have their own perceptual universes that are strange and fascinating. Jakob von Uexkull suggests that the unfamiliar world of an animal can be represented with a bubble to denote the self-world, or phenomenal world, of that animal. According to von Uexkull, perceptual and effector worlds form a closed unit, the "umwelt." "Figuratively speaking, each animal grasps its object with the two arms of a forceps: receptor and effector. With the first, it invests the object with perceptual meaning, with the second, operational meaning."

The animal world, its life-image, is what has meaning for the organism. It is a focus. The first principle of a life-image theory is that all animals, from the simple to complex, are "fitted to their unique worlds with equal completeness." A simple world corresponds to a simple animal; a well-articulated world to a complex animal. Von Uexkull implies that the human world is only one of many possibilities. Animals are not suboptimal beings relegated by evolution to second-rate habitats. Instead, they are optimally fitted to places that humans are not fitted to.

The theory of life-images is a basis for a new, genuinely nonanthropocentric metaphysics. Much earlier, some of the pre-Socratics developed a nonanthropocentric world view. Zeno the Stoic preached "life in agreement with nature" as the goal of ethics. Chrysippus added that as individual natures were parts of the nature of the whole, therefore, life was to be in accord with human nature as well as nature. Francis of Assisi tried to unite the compassion of Christianity and the animistic sense of union with the natural world. Natural processes take on an expression of significance of their own without reference to humanity. All beings have an "ultrahuman" (coined by Richard Jeffries in John Fowles) value of their own. St. Francis tried to depose man from his monarchy and set up a democracy of all God's creatures. The Taoists saw that humans are indistinguishable from other creatures; if they seem distinguishable, it is only because of their feelings of self importance. Lao Tzu turned the pyramid of human values upside down. He considered the laborer more successful than the aristocrat; cultivation of the inner life is more important than high status; physical enjoyment is more rewarding than constant acquisition. And, as there are more laborers than aristocrats, there is a net gain to the success of the community. Arne Naess offers a biospherical egalitarianism, where all beings have an equal right to life and fulfillment.

Nature is not anthropomorphic, in the image of man. Nor is it anthropocentric, centered around man. But, it is measured and valued by man, as, indeed, it is measured and valued by all beings. When humans evaluate ecological situations, preference is usually given to human values. But, there are other beings that are measuring their parts of habitats. There are other centers, and these centers are equals, brothers and sisters. (The terms derivative from brother and sister are more peripheral. The Latin terms for brother and sister yield 'fraternal' and 'cousin.' The Greek terms are twins, taken from the word for womb, and *delph*, on which 'delphinium' and 'dolphin' are based.)

Nature as Self

Our bodies contain the ashes of stars; human cell structure is shared with trees; we share our bodies with bacteria, fungus, insects, many of which are beneficial--and even those not considered beneficial may have positive effects on our health. As Lewis Thomas shows, our human bodies are living communities, hosting amoeba in the blood, mitochondria in the cells, bacteria in the intestines. We are connected to the largest and smallest beings.

In fact, humanity is embedded in the earth, according to Maurice Merleau-Ponty. From the oldest language we know, the Indo-european tongue, we took the word for earth and turned it into humus and human (*dhghem*=earth; *humanus* in Latin; human in English). Yet, the word for man was shaped into man image, world (Indo-European *wiros*=man; *weorold* in O. English; world in modem English). One word progresses from earth to human, the other from human to earth. We refer to the earth literally as world, 'man-image.' We cannot be any closer to the earth and its processes, since the parts are combined in us. We are indissolubly one with nature.

We have mistakenly concluded that our skin is the boundary to our selves. But, our intuition senses our interdependence with nature. We extend the boundaries of personality to other things and people. William James claimed that, beyond the body, the immediate family (father, mother, spouse, children) was part of the self. For Carl Jung, the 'Self' guided and integrated the whole of psychic life, conscious and unconscious. The concept of individuation, the process where a person discovers and evolves her Self, is central to Jung's psychology. This Jungian Self is awareness of our selves.

We participate in relationships in a field of relationships. Because we are in the field, the study of nature is, to some extent, the study of ourselves and our effects on the field. The individual self is not a skin encapsulated ego, but an organism / environment field. The organism is a point at which the field is focused.

Paul Shepard likens the human skin to a pond's surface. The skin's interpenetration ennobles and extends the self — the beauty and complexity of nature are continuous with ourselves. We know subjectively that we are not separate from the earth, that all other beings are necessary parts.

Perception of the body as landscape and of natural terrain as a body is as fundamental to psychology as it is to mythology. We depend completely on the natural environment, physically and psychologically. D.O. Hebb has conducted experiments that show the effects of a limited environment. Cut off from external stimuli, the mind becomes strange and distorted. Mental health can be related to the quality of the landscape, as Rene Dubos and others (e.g., John Passmore, Shepard, and Ian McHarg) have done. The external world is needed to keep us alive and sane. This world is composed of remote occurrences, on polar icecaps and distant stars, as well as immediate personal events. The individual is woven into the world.

If nature is a body, then it has vital organs. Certainly parts of nature function like organs, circulating nutrients and minerals and cleaning wastes. Nature is the body of our species. We can do without some of it, but not without all of it, as we can live without one kidney, much of a liver, or arms or legs. Human beings could sell their 'spare' organs if they chose. Not to sell them, in fact, is to forego an advantage of the resources of our bodies in an economic sense. Most people don't sell, however, because feeling whole and healthy is more important than the temporary income. There is an important parallel with nature.

Nature as Itself

G. Spencer Brown understands a much wider concept of self. In describing the conception of form, Brown notes that the self constructed in order "to see itself'. But, in order to do so, it must divide into one state that sees and another that is seen--it must become distinct from itself. In this sense, the world has divided and subdivided itself. Whenever another division is made, a self--Brown says a "universe"--comes into being. The skin of an organism only cuts off an inside from an outside. But, the skin is permeable.

The earth has innumerable modes of being that are not human modes. Our direct intuitions of nature tell us that the earth is infinitely strange; it is alien, even when gentle and beautiful. It seems often mysteriously impersonal, unconscious, immoral, hostile, awesome. J.B.S. Haldane recognized the strangeness of nature. "I have no doubt that in reality the future will be vastly more surprising than anything I can imagine. Now my own suspicion is that the universe is not only queerer than we suppose, but queerer than we can suppose." Perhaps the queerness results from sheer complexity. George Perkins Marsh believed that the equation of animal and vegetable life was "too complicated a problem for human intelligence to solve, and we can never know how wide a circle of disturbance we produce in the harmonies of nature..." Barry Commoner echoes them both: "not only is nature more complex than we think, but perhaps more complex than we can ever think." In its immense complexity, nature seems wholly other, nonhuman, ultrahuman. It seems distant. So it is feared as unfathomable and uncontrollable. Nature seems contradictory and sinister, shaped by death, which we fear. We fear to understand, to be compassionate. And, fear casts out love and, with love, goodness, beauty, truth, and intelligence. Until all that remains is fear of other beings and the unknown; fear of the smiling science and technology that take away more than is given; fear of fellow human beings, who are trying to regain what was taken.

But, love casts out fear. In the Upanishads it is written that "Who sees all beings in his own Self, and his own Self in all beings, loses all fear." As fears and unconscious motives are understood, the awareness of all feelings intensifies. Feelings that are dualistic at one level--fear and courage, pride and humility--are combined at a higher level. Unconditional love blends many feelings that cannot be understood at an intellectual level. Erich Fromm identifies four elements in loving: Care, the active concern for life and development; Responsibility, the desire to respond to others needs; Respect (meaning to look at), to recognize others' uniqueness; and Knowledge, combining objectivity with participation and intimate identification. These elements define a loving relationship. The inexhaustibility of a being or of relationships constitutes much of the nature of love. Human beings are compelled to seek other beings and love is the only approach. Seeking in their hearts with wisdom, the sages in the Rig Veda (X. 129) found that love was the first seed of the soul. Nature has evolved the seeds, as Alexander Pope understood. "beholde chain of Love/Combining all below and all above." Love exists in the conversation between human beings and other beings. Conversation is not limited to two individuals or to the present.

In the sense of living together, love is ethical (the word "symbiosis" means 'living together' from the Greek; ethics means 'doing together' from the Sanskrit). Abraham Maslow presents "love-knowledge" as unlimited. It is a path to objectivity with greater perception, which provides kinds of knowledge not available to nonlovers. (Maslow cites his work with monkeys. Konrad Lorenz, Michael W. Fox, George Schaller; and Jane Van Lowick-Goodall have found it to be true. A good teacher, parent, scientist, or friend functions this way.) Love creates an openness to experience, without judgment. Beings unfold. Love expands the awareness of self and other beings. Its intimacy permits distance. Its duration reaches future generations of beings. Love personalizes the universe, but keeps it free (the word 'free' is from the German, *freier*, meaning to love or to woo).

We cannot approach beings as they are through our personal and economic interests, but only on their own terms, in relation, through respect and love. Any other approach separates us from other beings and truncates our aesthetic responses with boundaries. The word animal means endowed with spirit (from the Latin *animus*). Our spirituality places sacredness in everything. We are part of the cycle, woven into a poetic, mythic unity. But, the unity may not be comfortable, and nature is not a father or mother or any entity of our wishing. It merely is. We are and dwell in metaphors in it These personal metaphors are as true as the metaphors of machines, and are a useful counterbalance to the scientific image of nature as objectified data.

Myths and metaphors are modes for conveying ecological wisdom; they are less concerned with survival than the survival value of a good fit. Myths provide equilibrium between self-restraint and self-expression, between self-protection and self-expansion. Myths also limit human cultures, so that other beings (brothers and sisters) can make their homes in their places (fathers and mothers lands) within the body of the earth. Wisdom cannot depend on perfect knowledge of other beings or places; such knowledge does not exist. Humans must act "as if" (in Vaihinger's words) they were wise, that is, circumspectly, with caution and respect, as if nature was our very self.

Chapter 45

Panethics
Ecological Rules for Living Together

Introduction
In its popular application, ethics is a set of rules governing the conduct of a group of people, based on traditional behavior of that group. Bioethics is an extension of the rules to interactions between medical practitioners and patients, based on medical and philosophical values. Both ethics and bioethics are limited cases of a pan ethics in an ecological context. An ethics abstracted from historical context, detached from other societies, and alienated from nature is academic, insular, and strange. Such ethics tend to depend entirely on local human utilitarian values. Philosophical systems are little better:

Kantian ethics is unsuitable for treating other beings in nature; deontological ethics has a weak subjective foundation in human conscience. The areas of concern of ethics and bioethics are not broad enough; their foundations are not deep enough. An ethics based on ecological knowledge, by comparison, places human behavior in vital social and biological communities in nature — birth, death, illness, and sex all take place within nature. The frame of reference of ethics is enlarged, leading to appropriate behaviors in a larger context. Human health is related to the health of ecosystems. This ethics, pan ethics, addresses animal rights and wilderness preservation, as well as human concerns.

Plain Ethics, Living Alone
The subject of ethics is human conduct, by definition. Ethics are almost exclusively centered on human action. Ethical theory attempts to give fundamental reasons why actions are right or wrong. Most ethical systems are based on utility and control.

Utility and Detachment
People in many modern societies are learning to be uncaring, unattached, and uninvolved, without really thinking about the effects on society and on the environment. Perhaps, it is a result of confusion or fear. We still fear nature because it is uncontrolled and unfathomable. We distance ourselves from what is uncontrolled or unknown. This detachment is the greatest threat to the welfare of nature. William James called for the moral equivalent of war against nature. But nature and civilization are not in an adversary relationship. Nature is not a threat. The ignorance of nature is a threat. And the false cosmology of industrial progress is a threat. The ambiguity of the word "nature" reflects the doubts and uncertainties that humans feel about nature. Nature is alien: other. As we have isolated ourselves in human artifacts, we seem to have lost touch and grown cold-hearted. Nature is separated by a wall of incomprehension. We see wilderness through a glass wall, as we see most all things. Through the thick, convex glass of

utilitarianism or the tinted glass of romanticism. Allowing wilderness a place to develop might remove the glass, as would recognition of our continuity with nature.

The attitude of distance toward beings excludes consideration of necessary information. Science accepts the sentience of plant life, but does not adjust its methods. Human knowledge grants emotions to animals, but uses them badly anyway. When we can control nature, for whatever purpose, the result is not always good. Our style of efficiency and proclamation of mass values destroys places. The environment of significant places becomes a flatscape. It is turned into uses. John Fowles warns that "We shall never fully understand nature (or ourselves) and certainly never respect it, until we dissociate the wild from the notion of usability — however innocent and harmless the use." This attitude of uselessness lies at the root of our hatred and indifference. The problem with utilitarian ethics is that it permits the use and exploitation of any natural object, including human beings. Based on a limited science, the ethic fails to value those beings and communities for which no use is known. But, the majority of the beings in nature have no human uses. Even ecologists cannot think of uses for many large birds and mammals. This makes a coldness in the heart of our coexistence with other species. The vivisection of the world depletes our ability to feel compassion for it.

Human Center or Human Measure
The first cosmologies, images of human worlds, were anthropomorphic. Human beings saw human forms in every form of nature. With Plato and Aristotle, nature became anthropocentric, it turned around a center, humanity. Humans were the most important beings, at least through the Middle Ages. By 15th century European and Arabic standards, the universe was a rational order. The human place was prominent, and human life had meaning and purpose. Then, the Copernican revolution transformed the universe from geocentric to centerless. The biological universe, however, was still considered to be a great chain of being. Humans resided between the beasts and angels — until Darwin linked them too closely with the beasts, and cosmology became less meaningful. Then the industrial, scientific revolution restored human importance in a cosmology. Nature is seen exclusively in an anthropocentric manner, as a human resource (especially by John Dewey). Even the Biosphere Reserves are justified according to anthropocentric use. To assume that evolution necessarily progressed to humans ascribes an anthropocentric purpose to nature. But for environmental effects, dinosaurs, birds, or whales could be the dominant species. So humans are not the unique end or goal. In fact, like new dinosaurs, humans are good competitors, suppressing other species and creating their own pseudo-species. All fields of study are trying to confirm that man makes himself, according to Paul Shepard, no matter how the world is made. Geography endorses economic determinism; history studies the rise of Promethean civilization; the arts separate abstract qualities from content; sociology encourages the theme that everything is possible; the sciences present value-free facts. Ideas are no longer connected. All aspects of life have become

interchangeable, including soil, water, and land. All concepts of natural seem to turn on the definition of human. The anthropocentric mistake has been the choice of an oppositional logic, instead of a synthetic one. The either/or complex misses the relationship between the terms.

Some of the pre-Socratics developed a nonanthropocentric world view. Zeno the Stoic preached "life in agreement with nature" as the goal of ethics. Chrysippus added that as individual natures were parts of the nature of the whole, therefore, life was to be in accord with human nature as well as nature. The universe is not anthropomorphic, in the image of man. Nor is it anthropocentric, centered around man. But it is measured and valued by man, as, indeed, it is measured and valued by all beings. When humans evaluate ecological situations, preference is usually given to human values. This is unavoidable. But there are other life-images that are measuring parts of habitats. There are other centers. The concept of reverence allows the center to be everywhere. This idea has reference in earlier philosophical ideas. The atomists advanced the idea of an infinite universe without center or edge. In the 15th century, Nicholas of Cusa argued that the universe was without edge or center — because its creator was infinite and without location. The universe is "a sphere of which the center is everywhere and the circumference is nowhere." The earth was not the center. The modern theory of containment reflects that attitude towards centers and edges — ontofugism, not centering, but scattering, framing. A genuinely nonanthropocentric metaphysics would be a fitting foundation for ecological ethics.

Reverence for Life
Albert Schweitzer believed that ethical thought had been developing since prehuman history and that it culminated in the principle of reverence for life. Schweitzer challenges us to plunge into the adventure of ethics: "Let it dare, then, to accept the thought that self-devotion must stretch out not simply to mankind but to all creation, and especially to all life in the world within reach of humanity. Let it rise to the conception that the relation of man to man is only an expression of the relation in which he stands to all being and to the world in general."

Our attitudes are grounded in a belief system that constitutes a particular world view. The system constitutes a coherent whole. With Schweitzer, the system began to shift toward a biocentric outlook. The concept of reverence (*Ehrfurcht*, meaning honor-fear in German) offered some respectability for nature through a proper attitude. Schweitzer proposed an ethics derived from Christian ethics (but really larger) that affirmed the world. Reverence for life sometimes conflicts with the Christian paradigm, however, which is just one particular manifestation of a reverential ethic, and Schweitzer's version could lead to an instrumental ethic. Furthermore, Schweitzer's reverence entailed a constant effort to make excruciating decisions. His attitude was a *noblesse oblige* toward lesser species, based on a Christian idealism, on the fear of death, and on a mistaken image of nature as brutish and dumb.

Through human evolution, the circle of responsibilities has widened, from the family, to tribe, nation, humanity, and now toward all life. Although

Schweitzer noted that the circle of knowledge was widening also, he felt the streams were divergent, that ethics had nothing to gain from understanding nature, and furthermore, that there was no hope of finding meaning in natural phenomena. His ethic was not based on ecological knowledge. His reverence for life principle acquires a new aspect when it is restored to ontologically and ecologically firm ground.

Ethics and Knowledge

The effectiveness of ethics depends on a knowledge of the way the world works. The Aztecs based an ethic on imperfect knowledge of the sun and suffered disastrously. The correction for ignorance of ecology is knowledge of ecology. The correction for nonacceptance of death is a generalized concept of reverence for life. Death is a part of life, but not a driving force. The reverence for life has to be based on an ecological ethic that understands the necessity of predation as well as altruism — killing as well as saving. Reverence for life must include awareness of natural laws. Wolves need deer and mice to survive, as much as those need wolves to be healthy. Human intervention into natural communities must be responsible, not sentimental. Knowledge by itself, however, merely permits a more efficient utility. Knowledge cannot be the sole basis of decision making. It is always incomplete and therefore cannot describe all aspects of the earth that bear on human life or environmental quality. Knowledge must be humane. Abraham Maslow saw the organism as having biological wisdom; it can be trusted as autonomous, self-governing and self-choosing. To treat organisms, and nature in general, science could shift to a taoistic approach, asking rather than telling, observing rather than manipulating; receptive and passive, not active and forceful; "nonintruding," and noncontrolling, not forcing. A taoistic approach stresses noninterfering observation rather than controlling manipulation; it is receptive rather than forceful. This is part of the paradox of duality; it is detached yet concerned; free yet committed; and independent yet responsible.

Deontological ethics

One particular kind of theory has been proposed as being appropriate for an ethics of nature. This theory, that holds the rightness of an action to be determined in whole or part by considerations other than results (as with utilitarian ones), is a deontological theory of moral obligation, or 'formalist' theory. Results may be irrelevant or relevant, but not sufficient.

W. D. Ross (in Rader 1964) tried to reach a compromise between the deontological doctrine of Kant and the ideal utilitarianism of G. E. Moore. The weakness of Kant was an excessive abstractness and formalism. Without an appeal to consequences, the formula of the imperatives is not plausible. Moral rules permit no exceptions. Utilitarianism, on the other hand, unduly simplifies our moral relations to fellows; we are possible beneficiaries to every act. Furthermore, utilitarian right is based too exclusively on future results. Duty must be equally retrospective. The *prima facie* (meaning 'at first sight' in Latin) duties of Ross are conditional. Every conflict must be resolved by weighing the respective claims of all such duties in that situation. Ross identifies such duties as the obligation "not to injure others," and "to do good to others." Deontological theories seek moral guidance from an inner faculty, usually the conscience. But the rules of the conscience are not really self-evident. Rader provides instances:

1. the self-evident moral rule turns out to be ambiguous, with different selves in different cultures (what is the difference between murder and killing for food?)
2. moral rules can conflict (do not lie and do not cause suffering, or do not steal and cause suffering)
3. the injunctions of conscience vary with time and place (such as attitudes toward infanticide or homosexuality).

The nonreflective conscience that deontological ethics depends on is custom, so ingrained as to be self-evident. Perhaps this is why newer social crimes, such as deliberate overcapitalization (theft), biased news reporting (lying), and saturation bombing (murder), are excused more readily. Deontological ethics maintains that the moral quality of acts depends on conformity to rules, rather than on goals and consequences, as does the teleological or utilitarian view. Deontological duties are stringent. Rightness is independent from goodness. In fact, duty should be done, even if it "detracts from the balance of values in the universe." Deontological ethics still depends on human conscience, without regard for sufficient knowledge.

A basic assumption in authoritarian and intuitional ethics is that moral acts can be evaluated as discrete, self-contained units that have good or bad characteristics. The modern taboo forbids confusing is and ought, fact and value. But they are linked inextricably. Moral judgments are more than subjective feelings. They derive from a consideration of facts. The fact-value dichotomy can be overcome with naturalism or with ecological knowledge. There is nothing that does not have some value.

Pan Ethics, Living Together

A new ethics starts from nature itself, not from an extension of the old anthropocentric ethics, with all its limited assumptions. It is based on ecological knowledge, grounded "in the breadth of being," in Hans Jonas' words, and founded on principles discovered in existence. Many ultrahuman cultures have standards (or codes) of behavior to regulate interactions. In birds and simple mammals, these rules may be very rigid and predictable. With increasing brain complexity, however, learning takes a larger role.

Animals are ethical already; they live together by rules; they respond to situations to other animals. Ultrahuman ethics are rules for living together. The human community is one of many communities that make up ecosystems. Human relationships embrace other beings as well as other humans. Human ethics describes a small part of the rules, perhaps the only self-conscious part.

An ecocentric view emphasizes human moral responsibility for vulnerable ecosystems and habitats. An ecocentric view recognizes the ends and means of all beings. An ecocentric ethics can place human ethics in proper perspective, in proper relationship with other living, interacting beings. Understanding ecological relationships permits the toleration of fluctuation, irregularity, uncertainty, diversity, spontaneity, flexibility, looseness, and limits. The basic premise of nature is interrelatedness. The interpenetration of boundaries makes humans less discrete, less alone. A shared biology establishes the fellowship of beings.

Ecological Ethics
Humanity is an integral part of food chain and part of an organic cycle of birth and death. Humans need to recognize that they automatically participate in everything, and that they cannot unparticipate by choice. Participation starts at the quantum level, through the ecological and cultural. Human nature does not find meaning in an absurd world, but discovers its structure through interaction with the ultrahuman order. Human identity exists partly in relation to nature; the destruction of one involves the other.

The word ethics is derived from the Greek word meaning 'custom' (Greek *ethos*, Sanskrit *svadha*, Latin *mores*, plural of *mos*, from *meare*.), which itself came from the Sanskrit word for one's 'own doing.' Since it was used in the plural, it meant 'doing together.' The word 'morality' comes from the Latin word for will of the people; the singular meant the 'will' of a person. It was probably derived from the verb 'to measure,' as to measure one's way, to go one's way. Morals means the 'way of going together.' Ethics means 'doing together,' which of course one does living together. And, in an anthropometric universe, it is entirely appropriate.

Ethics are assembled inductively, from experience in living in places. Because of the uncertainty of human actions, ethics has to encompass the far past and distant future. No one knew that when DDT killed mosquitoes, it would concentrate in the food chain to kill birds. Values are time dependent, and ecological time can be very long indeed. The futures we invent are viable only if compatible with constraints imposed by evolutionary past. An ethics that requires a long-range responsibility also requires a new humility, since technological power exceeds the ability to foresee its consequences. An ecological ethic recognizes the moral obligation to leave the world habitable for future generations.

Aldo Leopold proposed a conservation ethic, dealing with human relationships to land, plants and animals. The land ethic Leopold had in mind was a sense of ecological community between humanity and other species. "When we see land as community to which we belong, we will use it with love and respect." Such an ethic would change the human role from master of earth to plain member of it. Predators are members of the community; and no special interest group has the right to exterminate them for the sake of benefit for itself. This attitude is important for habitat protection. Leopold describes the extension of ethics as "actually a process in ecological evolution. Its sequences may be described in ecological as well as in philosophical terms. An ethic, ecologically, is a limitation on freedom of action in the struggle for existence. An ethic, philosophically, is a differentiation of social from anti-social conduct. These are two different definitions of one thing. The thing has its origin in the tendency of interdependent individuals or groups to evolve modes of cooperation."

The extension of ethics to animals and land is an ecological necessity.

Extended ethics defines a social conduct that is a mode of cooperation and, ultimately, symbiosis. Leopold argued that ethics are voluntary limitations of freedom, necessary in a complex world of which we remain incredibly ignorant. Ethics are developed in response to problems that arise from increasing knowledge. Science has phenomenally increased our knowledge of physical and biological processes. It has now become the basis of our moral code, but it cannot very long be a science divorced from feeling and art if that code is to help us survive. To do this science requires aesthetic perception as well as disciplined thinking and feeling. As there is a rational component to ethical judgments, so there is an intuitive and emotional one, also. An evolutionary ethic suggests that humans avoid tampering with complex evolved systems, not because they are good, but because they are the basis of life at this stage of development. Ecological ethics is situational because ecology is the study of changing systems. The morality of the act is determined by the current state of the system. Adaptive modes should conform to ecological patterns. An ecological ethics is based on attributes of ecosystems and human compliance with ecological laws. The aim of an ethic must be harmonious to the idea of the world's population of living beings.

Ethics is a series of rules for living together. Most sets of ethics make the rules easy to follow. They emphasize the differences (relativism) or similarities (absolutism) of human beings only; or of the individual or the group; or of good feeling, reason, or desire. But ethics has to confront the individual, embedded in a community, located in a bioregion, on earth. And the rules really are not as easy as human systems have presented. Schweitzer made them too difficult, with a constant valuing, but neither are they that difficult. An ontological ethics can be detailed only on a local level – even when it uses a global strategy.

Individualism can be tempered with the concept of common good, the good of the whole. The whole is diminished by individual loss, but the individual is crippled by the loss of the whole; wild children, for example, are never really human. Human good cannot be considered apart from the common good. We are in larger communities like an organ in a body. In *The Laws*, Plato has the Athenian say to a youth that all things are ordered with a view to the preservation of the whole, each portion contributes to the whole, and every other creature is for the sake of the whole. Ethics has expanded in wholes, from the family, to the human community, and to the ultrahuman community, on which all depend.

Natural Value
Humans have stripped the world of qualities and significance and claimed them for themselves. By valuing humans alone, value is made subjective, and ends are without value. The human perspective realizes only a small part of the spectrum of rich possibility of experience. Humans use an incomplete source of value for ultrahuman beings; the source is human need. Moral science is not so much concerned with good and bad or right and wrong as with fulfilling needs, such as survival, reproduction, or prestige. Abraham Maslow established a hierarchy of human needs, beginning with food and continuing through social acceptance to self-actualization. But

human needs are based on the health of the earth. Human needs extend to include a foundation of wilderness. Nature, which is self-supporting and self-managing, is the human life-support system. Human systems depend on natural ones, for recycling of wastes, water, and air. But, as human growth is logarithmic, so is human need, and need shapes facts, like kind and quality of resources. As Goethe recognized, all fact is theory, a blend of perception, imagination, and needs. Granting this, the need for wilderness is as much a fact as the need for food.

Wilderness is ecologically important, for it accounts for ninety percent of the energy trapped by photosynthesis from the sun; it is crucial in the global energy system. Trees and plants are good sources of energy. Energy and materials can be produced through photosynthesis. In fact, wilderness is the greatest producer of renewable sources of energy and materials. As habitat for incredible number of species, it is insurance against the dangers of simplified agricultural systems; it is a depository for genetic types. Biological species are essential to the maintenance of ecosystems. Wilderness is the source of evolutionary process.

Rich sensory and emotional experience can be derived from contact with the wilds. These values are not often mentioned by economists and planners. But values usually encode information having survival or prestige importance. Perhaps the most valuable thing is living time. This may be why humans value walking in the woods or observing the production of art. The value of wild nature is its independence and wildness.

Wilderness is the given primary reality, the matrix for life-images. If we can admit the independence of nature, that things continue in their own complex way, we may feel more reverence for them. Wilderness evolved without human help or interference, in equilibrium with the physical-chemical environment. We create human landscapes out of wilderness, from wilderness. The value of wilderness is that it cannot be reproduced. It can be apprehended only in nameless present by living senses. It can be known and entered in person, not behind glass.

We find that kind of knowing more in literature than in science. Whitehead turned to literature because it corrected the excess of objectivity that occurred in science. Poetry goes further and carries an expression of the organic character and value of nature. Value refers to the in-itselfness and for-itselfness of the process of realization. Besides having value for itself, it shares value with the rest of the universe. Everything has signification in the universe, says Whitehead: "Remembering the poetic rendering of our concrete experience, we see at once that the element of value, of being valuable, of having value, of being an end in itself, of being something which is for its own sake, must not be omitted in any account . . ."

Perhaps there is a hierarchy of value from simple forms to complex, corresponding to richness of experience, as John Cobb suggests. Although there is no hierarchy of being, there may be one of richness or value, depending on the frame of reference. There is value for each, and for the whole. And the whole may invert the value. Ecology can expand the narrow anthropocentric evaluation and see things from the perspective of the whole. Lovejoy's 'Great Chain of Being' traces the deductive order in classical nature from Greeks to German idealism. But Lamarck inverted the chain in his theory of transformism; mind is immanent and can determine transformations. Although the hypothesis of inherited characteristics was rejected by Darwin, who shared that hypothesis but denied mind as an explanatory principle, both Lamarck and Darwin inverted the value of life. By inverting the great chain of being, Lamarck escaped the directive that the perfect must precede the imperfect. The result of Elton's food chain was the realization that the bottom link — plants — is the most important.

Ethical values share the same fate. Francis of Assisi was the exception to the general attitude of Christianity: compassion to man only. He tried to unite the compassion of Christianity and the animistic sense of union with the natural world. Natural processes take on an expression of significance of their own without reference to humanity. All things have an ultrahuman value of their own. St. Francis tried to depose man from his monarchy and set up a democracy of all God's creatures. In parallel, the Taoists saw that we were indistinguishable from other creatures; if we seemed distinguishable, it was through our feelings of self-importance with our reason. Lao Tse turned pyramid of human values upside down. As there are more commoners than aristocrats, there is a net gain to the success of the community, according to Holmes Welch.

Ecological value starts at the broad base of the pyramid. Our ethics and legal system is species specific (Singer 1981), but it occurs in a larger, normative order. Ecological science has a very normative component. Even a set of genetic codes is normative; genes show what is valuable for a life-image. So value is morphic on all levels of being. A sort of natural relativity of frames of reference encompasses the smaller human system.

Perhaps we should not argue that things have value in the human system. Let us just respect the ultrahuman system. Bees have bee value; wolves have wolf value. Wolves are not efficient at binding nitrogen; neither are humans. Lichen are poor predators, but they break apart rock better than bighorns. The world would not be a better place without sharks, silverfish, cockroaches, rats, hyenas, or whales. Their existence has value; they have functions. From a functional point of view, all beings are equal. In an ecocentric perspective, all beings have intrinsic value and are equally important.

Every being has an intrinsic value, before any utilitarian value to humanity. Associations of plants and animals are just as unique as their components. The value of wild nature is its independence and wildness. Perhaps our astonishment contributes to its value. If we can admit the independence of nature, that things continue in their own complex way without human assistance, we may feel more respect. We can contemplate with admiration, and sense and appreciate as well as manipulate. The emergence of new moral attitudes depends on a more realistic philosophy of nature.

Reverence for Being

The recovery of implicit natural values can be expressed as a reverence for natural systems. Ervin Laszlo proposes a social ethic for the age of humanity, calling for reverence for the level-structure of the microhierarchy, including all systems on all levels, from atoms to an emerging planetary culture and ecology. "We can express the recovery of our implicit natural values in requesting a reverence for natural systems," he says. This reverence expresses the insight that humanity is in nature, a part of an embracing network of dynamic self-regulating and self-creating processes. Our thoughts and ideas are nature, as much as clouds and waves. Marveling at humanity is marveling at nature, the matrix from which we arose. Although there are others, humans are remarkable examples of dynamic order brought forth by the universe in its history.

Reverence must include all natural and artificial beings and things and fields, from atoms to weeds, computers to galaxies. Atoms, molecules, and organic cycles are parts of humanity. We must revere all the arrangements of earth stuff. The greatest human dignity follows from respectfulness of everything as meaningful as ourselves. Such a reverence treats all substances of the earth as precious, to be used carefully, if at all—and certainly not for a flood of mass-produced consumer items. It includes all human artifacts, manufactures, and societies. It promotes a society where individuals would live in close contact with natural support systems. It provides each with the aesthetic necessities of life to develop all capacities.

The basis of all value is being (the verb form). It is reality undistorted by human needs. Of course, humanity can still enhance nature with its presence, in a nonexploitable manner. The reverence for beings as they are is the law of noninterference. In nature, the law of noninterference means 'letting be' (Heidegger), 'letting alone' (Wilson), and 'not killing for pleasure' (Fox). Noninterference is not indifference, which is diffuse. It is caring. Noninterference will not lead to chaos, poverty, and stagnation. The technocratic vision strives for "life under control," but the earth is self-managing, productive, efficient, and orderly. Reverence can only be felt at the alienness of nature, not at its comfortable conquest. With so little wilderness left, however, formal rules may be necessary.

Rights

An ecological ethical model is not distorted by human needs and wants when it argues for the preservation of animals and habitats themselves, because they are, as they are. Paul Shepard says the argument is not new, and that its application is ambiguous because "unlimited rights" will conflict with human interest. But, there are two bad assumptions: that human interests are not ambiguous—they are—and that animals will be granted unlimited rights—they will not. Rights seem to follow the expansion of the sphere of ethics, as formal statements of intuitive knowledge. But codifying rights is more difficult, especially for philosophers, who tend to limit rights with a series of restrictions. For example, a contractual theory assumes a perfect detachment and a rational debate of rules. Animals and imbeciles are left out. Scott Lehman argues that natural objects are not the subjects of experience (certain animals excepted) and so cannot possess rights. He limits experience to mental states. (Perhaps he means nervous system states.) Some philosophers maintain that a right is a claim to something (Feinberg); others, that it is an entitlement. Richard Watson takes reciprocity as central to the general concepts of rights and duties; few animals and no natural objects have rights intrinsically. He mentions that some primates and mammals are moral entities because they are self conscious, have free will, understand principles, and intend to act accordingly. But the assumption of self-consciousness would rule out children and feeble-minded adults, as well as most beings. So, a larger definition of claim or reciprocity is needed, without regard for contracts and mutual duties.

When humanity was divided into citizens and slaves, there was less freedom. When it was divided into governors and governed, freedom was advanced by providing the governed with protection against the tyranny of governors. When people became self-governing, protection was needed against majority opinions, by distinguishing between the individual and society. New contraries — public and private — provide clarification of rights. Rights protect the interests of those holding rights. Natural rights are the rights of an underclass that has not been granted legal rights. These "natural rights" are used by minorities to legitimate their claims against controlling powers. Rights and obligations were first thought of in a political context consisting of customs and practices within and between states. In the 17th century they were thought of in a constitutional context, where forms of government were established to protect natural rights. Now they are thought of in a human context. Freedoms of — speech, worship — depend on institutional protection and are political rights. Freedoms from — want, fear — are extensions of economic rights. Freedoms for — pleasure, reproduction — are biological rights.

Humanity has taken its own opportunities. These opportunities have been codified for centuries as rights. Now, we must allow other beings equal opportunities. The interrelatedness of life dictates the interrelatedness of rights. And these rights are necessary to the integrity of the whole planet. Humanity developed in a community of animals and plants, as part of a clade on the same tree of life. The quality of human life has always depended on the quality of animal life. Animals have sensations and feelings, as important to them as ours are to us. Furthermore, the extension of rights to animals and plants does not deny any traditional human rights. Animals should be accorded higher moral regard and legal standing to reflect the intrinsic worth afforded by their existence and sentience. Welfare laws to conserve species and to guarantee humane treatment in research, transportation, and slaughter indicate a growing concern among people. A new ethic can keep animals free from human intervention, prejudice, or overuse. Animals should be preserved because they are as they are; their existence is moral justification. Their intrinsic worth is independent of the instrumental values imposed on them by humanity.

The strongest argument for rights is interrelatedness in communities. It is a basis for assigning rights to ultrahuman nature. Existence implies intrinsic worth. Garret Hardin considers interrelatedness, but interprets it narrowly. He considers rights as rules of competition; every right is a ploy in the struggle for existence, and every right implies an obligation to furnish it. This is good as far as it goes. But, life is more than competition; it involves cooperation and play. Rights are formal rules for living together. Society should be organized on the basis of functions, not rights. Ecological rights (customs) could be based on functions. It is foolish not to assign rights to animals, plants, and the earth because of contractual formalities. The reverence for all beings is concerned with the right functioning and right numbers in the right places, according to standards of health and quality of life.

Justice and Law

Socrates argued in *The Republic* against the conception of justice as giving every man his due, and proposed a definition of justice as every man performing his proper function. This proportion of reward to function in the community was named distributive justice by Aristotle. It describes the right to participate in the benefits of science and culture. If justice is a proportion set up in the community between men and goods, justice is also the restoration of the relationships of men and goods, when disturbed. Aristotle called this justice rectificatory. This has constituted the business of recent laws and courts. During this stage of universal rights, world order transfers the criterion from the nature of man to the community of men; both law and justice, obligations and rights, are reduced to equity. Thus, new nations demand to participate in a common justice, as opposed to the extension of natural rights. A principle of justice based on need can be extended to the ultrahuman community. To be sure, it needs to be altered to account for unconscious, interdependent beings. The right to use nature is a right to share. Current legislation on animal experimentation and protection implicitly recognizes the right to life and a healthy habitat.

One problem with the current legal system is that all ultrahuman beings are given the status of inferior human beings, legal incompetents, thus keeping humans in a guardian role. A new legal category is needed that would respect the existence, competence, and excellence of natural beings. Christopher Stone recognizes that the judicial system has granted rights to a variety of inanimate holders, trusts, corporations, and nations, for instance. The legal system already operates with fictions.

Formal law tends to seek guidance on normative issues from the general population, rather than from legal experts. People care for animals and wilderness. Natural rights are defined by positive laws and by negative restraints on behavior. Laws are needed to protect wilderness, now. Legal or religious action almost always precedes the general shift in conscience. The obligation to treat others equally includes an obligation to change human social patterns in the direction of equality.

Chapter 46

The Ecological Responsibilities of Corporations

Summary
The metaphor for a corporation used to be a simple mechanical model for turning resources into products. To be successful, a corporation had to grow and turn a profit continually. Unfortunately, the assumptions of the model were also simple and failed to consider human needs and natural cycles, causing great suffering and great disruption.

To be really successful, corporations need to adapt a more comprehensive model, one that reflects stability, cooperation, justice, and respect for nature. With an ecological model, the ecological responsibilities of corporations, to themselves, to nature, and to human communities, are described.

Introduction
A corporation is defined legally as an individual person, although one that is defined by law and exists only in contemplation of the law, hence it is artificial, invisible, and intangible. A corporation is not the stockholders or officers; it has its own entirely separate and distinct existence. This kind of artificial person was the invention of commercial interests; the first corporations in Britain were granted charters by special acts of parliament to provide services for the state. Most corporations have characteristic legal features: individuality, permanence, limited powers, continuing succession, action in the corporate name, limited liability of members, transferability of member's interests, and representative management. Charters have become easier to get, until now corporations can be formed under general law. Corporations have been adapted to meet most modern business and social needs. Real persons can do anything not prohibited by law; corporate actions are derived wholly from law and limited to the charter, although charters have become very comprehensive statements that allow a great variety of activities.

In the publication "Integrating the Enterprise" by Digital Equipment Corporation, it is stated that "Digital is a living organism," tending toward a state of dynamic equilibrium by adapting to circumstances as fast and economically as possible. This is a natural-enough metaphor, and it leads to interesting deductions.

Like an organism, each business is born, grows to a certain size, then matures and dies — perhaps a natural span is hundreds of years, like the Oxford University Press, for instance, or perhaps only a year like so many new businesses. Biologically, maturity marks the end of physical growth in humans, but not necessarily the end of emotional, intellectual, and social development. Like an organism, the corporation, unconsciously through its officers and managers, starts to act to preserve itself before completing its formative objectives, such as maximizing profit. Like an organism, a corporation lives in place and alters that place to some extent by living, although some of the larger ones risk destroying the mixed community of humans, animals, and plants (and all their associations) on which they depend. Every organism must fit in its community and environment. It must be integrated and limited — self-consciously so in the case of humans and corporations.

Corporations are unique organisms. Corporations are more than just groups of people without structure. Corporations are described by analogy with individuals. But, each has unique characteristics: the corporation has a right to property and free speech, but not a right to worship or vote. They can be longer lived than their human components.

1 *Problems with Corporate Organisms*

The old analogy of the corporation as a machine leads to bad assumptions: that everything is a resource; that resources are unlimited; that production must continue endlessly; that the corporation has to keep growing to survive; that the purpose of the state was to legitimize exploitation; that the purpose of humanity was to multiply, produce, and consume; and that the purpose of the universe was to supply human and corporate needs. The machine analogy also leads to false economic beliefs: that mass production is most efficient; that obsolescence is necessary for successful growth; that people's needs and wants are fulfilled by advertised products; and, that quality does not matter very much.

Some companies have started to work around these beliefs. Kodak and Head, for instance, have succeeded with custom production, high-quality, long-lived products that people did need. Other corporations are suffering problems as the results of bad assumptions and false beliefs. These problems include: overgrowth, with an increase in complexity and costs (many of them social); economic and ecological instability; social burdens (from pressures on family from relocation and powerlessness); misdirected effort on ill-conceived products; slack employee attitudes and performances (sickness, accidents, turnover, layoffs).

In the effort to control their problems, corporations have sought more control. The corporation tries to avoid being vulnerable to change and uncertainty, fluctuation, and market conditions by relying on planning and control. Corporations try to ensure stability by taking over the supplies of materials, controlling their subcontractors and the buyers of their products, controlling the work force with pay and incentives (as well as by cultivating identification), and managing demand by sales influence and advertising. Many corporations try to be flexible about resources, using what costs the least, say, cheap energy to replace expensive labor. Corporations, especially multinational ones, seek raw materials everywhere, in any nation, under any ocean. Corporations become international, mining, assembling, and selling in three different continents. Where developing countries were once regarded as sources of raw materials, they are now used as bases for manufacturing, and they are becoming growing markets.

This control is possible because corporations have acquired such great power. Large multinational corporations have great power to control national economies and ignore environmental laws, partly because of their historical promise of providing social services to national states. Unfortunately, as private good became identified with public good, corporations became less concerned with social service and more concerned with greater profit through greater technology. Greater technology lets the power to change overwhelm the power to see the consequences. So, social amenities, such as clean air and fresh water, are violated legally – after all, no right of contract or fair use of property has been breached. Corporations exercise enormous influence over our lives, probably more than governments or churches. Furthermore, the size of some companies means that their influence is felt like shock waves through societies and environments. Multinational corporations are a force in the business world and a major influence on world affairs. Size and complexity give them special power, not only financial but political. Power is supposed to evaporate in a purely competitive economy, but capitalist economies are not purely competitive; power accrues to corporations.

Usually the goals of a company are generously, nobly, and broadly stated as intentions to support the best interests of owners, managers, shareholders, employees, the public, and, lately, the earth. Many corporations pride themselves on their generosity and personnel standards and on their good corporate citizenship, although the public concept of good is being extended beyond traditional bounds as citizens become aware of the interactions of business, politics, and the environment. Even corporations that claim to make clean products in safe ways often seem to depend on "dirty corporations" for power and packaging, as well as for paper and materials.

Business as usual, with its inertial model of growth, could end in catastrophe for humanity and its environments. Industrial cultures, with their characteristics of simplification, naiveté, homogeneity, and incompleteness, turn wild landscapes into flatscapes, where variety disappears and significance is ignored for the comfortable standards of meaningless continuity. Rapid growth might precipitate a catastrophe sooner, while modest efforts at environmental protection and increased efficiency may only postpone catastrophe.

An organic model offers the best alternative, with its emphasis on energy efficiency and alternative sources, its major commitment to environmental protection and the internalization of environmental costs, and its change from growth to stable, sustained development. Such a model would provide organic assumptions and beliefs: that resources are limited, that human value is only part of ecosystem values, that humans are more than consumers and producers, and that the quality of life is more important than quantity of possessions.

Like organisms, corporations may have an optimal size and a home place; organisms that occur out of place are often called weeds, and organisms that grow too large monsters. Perhaps corporations should be limited in size and tied to one place.

Healthy organisms, even humans, are educated to take their place in a culture that limits their impact on the environment, proscribes their actions towards one another and towards others outside the culture, and trains them to reproduce themselves and renew the culture. Most corporations act like improperly socialized individuals who have not been taught how to take a proper place in society and how to be responsible for their actions; instead, corporations hide behind their legally limited liability.

Any organism, like robins to fermenting berries, can be addicted to certain things in certain circumstances. Corporations have become addicted to cheap energy and easy defense money. It is possible to create circumstances that limit or cure the addictions (for robins, the berries fall to the ground and get covered with snow).

Figure 7. Butterfly on Grandmother's Soul flowers (the author designed habitat for several corporations).

Ecosystem as Metaphor

Perhaps the metaphor itself is not adequate. Even organisms take a large part of their identity from their context, from the surrounding environment. It might be more productive, since corporations actually contain organisms, mostly human, to consider the corporation as an ecosystem. Ecosystems occur in a large diversity of sizes in nature, from a rotting log to one covering most of a watershed.

Definition. An ecosystem is defined as a biotic community and its nonliving environment functioning together. An ecosystem is also a self-organizing, chaotic system with emergent properties. Ecosystems develop in time. That is, a community develops by an reasonably orderly, directional process that involves changes in structure that results from community modification of the physical environment. Although the physical environment imposes limits and determines pattern and rate of change, the community controls the succession. The result is a relatively stable configuration characterized by a high biomass (or information content).

Characteristics – Energy. The "strategy" of ecosystem development is increased control of (or homeostasis or homeorhesis with) the physical environment — to protect itself from perturbations. There is a fundamental shift in energy flows, as increasing amounts are used for maintenance. The structure of a community changes: organic matter increase, inorganic nutrients are used internally (instead of being extrabiotic), biochemical and species diversities become high, and pattern diversity becomes well-organized.

As more and more energy is used for maintenance, the net community production approaches zero. Agriculture keeps an ecosystem immature in order to harvest the larger yield in immature systems. The mature system becomes more efficient, as it supports a larger biomass with the same amount of energy. The food chains become more weblike (dominated by detritus chains as opposed to linear grazing). Mineral cycles become closed and the nutrient exchange between organisms and the environment slows.

The life history of organisms undergoes change as well. Organisms tend to be larger (perhaps as a result of shift from inorganic to organic nutrients), with longer, more complex life cycles and narrower niche specialization. Population growth slows, with emphasis on the quality of life of organisms. Internal symbiosis becomes more developed, conserving nutrients and resisting perturbations.

Corporations as Ecosystems

Development in Time. Corporations have been evolving for hundreds of years. There have already been shifts in the meaning of corporations and in the forms of organization and style: From a stable product line to continuing innovation process, from a product based definition (shoes) to a process (information), from a single pyramid of organization to constellations of satellite concerns, from the static to the flexible, from product line management to networks and innovations (the management of change), and from stable forms to temporary. Mature systems have a greater ability to trap nutrients for cycling. Corporations have settled in an artificially maintained pioneer state, feeding on the extra productivity.

Energy Use. If corporations follow similar patterns over time as ecosystems, we would expect their "food chains" to become more complex, with most of the energy flow following detritus pathways. Optimizing material and energy use, reusing `wastes' as resources (food webs between companies, where wastes are products, not side-effects), and closing loops by recycling. Reciprocal adaptations between plants and animals, or between producers and consumers, leads to mechanisms that reduce grazing and increase feedback.

Low Productivity, High Stability. A mature ecosystem produces many things, most all of which are used by the system; wastes are broken down into component chemicals by microbes, while other resources, like nitrogen, are fixed to roots by other microbes. The tightening of the biogeochemical cycles is an important trend. Corporations sometimes parallel the aging of an ecosystem, from a pioneer state to a mature state.

Life History, Symbiosis. The longer lived the corporation, the less clear the divisions between private and public and economic and environmental concerns. Short-term individual concerns meld into long-range corporate and social concerns. The short-term pressures may seem immediate and irresistible, but the long-term goals cannot be ignored for long. We may need electric power or paper now, but the cost cannot be the destruction of the source of those needs. We need the social and environmental health and stability first and always. Partnership between unrelated species, say mycorrhizae and trees, becomes notable. Corporations would team up with other companies and social and environmental groups (the pattern could be industrial symbiosis). Pollution is the most limiting factor; possibly we could predict new corps arising to deal with pollution.

Differences. Although ecosystems can be long-lived — thousands of years for tropical rainforests — corporations can disband at any time. Furthermore, unlike ecosystems, corporations still seem to be bound inflexibly by the rule of two-year payback. This means that decisions are based on short-term return and not on long-term durability.

Furthermore, although large organisms are the case for mature ecosystems, the scale of many corporations is too large; the patterns are unsustainable — large institutions lose touch with their constituents, become self-absorbed and less responsible.

The differences between ecosystems and corporations allow for the possibility of changes. Human communities can redefine corporations and limit their impacts. They can change the charter of corporations for the benefit of the community. Corporate responsibility is more complex than a simple linear cost/benefit analysis. Using the metaphors of corporations as organisms and ecosystems, it is possible to outline a new set of responsibilities for corporations and a series of behaviors that human individuals and communities can do to integrate corporate behavior into the communities.

2. *Ecological Responsibilities of Corporate Organisms*
The public responsibilities of corporations, according to Harvard management, are to grow and prosper (thereby providing customer satisfaction, employment, taxes, and contributions to the economy) and to control their hazards. According to Milton Friedman, the only social responsibility of a corporation is to make money, by striving after profit as an efficient agent of production, although he admits that the corporation should conform to the rules and norms of society.

Profit making is a necessary part of business, but not the sole reason for business. The best business serves public goods as well as private interests. (This is similar to Ruth Benedict's original anthropological meaning of synergy as it applied to individuals. In secure, nonaggressive societies, an individual serves her own advantage as well as that of the group with the same act. The institution ensures mutual advantage; the acts are mutually reinforcing. High synergy institutions transcend the polarities of selfishness and altruism. Virtue pays because the rewards for selfishness coincide with benefit for the society. The social structure of low synergy cultures ensures opposition and counteraction; the advantage of one individual is a victory over another, as in a zero-sum game.) This is necessary for employees, since they have to feel like their work is meaningful and contributing to the public good. The path of production should therefore serve public good as well as profit. Environmental and social problems should get as much attention, because their part of the process, as sales, finance, and production.

Economic recession may bring a re-examination of values, not only by individuals who may have less material wealth, but by corporations that have emphasized growth. The public may insist that corporations consider social performance as well as strictly economic performance. The single economic purpose may only be the focus in a social ecological environment. Economic actions, such as where to build, who to relocate, hire, or dismiss, will be subjected to greater public scrutiny. Corporations will have to adapt to changes in standards. Business cannot assert primary self-interest at a cost to the public or environment. Corporations need to keep track of their environmental impacts. Many of the problems that corporations face are connected to the problems of the environment and society.

Every corporation depends on the stability of the environment and on the stability of social institutions. The environment provides air, water, and land, and provides renewing (both physical and psychological). Institutions, from sanitation, police, schools, churches, and community centers, provide a supporting network. As these institutions wobble or fail, corporations may have to subsidize or replace them to survive. Schooling for example, is often inadequate to provide literate, numerate, or ecolate workers. Police may not be able to provide secure conditions on corporate grounds for female workers. Public transportation to plants from town may not be available in enough volume.

A corporation has traditionally been seen as a morally neutral body, but even if it has only to conform to the nominal rules of society, it is already a moral agent. Corporations are no more neutral than other organisms. Etymologically, the word moral means simply "living together.' Sometimes even routine business (nonmoral and nonenvironmental) matters become deeper ethical conundrums about justice. Many areas of moral concern already are recognized: Worker safety, affirmative action, advertising truth, foreign investments, and harm to the consumer, public, and environment. Corporate responsibility occurs wherever the interests or rights of a person, society, or ecosystem are significantly affected by the actions of the corporation.

Responsibility can be understood in terms of costs and benefits, that is, through operations and their consequences rather than abstract behavior. Every action entails a gain and a cost (or profit and loss). Profits and losses are distributed privately, socially, or environmentally. Unfortunately, the modern system privatizes the gain and externalizes the loss (to the "commons," considered as a pool of "unowned resources," where in traditional societies, it was surrounded by rules for use). As long as this is possible, it is profitable to charge the cost to the environment. Externalizing costs works fine in an uncrowded world, where the costs are negligible and can be absorbed by natural processes. Resources were traditionally seen as free for the getting; air and water were seen as free sinks. Modern economies, embracing the notion that "nature is capital," draw on the accumulated "capital" of ecosystems for production. By ignoring the real cost of the capital, as well as the costs of natural services, such as nutrient recycling, soil building, and atmospheric renewal, these economies create a temporary wealth. Decisions regarding resources are made on short-term economic grounds and lead to material shortages and environmental degradation.

Similarly, labor was seen as minimum value. For example, the idea that "labor is a resource" implies that, like any common resource defined by industrial society, labor is cheap and can be used up. The real costs of free goods and externalities have had to be accounted for, yet—this often influences the selection of corporate priorities and growth. Furthermore, the production and distribution system for most corporations is linear (straight throughput) and not circular (complete recycling), although this is logical economically, given our frontier resource-use accounting. Major changes are occurring, though. The scale of civilization now makes externalization unfeasible. The costs of pollution and waste are being internalized; other inputs, such as labor and capital, are becoming more expensive. Corporations will have to internalize or be forced to internalize. With the internalization of costs (since the losses as well as benefits will accrues privately), the system will benefit from intrinsic responsibility.

Corporations need to work cooperatively to make sure the costs and benefits are extended equally throughout the system. They could start by sponsoring the rational use of rare resources through taxation. Influence the government to determine priorities for wilderness areas or special landscapes. Beautiful, fragile, unique, or endangered ecosystems and species must be protected at the expense of commercial activity.

Corporations have at least three large, ecological responsibilities.

2.1 To be Economically Healthy

The first responsibility of a corporation is the maintain its own health, to mature organically. limiting its size and impact to the locality.

- Create a department with ecological authority to envision long-range plans and impacts. Corporations need to react more quickly, to monitor their ecological and social environments for emerging patterns that determine their future. They need to anticipate and participate in the social and natural framework. A new department, with global, anticipatory functions could provide direction and continuity. Such a department could be justified in the same manner as military forces. Military expenditure is a nonproductive cost; its benefits are general and long-range, that is, it must discourage war in the next decade as well as in this one. Its scope of advice would include educational services as well as advertising, capital acquisitions as well as new products, plant engineering as well as security.
- Plan all foreseeable consequences of a product. Advanced technology permits the power to change to overwhelm the ability to foresee the consequences of change. Avoiding the opposite actions of intentions (the enantiodromia recognized by the Greeks as the operation of tragedy) is extremely difficult. Good intentions are not enough: Labor-saving devices may contribute to unemployment and social problems; foreign aid may result in starvation for more millions as local agriculture cannot compete; the environmental management of some species for sustainable yield causes population collapses.
- Determine the optimum corporate size. Limit the size of the

corporation. After a point, growth results in inefficiency and nonadaptibility. Development, on the other hand, can continue for hundreds or thousands of years. A smaller size could mean more flexibility and faster response to local conditions. Recognize material limits. The global economy is probably too large already to be supported by the natural systems of the planet. What is the upper limit to the economy of scale? Accept limits to growth based on materials and on nonrenewable or dangerous sources of energy. This should not limit development based on advancing technology and knowledge.

- Adjust corporate strategies to changing values. Smaller social and cultural groups have different and diverging values, so corporations are going to have to adjust to a diversity of values instead of to a monolithic standard. Now, the structure of power is disintegrating (with information replacing things as wealth). The knowledge-driven economy is more decentralized and customized. This moves us towards customization of production and away from mass production. Change the shape of the corporation to a framework coordinating separate divisions sharing information. Each could react much more quickly to market conditions.

- Work to delineate a new information model of production in which the stages of a process (capital, materials, workers, design, advertising, selling) are simultaneous and synthesized. The conception of the product is extended from design (even customer contributions and design of working conditions) to aftercare, including ecologically safe retirement and disposal (recycling). The notions of efficiency and productivity are changing. Innovation and computer technology shortens product life cycles. Production diversity is increasing. Convert the information model to an understanding model. Information is just data without appropriate structure. Provide a structure and material base for understanding through communication, education, and training.

- Enter partnerships with the employees. Address the optimum productivity of employees. For instance, government studies show that half-time employees are more efficient than full-time ones. So adjust the work force to fewer, more flexible hours (thus avoiding layoffs during the transition). Increase worker participation. What is extent of worker participation in management of workplace? New forms of ownership could mobilize workers in a more efficient and democratic economy. Productivity is declining, so is job satisfaction. Efficiency and productivity are often less important than use and appropriateness. Better pay and shorter work weeks have not compensated for lack of worker control. Offer more control. Streamline the organizational structure. Organizing workers hierarchically is costly. The best path for organization is lateral modularity, not bureaucratic hierarchy. The levels of management could be reduced drastically.

- Promote the principle of least effort, allowing the company to consume less, recycle, use longer, and avoid waste. Corporations could develop renewable energy sources. Reduce office costs through energy conservation plan. Use renewable energy sources. Corporations need to maximize recycling. Energy and materials can be used and reused, flowing through the system. Ship by the best transportation, probably rail. Cars are ecologically unacceptable forms of transport, yet companies intrinsically recognize them with large, free parking lots. Discourage commuting; encourage telecommuting or even alternate forms of transportation (bicycling, buses, and trains). Minimize wastes, for instance, by using permanent packaging. (Milk bottles and cola bottles can be reused forty or more times.) Conduct a complete series of audits, including an energy audit for every building, an environmental audit to determine negative impacts, from acid wastes or product disposal, and a problem audit, including inherited problems. Produce a comprehensive annual impact statement.

2.2 To Preserve the Health of Natural Communities

The second responsibility is to maintain the health of the natural communities — because environmental health is the basis for community health, and community health is the basis for economic health and worker health. The quality of life depends on the quality of the environment. If the environment is degraded to raise the quality of life, the effect will be very limited and never be self-sustaining. Fitting economic costs and needs to the limits of ecosystems and monitoring the economic process would reduce wastes and pressures on natural processes. The coupling of agricultural productivity to a solar budget, and the conscious restoration of degraded systems, would contribute to the health of ecosystems. Sufficient wilderness would allow the self-maintenance of global cycles. With the increase in security, wealth, and self-esteem, human populations could be dependent on ecosystem productivities and still be diverse and unique.

- Be accountable for ecological impacts. Corporations will be held more accountable for their technological impact. New technology will be more closely regulated. Corporations could anticipate this by favoring open appraisal of new technologies. By studying the potential consequences, physical, social, and ecological, as far as possible into the future, of its innovations in information technology, a corporation can gain credibility. Otherwise, it can wait and be forced by public and governmental pressure.
- Avoid interference with natural processes. Technological processes must be brought into balance with the cycles of the earth. They must not damage or degrade natural cycles. Avoid unnecessary harm. It may be appropriate to use trees or to compete with black bears for tree use, but it is never wise to destroy the ecosystem of trees and bears. Laws on pollution and noxious wastes have been notoriously lax and sometimes wrong-headed. Minimal acceptable tolerances are legal, yet people often prefer zero amounts of many substances. Minimal compliance with them is virtuous in comparison with many companies, but it would be better to lead to higher standards. Work toward setting zero-level goals. Do not dump exotic or dangerous wastes. Do not discharge quantities of `safe' wastes.

- Integrate loops and material flows; internalize cycles.
- Corporations maintain building and plants in thousands of locations, each requiring support. Convert to ecological grounds practices. Forgo economic development of key ecosystems, which are not available for human use. Consider adjusting economic pace to natural rates; do not cut trees, for instance faster than they grow. Consider minimizing use of ecosystem productivity to the net ecosystem productivity, rather than the gross productivity, especially as regards fisheries.
- Promote ecological design, which starts with questions. Is the product low-cost, aesthetically pleasing, and ecologically wise? Where does it fit in society? Ecological design, both responsible and socially responsible, must be radical, that is, rooted in a community in place. Membership in a place, in fact, leads to community. Corporations must become responsible members of the community. It would encourage an ecological approach to systems and processes in the whole environment, where the product, with its plant, engineers, and advertisers, is a link in a long biomorphic phylogenetic chain reaching from knotted ropes to surgical microchip memory implants. Ecological design has important characteristics for responsible technology: The products are designed by interdisciplinary teams considering all parameters and consequences; ecological sciences offer creative insights into design through a search for underlying organic principles; the product must be related to the particular environment, the tool is a link between human and environment.

2.3 To Preserve the Health of Human Communities
It is hard to protect communities when the way most business is done tends to disrupt community life. Because of its size, power, and intentions (for profit), the corporation should take higher risks than the surrounding communities. This will ensure the safety of products and wastes.

- Support the community. Work place isn't just collection of individuals. It is a number of groups. Group interaction can change attitudes. A working community can build mutual responsibility. Show proper behavior; learn community etiquette.
- Design the corporate structure and size for the community. Limit unnecessary movement or disruption. Plan the shape, size, and products of the corporation to fit the local community. Encourage self-reliance in communities. Communities can be self-reliant by producing enough food and shelter, by limiting their population to what can be produced, by using local products and raw materials (soil, minerals, plants), by using general and not specialized machines, by having multipurpose factories, by networking with other communities, and by doing without things that are not needed (bombs, food additives, or plastic bottles).
- Behave ethically. An ecological corporation could use corporate buying power to promote acceptable technologies and discourage unacceptable practices. Deal less with nuclear weapons contractors and more with solar energy companies. Deal less with one-shot paper

companies and more with recycling paper companies. Boycott paper companies involved in Rainforest destruction or old-growth forest destruction. Avoid banks that invest in anything that brings a high return (from third-world debt to Amazonian destruction and South African discrimination). Favor peace-oriented companies as business partners. And refuse to participate in work that is socially destructive.

- Participate in the economic and social functioning of the community. Economic development and social progress are necessary for the welfare of humanity, but must be conducted with environmental knowledge. The goal of economics and politics is to provide suitable and comfortable human habitations and meaningful activities. Human settlements must be planned and constructed within environmental constraints and according to ecological priorities. Work to preserve the structure of the natural and social communities. Corporations can encourage decentralization and restore schools, clinics, and shops to local communities. Offer cooperative control with the community. Change the pattern of ownership to reflect employee and community participation.

- Promote ecological education in a total context and interdependency. Encourage cultural traditions to stop letting social and spiritual needs be subverted by economic ends. Help lead the young into their adult responsibilities through training and participation (perhaps apprenticeship programs). Educate for appropriate ways to achieve wealth and well-being. Teach appreciation of services rendered by nature through flows and cycling. Point out the unexpectedness of consequences of even simple corporate interventions and innovations (positive feedback, biological concentration of poisons, and synergetic effects of simple new chemicals like CFCs). Trace the complex and reciprocal relations of soil, climate, vegetation, and human activity.

   Emphasize that a fixed set of ecological parameters in an ecosystem can not be maintained sustainably, because the system is dynamic and changing. For example, where do computers fit in schools? Children do not need computers to develop the powers of thought, but they do need an ecological curriculum where animals display greater powers of mind than computers or machines. The important technological advantages of a computer, word-processing, database searches, complex connections, and rapid computation, are not really needed before high school, unlike myths, languages, and physical activities.

- Implement community responsibility. In education, integrating business with humanities; the responsibility for the welfare of the citizens belongs in the community, as does education, safety, and the whole infrastructure. In management development programs and management as source of influence by setting goals, modifying structures, and introducing criteria for measuring progress. By the Board of Directors, the architects of responsibility and stewards of all resources. By the government, in its legislative, judicial, and regulatory functions. providing rules and permitting freedom.

3 Summary

The corporation, regardless of its legal definition, is a long-lived, collective, impersonal body. Yet, it has more physical, legal and moral power than any one individual. Its investment is long-term in actuality. Many stockholders keep their investments for decades or a life-time. They are not concerned about only one dividend. Like the corporate organism, they want the long-term outlook to be positive; they want to know that their investment is stable and that the quality of life it encourages or supports is continuous.

The complexity of environmental problems should not permit escape of responsibility. The context of corporate responsibility falls within the spectrum from individual responsibility to social responsibility (the designation of property or trading conventions — capitalism or communism). Perhaps that responsibility could be enforced if the entire earth were incorporated and concerned with maximizing its own values: healthy beings in living contexts. Certainly not having `free' services and resources would force corporations to internalize all costs of production.

In any case, there are strategies that a corporation could pursue to become ecologically minded. Instead of treating decisions as trade-offs, an ecological corporation could aim at a congruence of moral, economic, and ecological objectives. Responsibility could be manifested in organizational structure, manufacturing, and marketing practices, without departing from economic decision making.

Such a corporation could bring corporate research and development capacity to bear on the transition to a sustainable society. Where technologies play a role in the transition, companies can assume social responsibilities equal to their size and wealth. By commanding their vast resources, corporations can ease the transition to a sustainable society (which would actually meet their needs for stability). The model of corporate life needs to change, from dependence on continuous growth (of profits and waste), to be being based on stability, sustainability, cooperation, justice, and respect for nature.

CHAPTER 47

Incorporating the Earth: A Thought Experiment

Political systems are impotent to stop the massive interference in ecosystems by international corporations. The simplest and most direct way to give the earth a voice in the development of the earth by humanity is to incorporate the earth following international law. The entire planet, with its biochemical cycles and nonhuman communities, would become one legal body. Since corporations are human constructs, however, humans would have to represent ecosystems and their wealth of living organisms.

In early civilizations, the advancement of the state was expected to contribute to the welfare of its people. Corporations are recent devices created by states for public purposes. Most early American corporations, for example, were concerned with travel (turnpikes and inland waterways) or safety (fire insurance) — they resembled public agencies more than profit-seeking associations. In fact, the exclusive privileges and political power granted to corporations were based on the implicit promise of social services.

The association of economic development with national wealth allowed incorporation laws to be broadened. The corporation was given the constitutional rights of an individual. A corporation is a legal entity, independent from its founders, with its own rights, privileges, and liabilities. It is, however, required to obey laws and pay taxes; and it is accountable for its deeds in courts of law.

Unfortunately, as private good became identified with public good, corporations became larger, more acquisitive, and less concerned with social services. The quest for profit now has the effect of violating social amenities, such as clean air and clean water, instead of ensuring them. No responsibility is taken for environmental degradation since no right of contract or fair use of property has been breached.

Changes in societies, from rural to urban, from sparsely to densely populated, from culturally diverse to monotone, have transformed corporations and the societies themselves. Business corporations now provide the bulk of goods and services in many states. The scale of these corporations, the processes of production, and the size and needs of human populations, have altered and degraded many ecosystems and biogeochemical cycles.

Successful modern corporations create an identity based on their purpose in providing goods or services; they define their business in terms of profitability, growth rate, cash flow, and competitive position; they develop a corporate vision, with specific objectives and strategies, including long-term vision, collection of ideas and creative implementation, aggressive manufacturing, and reliable finance.

The purpose of a corporation often transcends simple financial gain — the corporation seeks to maintain its own existence, before profit. Financial objectives (sales, assets, profits) exist to sustain its existence. The goals that most motivate corporate managers are survival, independence, self-sufficiency, and self-fulfillment. Yet, these motives are consistent with the financial objectives of the corporation: to maximize corporate wealth. The responsibility of managers is to maximize the value of the company. Furthermore, because corporations are long-lived, that value should last a long time — a good reason for looking beyond the ten-year monetary horizon and the lives of its managers.

Although current wisdom (Milton Friedman et al.) holds that a corporation's only responsibility is to its stockholders, corporations are being pushed to include social purpose in their strategies, again. Alas, they are doing poorly at it. They do not know how much responsibility to take, or where to put limits, or whether to pursue policies that diminish their profits. Corporations have proved spotty in doing social and environmental good. It would be more appropriate to have them deal with the environment as a corporate entity concerned with maximizing its own values. Of course, that would mean no more "free" resources or environmental services.

The important advantages to incorporating the earth are the same as for incorporating a business.
1. Managerial flexibility: the stockholders are separate from managers; responsibilities are assigned by needs of the corporation.
2. Limited liability: the corporation borrows and repays. It shields its members from hazards to which they would otherwise be exposed.
3. Financial advantage: the ownership of assets can benefit stockholders and the corporation.
4. Tax advantage: investments in the good of the corporation may not be taxed by nations.
5. Estate planning and longevity: the corporation exists indefinitely beyond the lives of its participants.
6. Central management and representation: a large and complex business needs operational and managerial efficiency. Many of the participants have no direct voice in the operation — they must be represented.

The earth incorporated would focus on a core business: to ensure the integrity and continuity of life and all its connections and to secure the opportunity for development free from undue interference. It would operate to optimize values, like any good corporation, but the values would be ecosystem values (fungus values and earthworm values, as well as human values).

A temporary Board of Directors (the undersigned) would adopt bylaws, elect working officers, approve stock certificates, open accounts, and arrange a stockholders meeting. The stockholders would elect new directors, possibly from United Nations representatives or directly from elections, and decide on dividend declarations.

Stockholders, as citizens of independent nations, would turn over common and national property to the Earth Corporation, which would issue stock certificates to the stockholders. The corporation would allocate the purchase price of stock to capital at par value. Most of the shares — the percentage to be determined by the board as necessary to the operation of ecosystems — would be treasury shares. Anything more than par value would go to capital surplus, and only capital surplus could be distributed as dividends. Stockholders have the right to receive these dividends equitably, without resort to traditional distributions of wealth.

Stock certificates denote ownership of the corporation. Although the stockholders own the corporation, they do not own the property of the corporation, the earth, which is owned by the corporation itself. Stockholders, as individuals, groups, or nations, could make agreements about how business would be conducted, about what resources would be used or traded.

The elected board of directors would make decisions of distribution and limitation. Percentages would be deducted from the interest for the operation of the corporation and for equitable distribution to nations less favored by chance with biological or geological wealth. Furthermore, since the dividends would be distributed among people according to net ecosystem productivity and resource availability, no advantage would be gained by nations having large populations.

The basic functioning system would be considered capital, thus limiting the amount of human use of resources and probably the size of human populations. Interest would accrue in the form of net ecosystem productivity and diverted percentages of materials, such as gold or water.

The earth incorporated would solve the problem of having to value ecosystems in monetary or quantifiable terms; its systems would be untouchable capital. The human value of resources like copper, air, or water would be equated to the technological cost of recycling or producing them.

Raw material and energy are only two facets of the capital of a corporation — another is human ingenuity. Thus, human wealth would not be limited by restrictions on the availability of resources, but rather by a shortage of ingenuity.

An incorporated earth would be instrumental in conditioning international corporations to their social responsibility and in internalizing all costs. This corporation and governments could use traditional means, such as credit access, low interest rates, and setting priorities on equity issues, to evoke public interest in smaller and healthier human endeavors.

The suggested articles of incorporation are:

FIRST: The name of the corporation shall be The Earth, Incorporated.

SECOND: The purposes for which the corporation is formed shall include: The protection of functioning ecosystems and their living beings from destructive interference.
The conduct of inquiry into the operation of such systems and the role of humanity therein for scientific and educational purposes.
The taking of appropriate legal steps to carry out these purposes.
The maintenance of all real common property, including all lands, seas, and atmosphere, subject to the restrictions and limitations

hereinafter set forth, to use only the interest from income therein, reserving the principal thereof exclusively for the aforesaid purposes, it being intended that the corporation be organized and operated for preservational purposes and not for pecuniary profit.

The corporation is organized as a voice for nonhuman beings and systems. No part of the income of the corporation, if any, shall inure to the benefit of any trustee or officer of the corporation or to any private individual having an interest in the corporation (except for reasonable compensation) and no trustee or officer of the corporation or any private individual shall be entitled to share in the distribution of any of the assets of the corporation.

The corporation shall not be authorized to carry on propaganda, influence legislation, participate in any political campaigns, or discriminate against human cultures.

In furtherance of the foregoing purposes, the corporation shall have the following powers:

To accept and hold by gift or judicial order any real or personal property of whatever kind, nature, or description, wherever situated.

To sell, transfer, or dispose of the interest from any such property, but not the principal or any part thereof.

To make, accept, endorse, execute, and issue bonds, promissory notes, bills of exchange, and other obligations of the corporation for moneys borrowed for the purposes of the corporation.

To invest and reinvest its funds in stock, bonds, or in such other securities and property as its trustees shall deem advisable, subject to the limitations and conditions contained in any grant or gift.

In general, and subject to such limitations and conditions as are or may be subscribed by international law, to exercise such other powers which now are or hereafter may be conferred by international law upon a corporation organized for the purposes herein above set forth.

THIRD: The operations of the corporation are to be conducted on the surface of the earth but the operations of the corporation shall not be limited to such territory.

FOURTH: The principal office of the corporation is to be located in the United Nations, currently in the City of New York, State of New York, United States of America.

FIFTH: The number of directors, who shall be known as trustees, of the corporation shall be not less than 30 (a minimum number associated with major ecosystems), nor more than 3,300 (the number of independent cultures associated with biogeographical provinces and subprovinces).

SIXTH: The names and residences of the persons who shall be trustees
until the first annual meeting of the corporation, are:

C. J. Hagen, Seattle, Washington
L. G. Nieman, Viola, Idaho
V. L. Reason, Wilmington, Delaware
A. E. Wittbecker, Nashua, New Hampshire
M. H. Wolfe, Cambridge, Massachusetts

SEVENTH: All of the subscribers of this certificate are of full age; all of
them are residents of settled places on the Earth.

In witness whereof

   et alii

Author's Note: We all signed the original, but I have been unable to get
permissions from the others, two of whom have died and two of whom are
severely disabled from an automobile accident in Boston.

Chapter 48

Gandhian Nonviolence and Defending the Earth

Some groups of people, concerned with defending vacant lots, wilderness areas, ecosystems, and the earth as an organic body, have advocated using any means necessary. Earth First!, for instance, showed a lot of us arm-chair sympathizers that saving the earth requires more active participation than just letter-writing and circulating academic articles. They were among the first to put their bodies where their ideals were, and their stance has made some profiteers and environmental rapists more cautious. But, Earth First! made its wild reputation by monkeywrenching. Although the group has been effective so far, there is a possibility that its tactics may cause long-range difficulties in the form of violent retribution or the overreactive destruction of wilderness, as well as polarization inside other environmental groups. This article contrasts the tactics of Earth First! with a Gandhian nonviolence.

Dave Foreman of Earth First! has said,[1] "Monkeywrenching is nonviolent resistance to the destruction of natural diversity and wilderness. It is not directed towards harming human beings or other forms of life." Many Earth First! tactics, however, make this statement questionable. Is Earth First! really nonviolent, according to Gandhi's definition and practice of nonviolence?

Mohandas K. Gandhi characterized his ethics of group struggle by the Sanskrit word *ahimsa*, meaning "nonhurting" and "nonviolence." Ahimsa is a closely related set of prescriptions and descriptions. Gandhi said: "Ahimsa means avoiding injury to anything on earth in thought, word, or deed."[2] He adopted a wide interpretation of 'injury.' The subject 'anything' included all living beings and perhaps nonliving things.

Ahimsa is the absence of *himsa*, meaning hurting from the root *hins*, meaning hurt, a form of the root *han*, with a larger number of meanings, such as strike, kill, destroy, or dispel. Gandhi mostly had living beings in mind, but injury to nature or to natural processes, could come under the general principle of ahimsa.

Indeed, the concept of ahimsa is so wide that it could be interpreted to be an act of violence to abstain from efforts to prevent injurious acts, such as the exploitation of wilderness for profit. But, Gandhi referred specifically to destruction as part of sabotage as himsa, even if the things destroyed were not the property of anyone.

Gandhi also said:[3] "Ahimsa really means that you may not offend anybody, you may not harbour an uncharitable thought even in connection with one who may consider himself to be your enemy." Mental forms of injury include hurting people's feelings, their dignity, or their relationships — but the feelings and relationships must be positively valued, that is, it would not be himsa to hurt feelings of hatred, nor to save a victim from wrongdoing. Furthermore, some actions may be in accord with ahimsa if they are performed 'wholly unselfishly', although Gandhi does not accept the postulate of unselfishness as sufficient for the qualification of nonviolence. Most selfless terrorism, however, is still *not* nonviolent.

Mental forms of injury seem to occur in Earth First! campaigns, from name calling and humiliation to suggestive publications. By its verbal and physical stance, Earth First! seems to be violent towards many groups, including loggers and RVers. Foreman tacitly admits this when he makes the distinction between blockades (nonviolent civil disobedience) and monkeywrenching, while recommending that they not be combined in the same campaign.[4]

By its attitudes Earth First! polarizes most people. The Earth First! slogan is, after all, "No compromise in Defense of Mother Earth!" This really polarizes the opposition. One problem with opposition is its either/or character: "if you're not with us, you're against us." Positive action does not have to be bilateral or dualistic; it can transcend simple opposition and be positive without being adversarial. Gandhi was always willing to compromise on nonessentials. He characterized himself as a man of compromise because he was never sure that he was right. Compromise is an essential part of the nonviolent person, satyagraha; *ahimsa,* as unselfish love, demands compromise. There are principles that admit no compromise, and furthermore, if the compromise fails, the *satyagraha* is ready for "battle" (Gandhi's term).

Could Earth First! be truly nonviolent and still be effective? Could Earth First! be more visible and less destructive? Perhaps it could, with the adoption of a Gandhian campaign to defend wilderness. Before this campaign is presented, however, let us consider how violence may be a dangerous tactic for groups interested in preserving and protecting wilderness against an industry-dominated consumer public.

1. The hostage, wilderness, is very large and vulnerable to counter-attacks, as well as to swings in fad (which often determines how people vote to dispose of the object of the fad).
2. The destruction of materials, such as bulldozers, is usually illegal. Property is considered sacred in America and Norway, as well as in many industrial and consumer countries.
3. Violence tends to polarize opponents and some of the undecided against the long-range goals of a group, regardless of how well supported and argued.
4. Wilderness is symbolic of people's right to earn a living — even if the people in this case are loggers, drillers, or roadbuilders, who may destroy it in the process of using it.
5. Violence leads to escalation by opponents to protect their equipment and property. Violence leads to violent conflict as a *style* of opposition.

Most of the thousands of direct actions on behalf of the environment have been nonviolent in the Gandhian sense and some have been effective. The "hug the trees" movement in India, for instance, physically blocked excessive logging in the Himalayas. The Chipko (meaning "hug" or "cling to") movement started out to preserve trees by embracing them before axes could be used and has resulted in a ten-year ban on tree-felling in over 550 square miles of the Uttarakhand in India, a major source of timber and water power. The main goal is the "judicious use of trees," according to Chandi

Prasad Bhatt, the founder, and not complete preservation. The movement is pressing for a complete remaking of forest policy; they are also responsible for planting trees (over a million so far). One tenet of the movement is that the erosion of human values follows erosion of the land.

Both ecology and human ecology offer the best support for saving wilderness and humanity (which depends on wilderness). The principles from these sciences can inspire a series of hypotheses and norms (based on the Spinozan forms used by Arne Naess) that are the foundation of a Gandhian nonviolent campaign.[5] This series may also serve to clarify the aims of Earth First!

Hypotheses.
The hypotheses are segregated by common themes so that the derivations are more obvious and can be linked more easily. In this instance, hypotheses are presented from three broad levels, so that you can see the process of induction. This procedure is genetic rather than logical.

Third level hypotheses.
- Humans have the same genetic stock and the same basic interests as other species: food, welfare, self-actualization, love of place.
- Human make their worlds of facts from observations and theories colored by needs and wants, within perceptual and imaginative limits.
- Humans live together in human communities within biological communities in geographic places.
- Humans can communicate and do so to build cities and make art.
- Humans are capable of great trust and exhibit this in their use of automobiles, weapons, and money.
- Humans are capable of violence, for many reasons, from self-preservation to misunderstood symbols.
- Living in separate locations with differences in languages and cultures can lead to misunderstandings.
- Humans are autonomous beings and can choose their behavior.
- Humans can change their behavior.
- Nature is self-making and self-maintaining.
- Where nature is left alone, by definition wilderness, animals and plants reach their own balance.
- Animals and plants live in communities in geographic place.
- Animals and plants are autonomous beings.

Second Level Hypotheses.
- All beings (human and ultra) have long-range interests, especially in the continuation of their ecosystems.
- The presentation of facts, as uncolored by needs and wants as possible and within the limits of perception and imagination, increases the probability of understanding.
- Living together in the same community fosters cooperation and understanding on immediate, common goals.
- Working together on common projects increases cooperation.
- Constructive work is more binding; a constructive program is more

meaningful and generates trust and communication.
- Work abstracted from the community can cause its opposite (enantiodromia), that is, destruction instead of good.
- Misunderstandings can be corrected in peaceful confrontation.
- Constructive confrontation concentrates on faulty understanding in a situation and not on the opponent's action or personality.
- The focus on misunderstanding reduces violence.
- Humans are responsible for their own actions.
- Humans enjoy being with animals and plants in general and being in wilderness and using it for living and recreation.
- Human economies are based on wilderness, ultimately.
- Wilderness should be saved for many reasons.

First Level Hypotheses.
- The campaign to save wilderness should be constructive and positive and be addressed to issues and relationships, using unbiased facts. It should be well-advertised and simple, sticking to goals.
- The campaign should be nonviolent and given to the possibility of agreement, including compromise on some aspects.
- Violence against transgressors of wilderness may result in further violence against wilderness or perhaps further violence on the part of defenders to guarantee wilderness.
- The campaign should apply to a local area and address those who are knowledgeable in the local community.
- Responsible persons, acting in a group with an appropriate lead time, concerned with their own community, present an undeniable case.

From these hypotheses can be derived norms that regulate personal and group behavior in a nonviolent campaign.

Norms
Third Level Norms.
- Live in the community with your opponents!
- Formulate the essential, shared interests and try to cooperate on the basis of these interests!
- Refrain from provoking or humiliating your opponent!
- Seek personal contact with your opponent and his group and be available for meetings!
- Trust your opponent!
- Learn about wilderness!

Second Level Norms.
- Act as an autonomous, responsible person!
- Choose attitudes and actions that reduce conditions that lead to violence!
- Find common interests to build on!
- Present unbiased facts!
- Be flexible and ready to compromise!
- Be constructive! Suggest alternatives!
- Trust!

- Act nonviolently, peacefully, and responsibly!
- Defend wilderness!
- Don't stop!

For Earth First! (a norm perhaps?), the norm "Be constructive!" could take the form of alternative suggestions for loggers, such as private land management or unions or legislation to restrict (by tariffs) lumber sold below the real cost. The Marsh Institute, for instance, uses an educational tactic by offering to performing a consulting service for free on a small scale; the farmer and logger then can compare directly the results and the methods, such as pesticides versus integrated pest management to reduce aphids on a barley crop, or sanitation cuts versus single-tree trimming to eliminate dwarf mistletoe infestations. Then, there is defensive monkeywrenching, such as tree spiking—with signs notifying cedar thieves, and passive resistance, such as not informing hunters or biologists of the whereabouts of a bear. In general, logging and farming prices are unrealistically subsidized by free goods from nature, as well as government subsidies. Prices should be geared to costs, and tariffs and taxes used to balance trade discrepancies. The answer is not to fight over trees in wilderness areas or to give up wilderness as a one-time boost to timber interests. It is to attack the old, unreal ways of cutting, to attack unrealistic economic policies. The long-term strategy is to save most wilderness as completely as possible, but also to include the cost-benefit analyses of alternate plans for some areas. Sawmill workers have been complaining that wilderness removes real jobs from their grasps; this misperception needs to be exposed as the short-coming of a problem economy.

Anti-ecological decisions usually come after years of planning by government or industry bureaucracies, at extravagant costs. Therefore, it is best to seek out and address plans at the earliest stages, before momentum can carry the plan. This is hard to do when the bureaucracies resort to secrecy. One of the goals of Earth First! is exposure of destruction and the increase in public awareness. It might be more effective to team with reporters with cameras and send news flashes on destruction and illegal use than to destroy equipment.

These norms provide a consistent guide to reacting to dishonest or violent opponents in the struggle. For instance, if your opponent uses biased reporting, merely provide a factual presentation, with evidence, without resorting to provocation or name-calling. Although the stupidity or badness of opponents is not an issue, some people are stupid and bad, so it is a factor to be considered. The stupid and bad must be neutralized some way, by simple diversion through relatively harmless assignments.

Nonviolence is easily misunderstood. If your opponent respects violence in defense of property, and you misjudge your opponent and offer nonviolence, which is perceived as weakness, prompting him to violence, what should you do? Stay in the center of the issue and be active, but do not respond violently yourself. Most such opponents eventually recognize perseverance as an indicator of strength.

It is important to formulate one very clear, concrete, easily understandable goal for an action and alert the opponent to that goal as soon as possible. This is very hard to do when it is difficult to know who the opponent is and how to reach them. Any action can be part of a larger campaign, however, and this may be important for psychological reasons, since many actions are unsuccessful in reaching their goals, although that does not reduce their importance. The success of the campaign does not depend on the success of each single action. One side effect of actions is to attract the attention of the public, whom it is assumed would act rightly with knowledge.

A campaign may be part of a larger movement, such as one to eventually put 40 to 50 percent of the landscape (of North America or the planet) into special preserves. Such a movement may take many campaigns and a hundred years. But, the goal is important.

Wilderness campaigns are not really constructive in the sense of architectural or educational campaigns, because they are defensive. Furthermore, what is defended is, first, in the process of change, so it can not be saved as it is; second, it will always be vulnerable; third, it is an ambiguous concept misperceived by opponents; and fourth, it is ultrahuman. What is to be saved is the potential for evolution of uninhibited development of species and ecosystems.

Violence polarizes opposition. A stance of "no compromise" polarizes opposition. To turn your opponent into a supporter, compromise is far more effective than violence or coercion. We have to compromise. We don't know enough not to—we are not sure enough of the consequences of our actions, which often have the opposite effect of the one intended. Furthermore, compromise can be such that it satisfies the opponent's ego without giving up much, because it creates a state of cognitive dissonance, where a little token is enough to convince people that something is owed in return, a much bigger concession in this case, regardless of whether they know about how cognitive dissonance works.

The code of nonviolence, as presented by Gandhi, is not a rigid system. Exceptions are possible and, under some situations, even desirable. Arne Naess suggests that a small piece of a technical installation (a dam, for instance) could be destroyed in order to avoid the greater destruction of an area. Nevertheless, this violence is an exception and not a norm. Earth First!, by compromising on nonessential issues and using violence only as a warning, might increase its effectiveness. Either way, Earth First! already exemplifies an important, and neglected, aspect of Gandhi's philosophy: *you should follow your inner voice whatever the consequences*.

Author's Note: This article was rejected by *Earth First!* journal without explanation. Not radical enough? Too radical? You decide. It was subsequently published in *Pan Ecology* and has been reprinted several times.

Notes to Chapters

Chapter 1 Logical Fallacy and Computer Life

1. *Are Computers Alive? Evolution and New Life Forms*. Brighton: Harvester Press, 1983.
2. Ibid p. 170.
3. Ibid p. 25.
4. Ibid p. 108.
5. Ibid p. 52.
6. Ibid p. 27.
7. Ibid p. 109.
8. Ibid p. X.
9. Ibid p. 186.
10. Ibid p. 185.

Chapter 2 The First Self

1. New York: Simon and Schuster, 1984.
2. Ibid p. 306
3. Ibid p. 190.
4. Ibid p. 155.
5. *The Culture of Narcissism*. New York: Norton, 1978.
6. Turkle, p. 323.
7. Ibid p. 322.
8. Ibid p. 135.
9. *Person/Planet*. New York: Doubleday, 1978.

Chapter 3 Literacy and Computers

1. Vincent Rauzino, Conversation with an intelligent chaos, *Datamation* 28(5):123-138.
2. Ted Nelson, A new home for the mind, *Datamation* 28(3):168-181.
3. Daily Evergreen, Washington State University, May 10, 1984, p. 5.
4. EDUCOM News release, 1984.
5. Science and the Atari generation, *Science* 221(4611):1.
6. Pp 1-14 in *The Chronicle of Higher Education*
7. Silence is a commons, *Coevolution Quarterly* Winter 1983, pp. 5-9.
8. This term is from psychobiologist Stephen Rose
9. New age computer heresy, *Coevolution Quarterly* Summer 1983, pp. 98-100.
10. *Webster's Third International Dictionary*
11. Editorial, *BioScience* 33(8):479.
12 An ecolate view of the human predicament, pp. 49-71 in *Global Resources*. Baltimore: University Park Press.
13. *Democracy: Real and Deceptive*, Navajivan Publishing House
14. One energy slave equals 3000 Kcal of work per day
15. An evolutionist looks at computers, pp. 92-100 in *Naked Emperors*. Los Altos: Kaufmann.

Chapter 5 Wolves in Bulgaria

Spassov, N and G. Spiridonov. 1988. "Large Mammals of Bulgaria," in *Bulgaria's Biological Diversity*: Conversation Status and Needs Assessment, Vols I and II, ed. Curt Meine, Washington, DC: Biodiversity Support Program. P 470.

Spassov, N. et al. *Status and Conservation of the Wolf in the Southern Balkans* (Bulgaria, Macedonia and Albania). P. 5.

Editor. *The Sofia Echo*, Vol. 5 August 17-23, p 6.

Chapter 7 The Tragic Species

1. Ross, W., ed., *The Works of Aristotle*. Oxford: Clarendon, 1952, p. 677
2. Meeker, J. *The Comedy of Survival*. New York: Scribners, 1974
3. Hardin, Garrett. Tragedy of the Commons
4. *Complete Essays*. Stanford: Stanford University Press, 1958

Chapter 8 Reverence for Life

1. Schweitzer, Albert
2. *Religion in the Making,*
3. W. Brandon, *The Magic World*. New York: Morrow, 1971.
4. *The Philosophy of Civilization*
5. Fox, personal communication.
6. Albert Schweitzer, *Out of My Life and Thought*. p. 126.

Chapter 9 A Middle Way of Eating

1. Food for the Future
2. *Franju*
3. Fox, M. W. *Farm Animal Welfare and The Human Diet*
4. Cobb, J. B., Jr. *Is it too late? A theology of ecology*

Chapter 13 Aesthetic Education, Organic Dialectics, and Deep Ecology

1. In "The shallow and the deep, long-range ecology." *Inquiry*.
2. Theodore Roszak, *Where The Wasteland Ends*. Works by others as well.
3. see C. Gillispie, for instance
4. J. Winckelmann, 1787, "Reflections on the Painting and Sculpture of the Greeks."
5. Schiller's view of human civilization is presented in an essay in letter form, "On the Aesthetic Education of Man."
6. Eugene Odum

Chapter 15 Minimum Wilderness

Bergstraesser, A. 1962. *Goethe's Image of Man and Society*. Freiburg: Herder.

Birch, Charles, and Cobb, Jr., J. B. 1981. *The Liberation of Life*. Cambridge: Cambridge University Press.

Birch, Thomas H. 1982. "Man the Beneficiary?: A planetary perspective on the logic of wildland preservation." *International Dimensions of the Environmental Crisis*. R. Barrett, editor. Boulder: Westview Press.

Cheng, T.C. 1970. *Symbiosis*. New York: Pegasus.

Darwin, Charles. 1964. *The Origin of the Species*. New York: Collier.

Doxiadis, C. A. 1975. *Building Entopia*. New York: Norton.

Dubos, R. 1976. Symbiosis between the earth and humankind. *Science* 193:459-462.

—. 1980. *The Wooing of Earth*. New York: Charles Scribner's Sons.

Elton, Charles. 1966. *Animal Ecology*. New York: October House.

Evernden, N. 1981. *Out of Place*. unpublished manuscript.

Frome, M. 1974. Personal Communication.

Fowles, J. 1979. Seeing Nature Whole. *Harper's*. 259:49-56.

Hardin, Garrett 1977. *The Limits of Altruism*. Bloomington: Indiana University Press.

Hebb, D.O. 1958. Alice in Wonderland or psychology among the biological sciences. In *The Biological and Biochemical Bases of Behavior*. H. Harlow and C. Woolsley, eds. Madison: University of Wisconsin Press.

Holling, C.S. 1973. "Resilience and stability of ecological systems." *Annual Review of Ecology and Systematics*. R.F. Johnston et al., editors. 4:1-24.

Henberg, M. 1984. "Wilderness as playground." *Environmental Ethics* 6:253-263.

Krutch, J. 1970. *The Best Nature Writing of Joseph Wood Krutch*. New York: Pocket Books.

Lackner, S. 1984. *Peaceable Nature*. New York: Harper and Row.

Lehmann, Scott 1981. "Do Wildernesses Have Rights?" *Environmental Ethics* 3:129-146.

Leopold, Aldo. 1945. "The Green Lagoons." *American Forests*. 51:414.

—. 1949. *A Sand County Almanac. And Sketches of Here and There*. New York: Oxford University Press.

Lincicome, D.R. 1969. The Goodness of Parasitism: A New Hypothesis. Thomas C. Cheng, ed. *Aspects of the Biology of Symbiosis*. Baltimore: University Park Press.

Lovejoy, A.O. 1964. *The Great Chain of Being: A Study of the History of an Idea*. Cambridge: Harvard University Press.

Margalef, R. 1968. *Perspectives in Ecological Theory*. Chicago: University of Chicago Press.

Naess, A. 1972. The shallow and the deep, long-range ecology movement. A summary. *Inquiry*, 16: 95-100.

Odum, E. 1971. *Fundamentals of Ecology*. 3rd ed. Philadelphia: Saunders.

Rodman, J. 1977a. "The Liberation of Nature?" *Inquiry* 20:83-145.

Rolston, III, H. 1983. "Values Gone Wild." *Inquiry* 26:181-207.

Schaller, G.B. 1972. *The Serengeti Lion*. Chicago: University of Chicago Press.

Shepard, P. 1973. The Tender Carnivore and the Sacred Game. New York: Scribner's Sons.

—. 1974. Animal rights and human rites. *The North American Review* Winter, p.35.

—. 1978. *Thinking Animals*. New York: Viking Press.

Singer, Peter. 1981. *The Expanding Circle: Ethics and Sociobiology*. New York: Farrar, Strauss & Giroux.

Smith, P.M., and Watson, R.A. 1979. "New Wilderness Boundaries." *Environmental Ethics* 1:61-64.

Snyder, Gary. *The Coevolution Quarterly*, Fall 1983, p. 14.

Soleri, Paolo. 1983. *The Food Chain* (pamphlet).

Thomas, Keith. 1983. *Man and the Natural World*. New York: Pantheon.

Uexkull, J. von. 1957. A Stroll Through the World of Animals and Men. *Instinctive Behavior*. C. Schiller, ed. New York: International Universities Press Inc.

Waddington, C.H. 1960. *The Nature of Life*. New York: Atheneum.

Whitehead, A.N. 1967. *Science and the Modern World*. New York: Free Press. P. 136.

—. 1958. *The Function of Reason*. Boston: Beacon Press. .

Wittbecker, A. E. 1983a. "An optimum global population based on NEP." Fargo: Contributed paper, Ecol. Soc. Am.

Wittbecker, A. E. 1983b. "Ecology, mythology, and holopoetic culture." Montreal: Proceedings XVII World Congress Phil.

Wittbecker, A. E. 1984. "The law of noninterference in ecology and politics." *One Earth, Many Worlds*. Wilmington: Mozart & Reason Wolfe.

Wolfe, L. M. 1945. *Son of the Wilderness: The Life of J. Muir*. New York: Knopf.

Chapter 16 *Maximum Populations*

Anderson, Douglas C., Robert S. Hoffman, and Kenneth B. Armitage. 1979. "Above-ground productivity and floristic structure of a high subalpine herbaceous meadow." *Arctic and Alpine Research* 11(4):467-476.

Blum, Udo, Ernest D. Seneca, and Linda M. Stroud. 1978. "Photosynthesis and respiration of *Spartina* and *Juncus* salt marshes in North Carolina: Some models." *Estuaries* 1(4):228-238.

Brown, Lester. 1979. "Crossing the threshold? — Pressures on earth's biological systems." *Environment* 21(8):12-37.

Burlingh, P. et al. 1975. *Computation of the Absolute Maximum Food Production of the World* Wageningen, Netherlands: Agriculture University.

Clark, Colin. 1958. "World Population." *Nature* 181:1235-1236.

Clark, Colin. 1967. *Population Growth and Land Use*. London: Macmillan.

DeWit, K. 1967. Incomplete reference.

Dobben, W.H. van, and R.H. Lowe-McConnell, eds. 1975. *Unifying Concepts of Ecology*. Report of the Plenary Sessions of the First International Congress of Ecology. The Hague.

Eyre, Samuel. 1978. *The Real Wealth of Nations*. London: E. Arnold.

Gabel, Medard. 1979. *HO-PING: Food for Everyone*. Garden City, NY: Anchor Press/Doubleday.

Golley, Frank B., K. Petrusewicz, and L. Ryszkowski, eds. 1975. *Small Mammals: Their Productivity and Population Dynamics*. New York: Cambridge University Press.

Hulet, H.R. 1970. "Optimum world population." *Bioscience* 20(3):160-161.

Kozlovsky, Daniel G. 1974. *An Ecological and Evolutionary Ethic* New York: Prentice-Hall.

Kumar, Arun. 1975. "Variety, standing crop and net community productivity of the vegetation on a hard ground and stabilised dunes near Pilani, Rajasthan." *Annals of Arid Zone* 14(2):124-134.

Lappe, Frances Moore and Joseph Collins. 1979. *Food First: Beyond the Myth of*

Scarcity New York: Ballantine.

Lieth, Helmut F. H., ed. *Patterns of Primary Production in the Biosphere.* Stroudsburg, PA: Dowden, Hutchinson & Ross, Inc.

Lieth, Helmut F H. and Robert Whittaker, eds. 1975. *Primary Productivity of the Biosphere* New York: Springer-Verlag.

Naik, M.L. and G.P. Mishra. 1977. "Ecological studies of some grasslands at Ambikapur. III. Net community productivity and efficiency of the grasslands." *Tropical Ecology* 18:52-59.

Nigam, B.C., M.C. Joshi, and Arun Kumar. 1977. "Effects of herbage removal on summer season biomass and primary net community productivity of three sites around Pilani." *Tropical Ecology* 18:184-192.

Odum, Eugene P. 1970. "Optimum population and environment: A Georgian microcosm." *Current History* 58:355-366.

Odum, Eugene P. 1971. *Fundamentals of Ecology* . 3rd Edition. Philadelphia: Wm. B. Saunders.

Odum, Eugene P., Clyde E. Connell, and Leslie B. Davenport. 1965. "Population energy flow of three primary consumer components of old-field ecosystems." *Ecology* 43(1):88-96.

Odum, Howard T. 1957. "Trophic structure and productivity of Silver Springs, Florida." *Ecological Monographs* 27(1).

Odum, Howard T. and Elisabeth C. Odum. 1981. *Energy Basis for Man and Nature.* New York: McGraw Hill.

Ovington, J.D. 1961. "Some aspects of energy flow in plantations of *Pinus sylvestris L.*" *Annals of Botany, N.S.* 25(27):12-17.

Ovington, J.D., Dale Heitkamp, and Donald Lawrence. 1963. "Plant biomass and productivity of prairies, savanna, oakwood, and maize field ecosystems in central Minnesota." *Ecology* 41(1):52-65.

Pirie, N. W. 1976. *Food Resources.* London: Pelican Books.

Rodin, L.E., N.I. Bazilevich, and N.N. Rozov. 1975. "Productivity of the world's main ecosystems." IN: *Productivity of World Ecosystems.* Washington, D.C.: National Academy of Sciences.

Singer, S. Fred, ed. 1971. *Is There an Optimum Level of Population?* New York: McGraw-Hill.

Tansley, A.G. 1935. The use and abuse of vegetational concepts and terms. *Ecology* 16:284-307.

Trivedi, B.K. and G.P. Mishra. 1979. "Seasonal variations in species composition, plant biomass, and net community productivity of two grasslands in *Sehima-Dichanthium* cover type." *Tropical Ecology* 20(1):114-125.

Webb, Warren L., William K. Lauenroth, Stan R. Szarek, and Russell S. Kinerson. 1983. "Primary production and abiotic controls in forests, grasslands, and desert ecosystems in the United States." *Ecology* 64(1):134-151.

Westing, Arthur H. 1981. "A world in balance." *Environmental Conservation* 8(3):177-183.

Whittaker, R.H., F.H. Bormann, G.E. Likens, and T.G. Siccama. 1974. "The Hubbard Brook ecosystem study: Forest biomass and production."

Ecological Monographs 44:233-252.

Wittbecker, A.E. 1976. *Eutopias: Making Good Places Ecologically* (unpublished manuscript).

Woodwell, G. M. and Robert Whittaker. 1968. "Primary production in terrestrial ecosystems." *American Zoologist* 8:19-30.

Chapter 17 Ecological Design and Planning

Bateson, Gregory. 1987. *Steps to an Ecology of Mind*. Northvale, NJ: Jason Aronson Inc.

Bellah, Robert N., Richard Madsen, William M. Sullivan, Ann Swidler, and Steven M. Tipton. 1991. *The Good Society*. New York: Alfred A. Knopf.

Borgstrom, George. 1965. *The Hungry Planet*. New York: Macmillan Co.

Brown, Lester. 1979. "Crossing the threshold? Pressures on earth's biological systems." *Environment* 21(8):12-37.

Bryan and McClaughry. 1991. *The Vermont papers*. Incomplete reference

Burlingh, P. et al. 1975. *Computation of the Absolute Maximum Food Production of the World*. Wageningen, Netherlands: Agriculture University.

Calhoun, J. B. 1962. "Population density and social pathology." *Scientific American* 206(2):1399-1408.

Carson, Rachel. 1962. *Silent Spring*. Boston: Houghton Mifflin.

Daly, H. and Cobb, J. B., Jr. 1989. *For the Common Good*. Boston: Beacon Press.

Dobben, W. H. van, and R. H. Lowe-McConnell, eds. 1975. Unifying Concepts of Ecology. *Report of the Plenary Sessions of the First International Congress of Ecology*. The Hague.

Drucker, Peter. 1990. *The New Realities*. New York: Dutton.

Eyre, Samuel. 1978. *The Real Wealth of Nations*. London: E. Arnold.

Frankel, Otto. 1975. *Crop Genetic Resources for Today and Tomorrow*. O. Frankel and J. Hawkes, eds. New York: Cambridge University Press.

Fuller, R. Buckminster. 1981. *Critical Path*. New York: St. Martin's Press.

Gabel, Medard. 1979. *HO-PING: Food for Everyone*. Garden City, NY: Anchor Press.

Golley, Frank B., K. Petrusewicz, and L. Ryszkowski, eds. 1975. *Small Mammals: Their Productivity and Population Dynamics*. New York: Cambridge University Press.

Hulet, H.R. 1970. "Optimum world population." *Bioscience* 20(3):160-161.

Kohr, Leopold. 1957. *The Breakdown of Nations*. New York: E. P. Dutton.

Kozlovsky, Daniel G. 1974. *An Ecological and Evolutionary Ethic*. New York: Prentice-Hall.

Laszlo, Ervin et al. 1977. *Goals for Mankind*. New York: E. P. Dutton.

Leopold, Aldo. 1949. *A Sand County Almanac*. New York: Oxford University Press.

Liebig, J. von. 1840. *Chemistry in its Application to Agriculture and Physiology*. London: Taylor and Walton. (cited in Odum 1970)

Lieth, Helmut F. H., ed. *Patterns of Primary Production in the Biosphere*. Stroudsburg, PA: Dowden, Hutchinson & Ross, Inc.

Lieth, Helmut F. H. and Robert Whittaker, eds. 1975. *Primary Productivity of the Biosphere*. New York: Springer-Verlag.

Mander, Jerry. 1991. *In the Absence of the Sacred*. San Francisco: Sierra Club Books.

McHarg, Ian. 1969. *Design with Nature*. Garden City: Natural History Press.

McLuhan, Marshall. 1964. *Understanding Media*. New York: McGraw Hill.

Mumford, Lewis. 1956. *The Transformation of Man*. New York: Harper and Row.

Odum, Eugene P. 1970. "Optimum population and environment: A Georgian microcosm." *Current History* 58:355-366.

Odum, Eugene P. 1971. *Fundamentals of Ecology*. 3rd Edition. Philadelphia: Wm. B. Saunders.

Odum, Eugene P., Clyde E. Connell, and Leslie B. Davenport. 1965. "Population energy flow of three primary consumer components of old-field ecosystems." *Ecology* 43(1):88-96.

Odum, Howard T. and Elisabeth C. Odum. 1981. *Energy Basis for Man and Nature*. New York: McGraw Hill.

Reichel-Dolmatoff, G. 1977. "Cosmology as ecological analysis." *The Ecologist* 7:4-11.

Sahlins, Marshall. 1972. *Stone Age Economics*. Chicago: Aldine Publishing.

Sale, Kirkpatrick. 1980. *Human Scale*. New York: Coward, McCann and Geoghegan.

Schumacher, E. F. 1973. *Small is Beautiful*. New York: Harper and Row.

Singer, S. Fred, ed. 1971. *Is There an Optimum Level of Population?* New York: McGraw-Hill.

Smith, Anthony D. 1991. *The Ethnic Origins of Nations*. Cambridge: Blackwell.

Soule, MI and Wilcox, B. A., eds. 1980. *Conservation Biology: An Evolutionary-Ecological Perspective*. Sunderland, MA: Sinauer Associates.

Tainter, Joseph A. 1990. *The Collapse of Complex Societies*. Cambridge: Cambridge University Press.

Tansley, A. G. 1935. "The use and abuse of vegetational concepts and terms," *Ecology* 16:284-307.

Webb, Warren L., William K. Lauenroth, Stan R. Szarek, and Russell S. Kinerson. 1983. "Primary production and abiotic controls in forests, grasslands, and desert ecosystems in the United States." *Ecology* 64(1):134-151.

Westing, Arthur H. 1981. "A world in balance." *Environmental Conservation* 8(3):177-183.

Wittbecker, A. E. 1970. *Aesthetic Space and Human Limits* (Newark: University of Delaware).

Wittbecker, A. E. 1976. *Eutopias: Making Good Places Ecologically* (unpublished manuscript).

Wittbecker, A. E. 1981. "The fifty percent rule in ecology and economics." *Proceedings of the G. P. Marsh Institute* 5:4-8.

Wittbecker, A. E. 1983. "An optimum human population based on NCP." Ecological Society of America annual meeting, Fargo.

Wittbecker, A. E. 1986. "Palouse Ecosystem Restoration." 4th International Congress of Ecology, Syracuse.

Wittbecker, A. E. 1991. "Recognizing primary cultures and independent nations and creating a framework for them." *Pan Ecology* 6(1):1-12.

Woodwell, G. M. and Robert Whittaker. 1968. "Primary production in terrestrial ecosystems." *American Zoologist* 8:19-30.

Zadeh, L. A. 1965. "Fuzzy Sets." *Information and Control* Vol. 8:338-353.

Chapter 27. Revolutionary Ecology

1. William Gibson, "Ecology and Justice", *Wilderness* (Summer 1986):52-56.
2. Nathan Hare, "Black Ecology", *The Black Scholar* 1(April 1970):2-8.
3. William Tucker, "Is Nature Too Good for Us?", *Harper's* (March 1982).

Chapter 34. Ecological Forestry as a Relativistic & Crisis Science

Franklin, Jerry and C. T. Dyrness. 1988. *Natural Vegetation of Oregon and Washington*. Corvallis: Oregon State University Press.

Harris, Larry. 1984. *The Fragmented Forest: Island Biogeography Theory and the Preservation of Biotic Diversity*. Chicago: UC Press.

Perry, D. A. 1995. *Forest Ecosystems*. Baltimore: Johns Hopkins University Press.

Smith, David. 1986. *The Practice of Silviculture*. New York: John Wiley.

Toumey, James W. 1947. *Foundations of Silviculture Upon an Ecological Basis*. Second ed. rev. C. F. Korstian. John Wiley & Sons, New York.

Chapter 35. Gigatrends in Ecology & Forestry

Naisbitt, John. 1984. *Megatrends: Ten New Directions Transforming Our Lives*. New York: Warner Books.

Chapter 37 Radical Ecology

1. *Out of Place*, unpublished manuscript (1981).
2. See E. Odum. "The emergence of ecology as a new integrative discipline." *Science* 195(4284):1289 – 1293 (1977).
3. "Introduction: Ecology and man – A viewpoint." *The Subversive Science* (Houghton Mifflin, New York, 1969).
4. "Ecology – A subversive subject." *Bioscience* 14(7):11 (1964).
5. *Where the Wasteland Ends* (Harper & Row, New York, 1972).
6. Wittbecker, *One Earth, Many Worlds: An Essay on Holopoetic Cosmology*. Ph.D. dissertation (1983).
7. *Eco-Philosophy* (Boston, Marion Boyars, 1981). Also, David Klein, "Ecophilosophy," invited paper (1972).
8. "The shallow and the deep long-range ecology movement: A summary." *Inquiry* 16:95 – 100 (1973).
9. "Theory and practice in the environmental movement: Notes towards an ecology of experience." Pp. 45 – 56 IN: *The Search for Absolute Values in a Changing World* (International Cultural Foundation Press, New York, 1977).
10. Theodore Roszak, op. cit.
11. A. Maslow, "Towards a humanistic biology," *The Farther Reaches of Human Nature* (Viking Press, New York, 1971). 12. See Daniel Kozlovski. *An Ecological and Evolutionary Ethic* (Prentice-Hall, New York, 1974).
13. *A Sand County Almanac And Sketches Here and There* (Oxford University Press, New York, 1968).

14. See John Fowles, "Seeing Nature Whole." *Harper's* (November, 1979).
15. *Ecology and the Politics of Scarcity* (Freeman, San Francisco, 1977).
16. *Reflections on the Paintings and Sculpture of the Greeks*, J. Winckelmann, 1787
17. See Charles Birch and John Cobb, *The Liberation of Life: From the Cell to the Community* (Cambridge University Press, London, 1981) and John Rodman, "The liberation of nature," *Inquiry* 20:83—145 (1982).

Chapter 40 Deep Anthropology
1. Arne Naess, "The shallow and the deep, long-range ecology movement: A summary." *Inquiry*. 16:95-100, 1973.
2. in "A critique of anti-anthropocentric biocentrism." *Environmental Ethics*. Vol 5: 245-256. And "A note on deep ecology." *Environmental Ethics*. 6:377-379.
3. in "Deeper than deep ecology: The eco-feminist connection." *Environmental Ethics*. 6: 339-345.
4. in "The dogma of anti-anthropocentrism and ecophilosophy." *Environmental Ethics*. Vol. 6: 283-288.)
5. The word archaic is used as a comprehensive term for aboriginal and nonindustrial.
6. from Richard Jefferies. In: John Fowles, "Seeing Nature Whole." *Harper's* 259:49-56, 1979.)
7. This is one meaning of 'thinking like a mountain.' It is not 'mountaining like a mountain,' as Skolimowski argues. It is remaining human while acting as if one had the identity of a mountain. .
8. "Cosmology as ecological analysis." *The Ecologist* 7: 4-11, 1981.
9. Roy Rappaport.
10. R. H. Dicke, "Dirac's Cosmology and Mach's Principle." *Nature* (1961) 192: 440.
11. *Eco-Philosophy*. Boston: Marion Boyars, passim. p. 73.
12. *North American Review*.
13. J.G. Neihardt, *Black Elk Speaks*. New York: Pocket Books, 1959.
14. *New Paths in Biology*. New York: Harper & Row, 1964.
15. "A stroll through the world of animals and men." *Instinctive Behavior*. C. Schiller, ed. International Universities Press, New York, 1957.
16. Some of the pre-Socratics developed a nonanthropocentric world view. Zeno the Stoic preached "life in agreement with nature" as the goal of ethics. Chrysippus added that as individual natures were parts of the nature of the whole, therefore, life was to be in accord with human nature as well as nature. Francis of Assisi tried to unite the compassion of Christianity and the animistic sense of union with the natural world. A scientific argument supports these insights.
17. This idea has reference in earlier philosophical ideas. The atomists advanced the idea of an infinite universe without center or edge. In the 15th century, Nicholas of Cusa argued that the universe was without edge or center—because its creator was infinite and without location. The universe is "a sphere of which the center is everywhere and the circumference is nowhere." The earth was not the center. The modern

theory of containment reflects that attitude towards centers and edges. This is ontofugal, not centering, but scattering, framing.

18. Latin *mores*, plural of *mos*, from *meare*. Ethics means 'doing together,' which of course one does living together. The word symbiosis comes from the Greek meaning 'living with.'

19. *The Primacy of Perception*. James Edie, ed. Evanston: Northwestern University Press, 1964.

20. Skolimowski admits that, faced with a choice between mosquito and human, he would take the life of the mosquito. But what if there were five billion humans and fifty mosquitos? Which life would be more valuable, then? Which would he take? I would save the mosquitos first. Garrett Hardin once remarked that in view of their relative numbers, he would, if forced to choose, support the existence of one redwood tree over one human baby. "Destroying wildlife in the people's name." *Defenders* 56:22, 1981.

21. Michael W. Fox, personal communication.

22. "A defence of the deep ecology movement." *Environmental Ethics*. 6:265-270.

23. from the Greek word "*logos*."

24. Watson has a thermodynamic concept of ecological balance to dismiss the barren plateaus of the Mediterranean as just as balanced ecologically as the forests they replaced. It is balanced at the lowest level of productivity.

25. Only Paul Shepard's techno-cynegetics actually makes this proposal in *The Tender Carnivore and the Sacred Game*. New York: Scribner's, 1973.

26. Alan Wittbecker, *Eutopias: Towards the Biological Definition of Good Places*. Wilmington: Mozart & Reason Wolfe (in review).

Chapter 41. Metaphysical Implications

1. J. Baird Callicott, "The Metaphysical Implications of Ecology," *Environmental Ethics* 8 (1986):301-316.

2. Ibid., p. 307.

3. Albert Einstein, *Essays in Science* (New York: Wisdom Library, 1934), p. 68.

4. Alan Wittbecker, *The Poetic Archaeology of the Flesh*. (Wilmington: Mozart & Reason Wolfe, 1976).

5. Ibid, pp. 301, 310–311.

6. see E. Wilson's experiments on small islands.

7. The scientific description of the concept is still unsatisfactory, as it was to Weiss.

8. Ibid, p. 312.

9. see Wittbecker (1976).

10. see D.R. Brooks and E. O. Wiley, *Evolution as Entropy: Toward a Unified Theory of Biology* (Chicago: University of Chicago Press, 1986). Also see Jeremy Rifkin, *Entropy* (New York, Viking, 1980) and Nicholas Georgescu-Rogen, *The Entropy Law and the Economic Process* (Cambridge: Harvard University Press, 1971), for applications to cosmology and economics.

11. Harold J. Morowitz, "Biology as a Cosmological Science," *Main Currents in Modern Thought* 28 (1972): 156.
12. the other type of organization is nonequilibrium stationary state, such as a solar system. Ilya Prigogine, From Being to Becoming. (San Francisco: Freeman, 1980).
13. see Edward Goldsmith. "Superscience—Its Mythology and Legitimisation." *Ecologist* (1981):228-241.
14. Ibid., p. 310.
15. Ramon Margalef, *Perspectives in Ecological Theory* (Chicago: University of Chicago Press, 1968).
16. in Eugene Odum, *Fundamentals of Ecology* (Philadelphia: W. B. Saunders Co., 1971).
17. It is not necessary to associate scientific holism with Hindu metaphysics. A more productive comparison is with contemporary Gestalt thinking and with the Gestalt psychology of W. Kohler (1913-1940), which emphasized the importance of wholes and tried to identify the organizing principles of perception in terms of wholes.
18. Arthur Koestler, *Janus: A Summing Up*. (New York: Random House, 1978).
19. among them: R.V. O'Neill, D.L. DeAngelis, J.B. Waide, T.F.H. Allen, T.W. Hoekstra, T.B. Starr, and L. Johnson.
20. see Francisco Varela, *Principles of Biological Autonomy* (New York: North Holland, 1979).
21. Ibid., p. 312.
22. McIntosh, *The Background of Ecology* (1985), p. 40. Waddington sees physical laws as merely constraints for living organisms.
23. Ibid., p. 315.

Chapter 42. *Metaphysical Principles Based on the Foundation of Ecology*

1. Living systems do not display homeostasis (constant change) so much as a particular course of change in time (homeorhesis). The course is stable, not constant, according to Conrad Waddington, *The Evolution of an Evolutionist* (Ithaca: Cornell University Press, 1975). Changes to a system are symbolized by trajectories in a multidimensional phase space (or landscape).
2. J. Von Uexkull, "A stroll through the world of animals and men," In *Instinctive Behavior*, C. Schiller, ed. (New York: International Universities Press, 1957).
3. John Rodman. Incomplete reference. Inquiry (1976).
4. Alfred North Whitehead, *The Function of Reason* (Boston: Beacon Press, 1958), p. 8.
5. Parasitism and predation are considered similar, as are protocooperation and mutualism. The significant difference is the necessity of the latter arrangements, that is, the organisms in predation and mutualism need each other.
6. Michael W. Fox. Personal communication, 1980.
7. in Begon, M., J. Harper, and C. Townsend. *Ecology: Individual, Population, and Community*. (Sunderland, MA: Sinauer Associates, 1986.).
8. Steven Stanley, *The New Evolutionary Timetable: Fossils, Genes and the Origin*

of Species. (New York: Basic Books, 1981).

9. Neil Evernden, *Out of Place* (unpublished manuscript, 1981).

10. Maurice Merleau-Ponty, *The Visible and the Invisible*, trans. A. Lingis (Evanston, IL: Northwestern University Press, 1968).

11. Whitehead might be a more appropriate source for a discussion of internal relations than Hegel or others. His cosmology is more ecological. See John B. Cobb, Jr. Is *It Too Late? A Theology of Ecology* (New York: Glencoe, 1971).

12. J. Baird Callicott, "The metaphysical implications of ecology," *Environmental Ethics* 8 (1986): 312.

13. Insects and plants for instance, often mutually adapt. Paul Ehrlich and Peter Raven refer to this mutual adaptation as coevolution, the emergence of a highly ordered complexity to full structuration. Insects may evolve to fertilize just one species of flower, as in the case of a wasp and orchid. The ground orchid in Australia (*Cryptostylus leptochilla*) has flowers that look like, and may smell like, a female wasp (*Lisopimpla semipunctata*). The orchid blooms at a time when the male wasp emerges from its pupal case. The male mates with, and pollinates, the orchid. Later, when the female emerges from the soil, he mates with her with more experience.

14. Paul Shepard emphasized this, but Ervin Laszlo made the idea an important part of his system ethics.

15. McIntosh, *The Background of Ecology* (1985), p.40. Waddington sees laws as constraints for living organisms.

16. Garrett Hardin (1960) states the competitive exclusion principle as: Complete competitors cannot coexist. Niches must be different for species. Krebs states that the fundamental niche of a species has an "infinite number of dimensions," making complete determination impossible. Another difficulty in definition is the assumption that environmental variables can be ordered linearly and measured. Furthermore, competition is dynamic, whereas models freeze single theoretical instants.

17. Succession appears to be a process of self-organization in a cybernetic system at the ecosystem level. It is primary for Eugene Odum.

18. Callicott (p. 307) states that "the producers must be many times more numerous than consumers" as an example of common structures. As a general rule, there is a decline in biomass with each increase in trophic level, as specified by energy flows. But, there can be a greater weight of consumers than producers if the turnover in producers is higher than consumers, for example, where fish feed on photoplankton. Nutrients are considered to cycle, but not energy.

19. Diversity incorporates species richness (how many different kinds) as well as a measure of abundance (how many of each). Other aspects of diversity, e.g., life cycles, are less often considered.

20. Charles Elton, *The Ecology of Invasions by Animals and Plants* (London: Methuen, 1958).

21. R. M. May, *Complexity and Stability in Model Ecosystems* (Princeton:

Princeton University Press, 1973).

22. S. J. McNaughton (1977), Eugene Odum (1971).
23. Arne Naess, "The shallow and the deep, long-range ecology movement. A summary," *Inquiry* 16 (1973): 98-104. Naess's structure for norms and hypotheses is used throughout.
24. The archetype of these sentences appears to be in the Bible, when God says, "Let there be light!" The function is meaningless in terms of information theory, as there were no receivers at the time, but it is meaningful in terms of expression.
25. This list also incorporates Michael Soule's norms, with slight changes.
26. Hobart Smith, editorial, *Bioscience* (1984).
27. in Peter Singer, *The Expanding Circle: Ethics and Sociobiology* (New York: Farrar, Strauss & Giroux, 1981).
28. R. H. Whitacker, "On the broad classification of organisms," *Quarterly Review of Biology* 34 (1959):210-226. See also Lynn Margulis and K. Schwartz (1981).
29. Interference is not the same as exploitation, which is the normal use of a resource or species by another species; exploitation has a rejuvenating effect. Interference is not general competition, either; as it is used here, it is destruction without gain.
30. Darwin admitted that he would rather have used the term "natural preservation" instead of "natural selection," in G. Stent, *Paradoxes of Progress* (San Francisco: Freeman, 1978).

Chapter 43. Ecophilia
(see general bibliography)

Chapter 44. Nature as Self (in order cited)

Portmann, Adolf. 1964. New Paths in Biology. Harper and Row, New York.
Fox, Michael W. 1980. Returning to Eden: Animal Rights and Human Responsibilities. Viking, New York.
Snyder, Gary. 1969. Earth Household. New Directions, New York, p. 105.
Lovelock, James 1979. Gaia. Oxford University Press, New York.
Saint Francis. The Writings of Saint Francis of Assisi. Transl. B. Fahy. Burnes and Oates, London, p. 130.
von Uexkull, J. 1957. "A stroll through the worlds of animals and men," Instinctive behavior, C. Schiller, ed. International Universities Press, New York.
Naess, Arne, 1972. "The shallow and the deep, long-range ecology movement," Inquiry 16:95-100.
Fowles, John. 1979. "Seeing nature whole," Harper's 259:49-56.
Welch, Holmes. 1966. Taoism. Revised edition, Beacon press, Boston.
Thomas, Lewis. 1973. The Lives of a Cell. Bantam, Toronto.
Merleau-Ponty, Maurice. 1968. The Visible and the Invisible. A. Lingis, transl. Northwestern University Press, Chicago.
Brown, G. S. 1972. Laws of Form. Bantam, New York.
Haldane, J. B. S. 1927. Possible Worlds and Other Papers. Chatto and Windus, London.

Marsh, G. P. 1964. Man and Nature. Belknap, Cambridge.

Commoner, Barry. 1971. The Closing Circle. Knopf, New York.

Upanashads. 1966. R. C. Zachner, ed. Dutton, New York.

Fromm, Eric. 1956. The Art of Loving. Harper and Row, New York.

Pope, Alexander. 1966. "An essay on man," Poetical Works. Oxford University Press, New York. p. 259.

Vaihinger, Hans. 1949. The Philosophy of As If. C. K. Ogden, transl. 2nd ed. Routledge & Kegan Paul, London.

Chapter 48. Nonviolence

1. Dave Foreman, ed. *Ecodefense: A Field Guide to Monkey-wrenching* (Tucson: Earth First! Books, 1985, p. 10).

2. *Harajan*, 7.9.1935, p.234.

3. R. Prabhu and U. Rao, eds. *The Mind of Mahatma Gandhi* (Madras: Oxford University Press, 1946).

4. Foreman (1985, p. 11).

5. Arne Naess. *Gandhi and Group Conflict* (Oslo: Universitets forlaget, 1974).

Index

During a brief career in astrophysics and astronomy at the University of Arizona, where he worked on mathematical models of stars and on spectrometric analysis, Alan Wittbecker spent his daylight hours climbing trees and trying to track mountain lions. He shared a trailer with a mouse, cockroach, squirrel, and bat (but did all the cooking himself).

Encouraged by research cuts to pursue a different direction, Wittbecker went to graduate school in psychology, anthropology, philosophy, and ecology (his degrees are in these fields). As a graduate student in 1970, he was a cofounder of the G. P. Marsh Institute for Research in Ecology, where he worked for 22 years, including 3 as Director by rotation. When projects were sparse, he worked in other occupations, such as ditchdigger, janitor, gardener, diving coach, gymnastics teacher, artists model, librarian, systems engineer, editor, graphic artist, typesetter, forester, and math instructor.

In 1976, with three partners, Wittbecker cofounded Nieman Ryan Community Designs, specializing in private and urban local landscape design — but, also designing books, posters, journals, packages, landscapes, and buildings. In 1992, he founded SynGeo ArchiGraph, a firm specializing in global and regional ecological designs; he created designs for several bioregions, as well as international frameworks. A year later he set up the educational program for the new Ecoforestry Institute, becoming an Instructor in 1994, journal Editor in 1995, and Director in 1997. He has worked on public and private forests from British Columbia to California, and on wildlife projects, from Siberia to Norway.

A veteran of the US Air Force, Wittbecker is also a returned Peace Corps Volunteer from Bulgaria, where he monitored wolves in the Central Balkan Mountains. He has written newspaper columns and articles on ecology. He is the author of eight books, including *The Poetic Archaeology of the Flesh: An Investigation into the Phenomenology and Ecology of Being*, and over 100 articles.

Author's Notes

Many of these essays are taken from public talks, research discussions and class interactions, as well as from newspaper columns and articles, and thus are less stilted than my usual academic style. I have resisted the impulse to adorn them with bombastic neologisms, at the risk of having them seem informal.

To make up for the loss of trees and their services, as a result of my use of paper in these books, I have planted over nine thousand trees, during a period of twenty years, at the Altazor Forest in Idaho. More plantings are planned in forests in Oregon, Massachusetts, Florida, and Virginia.

Third nursery at Altazor Forest in Idaho, 1987

Colophon

Type: Palatino (designed by Hermann Zapf in 1948
   at Stempel AG)
Display Type: Palatino
Book Design: Rian Garcia Calusa Designs
Cover Design: Rian Garcia Calusa
Photographs & Graphics: Alan Wittbecker
Author Photo: Mike Barnes
Editing: J. Garcia B. of Rian Garcia Calusa
Hardware: Macintosh G5
Software: Adobe InDesign & Acrobat
Furious Charge & Entertainment: Pippi Frog
Spiritual & Material Support: Precious Woulfe